MACHINING AND FORMING TECHNOLOGIES

MACHINING AND FORMING TECHNOLOGIES

VOLUME 2

Machining and Forming Technologies

Additional books in this series can be found on Nova's website under the Series tab.

Additional E-books in this series can be found on Nova's website under the E-books tab.

MACHINING AND FORMING TECHNOLOGIES

MACHINING AND FORMING TECHNOLOGIES

VOLUME 2

J. PAULO DAVIM
EDITOR

Nova Science Publishers, Inc.
New York

LIBRARY OF CONGRESS CATALOGING-IN-PUBLICATION DATA

ISSN: 2157-250X

ISBN: 978-1-61122-820-5

Published by Nova Science Publishers, Inc. † New York

CONTENTS

PREFACE

Machining and forming technologies are two of the most important manufacturing processes, which includes a large group of manufacturing processes in which plastic deformation and others techniques are used to change the shape of workpieces. This book presents and discusses important aspects of machining and forming technologies.

Chapter 1 - The Die Sinking–Electro Chemical Spark Machining (DS–ECSM) process has been successfully applied for cavity machining in electrically non-conducting materials such as glass, quartz, ceramics and composites using non-rotating tool electrode. In this paper, a thermal finite element model has been developed to estimate the temperature distribution within the heat affected zone of single spark in the workpiece accounting temperature dependent properties of workpiece and same is used for the determination of material removal rate (MRR) and average surface roughness (ASR) in DS–ECSM. The computational results are verified by comparing them with the experimental results available in literature. It has been found that surface finish improves with the increase in spark radius and electrolyte concentration above a critical value (C = 10%). On the other hand MRR improves with the increase of energy partition and duty factor at the cost of deterioration in surface finish.

Chapter 2 - Surface roughness (Ra) is a significant upshot in the manufacturing process and it materializes a major part in the manufacturing system. It depends on different machining parameters and its prediction and control is a query to the researchers. In this study, a regression model and two artificial neural networks (ANNs) namely: Back propagation and radial basis function, were developed to predict surface roughness in electrical discharge machined surfaces.

In the development of predictive models, machining parameters of discharge current (Ip), pulse duration (Ton) and duty fraction (τ) were considered as model variables with a constant voltage 50 volt. For this reason, extensive experiments were carried out in order to collect surface roughness dataset. The developed models are validated with a new set of experimental data, and predictive behavior of models is compared, subsequently relative advantages of each model are analyzed.

Analysis of variance (ANOVA) and F-test were used to check the validity of regression model and to determine the significant parameter affecting the surface roughness. The statistical analysis exemplify that the Ip, Ton and τ were the factors in sequence have significant influence the on surface roughness. The microstructures of machined surfaces were also studied by scanning electron microscopy (SEM). The SEM investigations revealed that EDMed produced were increased significantly with Ip.

Chapter 3 - In this study, wire electrical discharge machining of aluminium matrix composites has been experimentally investigated. Three work materials viz. $A359/SiC_p/10\%$, $A359/SiC_p/20\%$ and $A359/SiC_p/30\%$ have been machined at various combinations of machining parameters such as pulse on-time, off-time, wire speed and wire tension. Empirical models have been developed to predict the cutting rate, surface roughness and kerf width of the machined composites. The models were developed by using the technique of response surface methodology. The constants and coefficients of these models were determined by employing the results of the experiments. The experiments were conducted by using carefully planned response surface design matrix (central composite design). The model predictions are in good agreement with the experimental results. The R^2 values of the proposed cutting rate, surface finish and kerf width models are above 0.99, 0.99 and 0.98 respectively. Volume fraction of ceramic particle and pulse on-time were found to influence cutting rate and surface finish significantly. Kerf width was significantly affected by volume fraction, pulse on-time, off-time, wire speed, and wire tension.

Chapter 4 - Finish-turning of hardened steel to grinding-like accuracies requires proper selection of cutting tools and process parameters. In the present work, the bearing steel AISI 52100 hardened to 55 HRC is finish-turned using cBN inserts, and performance measures such as tool wear, surface finish, cutting tool temperature and vibration are investigated. For the first part of the study, experiments have been carried out employing orthogonal array in the design of experiments.

Significance of process parameters and their influence on performance measures are brought out by analysis of variance and response equations. Optimal conditions are identified for each of the performance measures. The second part of the investigation deals with performance of cBN under the optimal conditions. In such optimal conditions, the cutting tool may take long time to wear out and reach a failure stage. To simulate the condition of tool wear, cutting tools with different flank wear are obtained using aggressive cutting conditions.

Results obtained using such cutting tools with a pre-measured flank wear are presented and discussed in this paper.

Chapter 5 - Hard machining is a competitive finishing operation in precision manufacturing. Hard machining applications have been remaining slow due to the existence of surface damage such as chatter marks. The chatter marks on a machined component surface could be found only after machining. This post-process scenario imposes a potential danger to subsequent product performance such as fatigue life. Therefore, real-time monitoring of hard machining operation has significant economical and durability importance. Hard turning creates high pulsating pressures on the insert seat and shim. The shim acts as a shock absorber for the insert and workpiece. The shim protects both insert and tool holder from damage due to high cutting forces. In this research, the shim material is replaced with different damping materials, and cutting tests were conducted to find out the improvement in surface finish. The improvement in quality of surface finish is investigated by real time observations using the vibration data collector and MCMe software. The experimental results show that RMS and frequency shift may be used to monitor the surface integrity factors. The surface finish of the workpiece is significantly improved after replacing Carbide shim with Brass shim.

Chapter 6 - Cutting forces affecting cutting tool in metal cutting are one of the important parameters necessarily to know tool failure and to optimize the cutting parameters. It is known that flank wear describing the tool life is formed with mechanical and thermal interaction between cutting tool and workpiece. In present study, the cutting stress analyses were carried out from two aspects. Firstly, changing in cutting stress distribution with wear band width is determined with modeling of between 0.3 mm and 0.9 mm wear band values by finite element method using Ansys 10.0 programme. Secondly, the effect of the some important cutting parameters such as cutting depth, feed rate, and approach angle on the cutting stress was performed in present analyses. The results showed that stresses increase to parallel an increase with flank wear value. The cutting depth has more a significant effect on the tool stresses than feed rate parameter in respect of von-Mises stresses. Also, cutting stresses decreased with an increase in the approach angle.

Chapter 7 - In the present work an attempt has been made to predict surface roughness and material removal rate (MRR) in CNC end milling process considering spindle speed, feed rate and depth of cut as machining parameters. The experimental work has focused on determination of optimum process environment for achieving expected quality as well as high productivity by ensuring minimization of surface roughness and maximization of material removal rate. Optimization of the process parameters for the desired surface roughness and material removal rate has been carried out by utility based Taguchi approach. The aforesaid method has been proved very helpful to optimize multiple performance responses for continuous quality and productivity improvement of the process/product.

Chapter 8 - Because of excellent creep properties near alpha titanium alloys are used as aeroengine components. Near alpha titanium alloy IMI 834 is one such material that is used as high pressure compressor discs, shafts and blades in aeroengines. This article deals with how dynamic materials modeling (DMM) has been used to understand the high temperature deformation behaviour of the material and to identify the optimum thermomechanical processing parameters i.e., temperature, strain rate and strain. Based on DMM, a two stage – two step process was designed to near isothermally forge 434 mm diameter discs to thier near net shape. Finite element analysis was used to evaluate the process and die design. The results obtained are presented and discussed in the paper.

Chapter 9 - This paper reports an investigation of micro texture in the cast condition and in the warm rolled conditions (50% size reduced) of nitrogen alloyed duplex stainless steel. The micro texture measurements were carried out using electron backscattered diffraction (EBSD) system. Grain orientations, Grain size distributions and Misorientation angles were also reported for the dual phases, ferrite (alpha) and austenite (gamma) under both the conditions of before heat treatment and after heat treatment. The results reveal that there is a definite improvement in the mechanical properties after reducing the specimen size to 50% by warm rolling, followed by heat treatment (solutionizing). This improvement in the mechanical property is correlated with the volume of low angle grain boundaries (EBSD analysis) and with the cleavage facets (SEM analysis).

Chapter 10 - This paper reviews the challenges during drilling of composite, titanium, and aluminium stacks. The machinability aspects of these three material stacks commonly used in aerospace applications are reviewed individually based on influence of drill material, drill geometries, and drilling parameters. Composite material requires high speed and low feed rate, drilling titanium requires low speed and high feed rate whereas drilling aluminium requires a balance between speed to feed rate. The review reports critical results and gap in the literature for composite, titanium and aluminium. Drilling of multimaterial stacks require a compromise between many parameters. The problems and solutions related to drilling of multimaterial stacks are discussed and the directions in which the research on drilling of multimaterials may be carried out are suggested in this paper.

Chapter 11 - The paper outlines the development of 3D finite element model for drilling of AISI 1010 steel using commercial FE code MSC Marc Mentat. Thrust force, Torque and temperature distribution in the cutting region are calculated for selected cutting conditions. For experimental measurement of thrust force and torque, a strain gauge based three component dynamometer is used. Standard k-type thermo couple is used for the measurement of temperature at the tool-workpiece interaction zone during drilling process which is inserted in to the coolant hole of a drill bit. The effect of lubrication on thrust force and torque is also studied. To study the effect of lubrication on thrust force and torque two contact conditions, contact with friction and contact without friction are used in finite element analysis of drilling process.

Chapter 12 - In the present work, principal component has been fused with artificial neural network (ANN) for prediction of drill wear so as to reduce the redundancy of the network. Large numbers of experiments have been conducted and sensor signals have been acquired using data acquisition system. Cutting forces, torque, vibrations along with other process parameters such as spindle speed, feed rate, drill diameter, chip thickness and surface roughness have been used as input signals to train a standard back propagation neural network (BPNN) with flank wear as output. These input parameters with different combination of inputs have been tried and it has been observed that inclusion of chip thickness and surface roughness leads to a better trained network and hence a better prediction of wear. Principal component of these input parameters has been derived thereafter and has been used to predict the flank wear using BPNN.

Chapter 13 - Milling is used to manufacture a large number of components. Studying the thermomechanics of the milling process by means of purely experimental approach is difficult and time consuming, especially when accounting for various parameters, e.g. tool geometry, material's properties and cutting conditions etc. In the present study fully thermomechanically 2D coupled finite element analysis of slot milling operation is performed. The cutting forces, temperature distribution in the workpiece and in the cutting tool are calculated for selected cutting conditions. For the measurement of cutting forces a strain gauges-based three-component dynamometer was designed and constructed. The cutting forces measured for the case of dry slot milling were compared with results of the finite element analysis. The effect of friction on cutting forces was also studied with two contact conditions.

Chapter 14 - The objective of the present study is to develop a multi-layer feed forward neural network model to predict fractal dimension in CNC milling of brass material. Experiments are carried out using full factorial design and spindle speed, feed rate and depth of cut are considered as cutting parameters. These cutting parameters are used as the input neurons and corresponding fractal dimension is considered as the output of the network. Neural networks are trained with Levenberg-Marquardt algorithm. Minimum mean squared error is considered as the criterion for selecting the best network. It is observed that the network with 3-16-1 architecture gives the best performance. The network predicted and experimental fractal dimension are very close to each other. Mean absolute percentage errors are 0.54% and 1.26% for training and testing pattern respectively.

Chapter 15 - In this study, the effect of cutting parameters on the wire electrode wear in wire electrical discharge machining (WEDM) was investigated experimentally. The experiments were carried out utilizing brass wire of 250 μm diameter on E0300 alloy steel. Experimentally it is found that wire feed rate (WF) is the most significant parameter for wire crater depth (d, μm). Wire tension (WT) is the most significant parameter for both the performance criteria e.g. electrode wear rate (EWR, mm3/min) and surface finish (Ra, μm). Mathematical models for WEDM parameters are developed utilizing the results of various machining performance characteristics. The confirmation test results show 37.52% improvement of surface finish when the machining test was conducted with optimum parametric setting as compared to the initial parametric setting..

Chapter 16 - The paper reports the optimization of process parameters in powder mixed EDM of HSS tool steel using Taguchi based grey relational analysis. Eighteen experimental runs based on orthogonal array were performed. The material removal rate (MRR), relative wear ratio (RWR), and the surface roughness (SR) of the machined specimen were the response variables. By analyzing the grey relational grade, the degree of influence for each controllable process parameters onto individual response variables was estimated. Analysis of variance (ANOVA) of the grey relational grade shows that the pulse current is the most significant parameter influencing PMEDM process according to the weighted sum grade of the MRR, RWR and SR.

Chapter 17 - Aluminum Matrix Composites (AMC) are difficult to machine due to the presence of hard and brittle ceramic reinforcements. Electro Chemical Machining (ECM) is an important process for machining such materials. In this study four factors: electrolyte concentration, electrolyte flow rate, applied voltage and tool feed rate of ECM, were chosen as variables in order to study the process performance in terms of Metal Removal Rate (MRR) and Surface Roughness (Ra).The Response Surface Methodology (RSM) is used to identified the most influential parameters for maximizing the MRR and minimizing the Ra. This paper attempts to establish a comprehensive mathematical model for correlating the interactive and higher-order influences of various machining parameters through RSM. Analysis of variance (ANOVA) has been performed to verify the fit and adequacy of the developed mathematical models.

Chapter 18 - The elastic-plastic deformation region plays an important role at structural materials operating under low load, e.g. in case of fatigue or continuous load. The elastic-plastic deformation is the most inhomogeneous part of the whole deformation process. The plastic deformation begins in a microstructure which could be described as the softest realizable structure with statistically disordered crystal orientation. The extremely wide deformation options of this structure determine the properties of the intermediate deformation region. These are mainly the elastic and plastic anisotropy: the former provides zero to three times bigger possibility for the start of the plastic deformation, at the latter this value is between 0.23 and 0.5. In this chapter we show that – independently from the accepted or not accepted assumptions of the literature – the start of the plastic deformation always begins in only those few crystals, which have the optimum orientation according to the theoretical assumption. Increasing the stress the plastic deformation extends continuously to the wider fraction of volume. The rate of this phenomenon – till it is extended to all of the crystals – is influenced by the direction dependent Young-modulus.

The effects of the anisotropy have to be taken into account in plastic deformation, especially at small deformation rates. In case of fatigue tests, some crystals can take larger part in deformation caused by the cyclic load than other crystals do. Similarly to this, only some of the crystals take part in deformation during creep tests, if the load is smaller than a certain value.

Versions of these chapters were also published in *Journal of Machining and Forming Technologies,* Volume 2, Numbers 1-4, edited by J. Paulo Davim, published by Nova Science Publishers, Inc. They were submitted for appropriate modifications in an effort to encourage wider dissemination of research.

In: Machining and Forming Technologies, Volume 2
Editor: J. Paulo Davim
ISBN: 978-1-61122-820-5

Chapter 1

THERMAL MODELLING OF MATERIAL REMOVAL RATE AND SURFACE ROUGHNESS IN DIE SINKING–ELECTRO CHEMICAL SPARK MACHINING PROCESS

Mohan Charan Panda and Vinod Yadava*
Mechanical Engineering Department,
Motilal Nehru National Institute of Technology,
Allahabad (U.P.) – 211004, India

ABSTRACT

The Die Sinking–Electro Chemical Spark Machining (DS–ECSM) process has been successfully applied for cavity machining in electrically non-conducting materials such as glass, quartz, ceramics and composites using non-rotating tool electrode. In this paper, a thermal finite element model has been developed to estimate the temperature distribution within the heat affected zone of single spark in the workpiece accounting temperature dependent properties of workpiece and same is used for the determination of material removal rate (MRR) and average surface roughness (ASR) in DS–ECSM. The computational results are verified by comparing them with the experimental results available in literature. It has been found that surface finish improves with the increase in spark radius and electrolyte concentration above a critical value (C = 10%). On the other hand MRR improves with the increase of energy partition and duty factor at the cost of deterioration in surface finish.

Keywords: Sinking-Electro Chemical Spark Machining (S-ECSM), Material Removal Rate (MRR), Average Surface Roughness (ASR), Finite Element Method (FEM).

---

* Corresponding Author, Email: vinody@mnnit.ac.in

1. INTRODUCTION

Electrically non-conducting advanced ceramic materials, glass, quartz, composites, etc. are gaining industrial acceptance, mainly due to the way in which technological limitations can be overcome by the use of these non-metallic materials. But manufacturing engineers are facing new challenges to shape and size these electrically non-conducting materials economically and efficiently. Machining of these materials by conventional methods is a serious problem yet to be resolved. Ultrasonic Machining (USM), Abrasive Jet Machining (AJM), Laser Beam Machining (LBM), and Electron Beam Machining (EBM) are some of the advanced machining processes that can be used for machining these materials, but surface quality and dimensional accuracy of the machined surfaces are the major concern. The development of the Electro-Chemical Spark Machining (ECSM) is the outcome of machining requirements of such materials.

ECSM is one of the hybrid machining processes (HMPs), combining the features of ECM and EDM, used for machining non-conducting materials. ECSM has been found to machine features in four different ways such as Die Sinking–ECSM, Hole Drilling–ECSM, Wire Cutting–ECSM and Contour Milling–ECSM. DS–ECSM operation usually involves machining of cavity using non-rotating tool electrode where as in Hole Drilling–ECSM a rotating tool electrode is used with the main focus on surface quality of side wall of the hole. Wire Cutting–ECSM is capable to do slicing large volume and complex shape to non-conducting material without the need of full form tool electrode. In Contour Milling–ECSM, a simple shape tool electrode is used to produce 3-D cavity by adapting a movement strategy similar to conventional milling.

ECSM based on electro-chemical discharge (ECD) phenomena, was presented for the first time in 1968 by Kurafuji as "electrochemical discharge drilling" for micro holes in glass (Kurafuji and Suda, 1968). This process is also referred as Electrochemical Arc Machining (ECAM), Electrochemical Discharge Machining (ECDM), and Spark Assisted Chemical Engraving (SACE) (Wuthrich and Fascio, 2005). Machining with ECD is a complex process influenced by several parameters. Until today it is not yet clear which parameters control mainly the machining. Cook et al. (1973) studied about the influence of several parameters, like electrolyte properties, applied voltage and others on the material removal rate (MRR) during DS–ECSM. Bhattacharyya et al. (1999) studied the machining of aluminium oxide (Al_2O_3) ceramic using DS–ECSM under varying process parametric conditions such as applied voltage and electrolyte concentration. Sarkar et al. (2006) developed a second order, non-linear mathematical model for determination of MRR, radial overcut (ROC), and thickness of heat affected zone (HAZ), as a function of input parameters such as applied voltage, electrolyte concentration and inter-electrode gap, during DS–ECSM operation on silicon nitride. Tandon et al. (1990) carried out experimental study of DS–ECSM on Kevlar-fiber-epoxy and glass-fiber-epoxy composites to find the effect of change in voltage and specific conductance of the electrolyte on the MRR, TWR and ROC. Gautam and Jain (1998) carried out experiments with the different tool kinematics in electro-chemical spark drilling (ECSD) to enhance the process capabilities. They also carried out the parametric study for observing the effect of tool rotational speed, depth of cut and tool eccentricity on MRR and Material Surface Quality (MSQ).

Basak and Ghosh (1997) have developed an analytical model for the estimation of MRR, but the basic theory of spark generation proposed by them does not match with the actual situation. Jain et al. (1999) computed material removal rate during DS–ECSM using Finite Element Method (FEM). In their model random number generation scheme to locate the spark over the workpiece has been used. They assumed the nature of the spark as prismatic column with square cross-section, which is far from real situation. Bhondwe et al. (2006) developed a FEM model for computation of MRR from two types of material, soda lime glass and alumina during DS–ECSM. Parametric study was also performed to study the effect of electrolyte concentration, duty factor and energy partition on MRR.

From the above literature survey, it can be concluded that very little effort has been made towards modeling of DS–ECSM process. Most of the work on this process is experimental and consists of the study of process principle and effects of various process parameters on the MRR. It has also been observed that no theoretical/numerical study has been reported for the prediction of average surface roughness (ASR) during DS–ECSM. Present work is an attempt in this direction.

In this paper, an axisymmetric finite element transient thermal model has been developed to estimate the temperature field due to Gaussian distributed input heat flux of a single spark. The estimated temperature field is further post processed for calculating MRR and ASR during DS–ECSM. Computational experiments have also been performed to determine energy partition and spark radius. Particular attention has been paid to study the effects of energy partition, duty factor, spark radius, and electrolyte concentration on MRR and ASR during DS–ECSM of silicon nitride workpiece. The present developed model has few similarities with the model developed by Bhondwe et al. (2006) but entirely different approach. The difference of the present model with the previous model of Bhondwe et al. (2006) is that they developed the model for calculation of MRR only, considering critical current and critical voltage for calculation of input heat flux, whereas the present model is developed for the calculation of MRR and ASR both, considering supply voltage and circuit current for calculation of input heat flux as well as temperature dependent material properties of the workpiece.

2. Modeling for Temperature Field in Heat Affected Zone of a Single Spark

The configuration of a DS–ECSM setup is shown in Figure 1. In ECSM the hydrogen gas bubbles are formed at the tip of the tool electrode. The coalescence of bubbles forms a gas film around the tip of the tool electrode when the supply voltage reaches a critical value. This gas film isolates the tool electrode from the electrolyte. Electrical discharges in the form of sparks take place in this gas film and machining of material is possible if the workpiece material is kept in the vicinity of the sparking zone. The material removal in DS–ECSM is mainly caused by the melting and vaporization due to heat generated by the spark. It is considered as a thermal phenomenon (Wuthrich and Fascio, 2005). For the modeling of temperature field in the heat affected zone of a single spark, following simplifying assumptions are made:

(i) Workpiece material is homogeneous and isotropic.

(ii) Only a fraction of total spark energy of a single spark is dissipated as heat into the workpiece. Remaining heat is assumed to be distributed between the tool and the electrolyte.

(iii) At a time only one spark is produced at the workpiece top surface (single Spark phenomenon) (Kulkarni et al., 2002) and the duration of spark is same for all sparks. The zone of influence due to single spark is assumed to be axisymmetric in nature $\left(\dfrac{\partial T}{\partial \theta}=0\right)$.

(iv) The boundaries (B_2 and B_3) of the domain (Figure 2), away from the heat flux are chosen at such a distance that there is no heat transfer across those boundaries i.e. $\dfrac{\partial T}{\partial n}=0$.

(v) Shape of heat flux is assumed to be *Gaussian* distributed. From the experimental studies of Kulkarni et al. (2002) for single spark, the crater shape was observed as dome shape (part of sphere). So reflecting the shape of crater, the nature of the heat flux can safely be approximated as *Gaussian*.

(vi) Ejection efficiency is assumed to be 100%. Also, there is no deposition of recast layer on the machined surface.

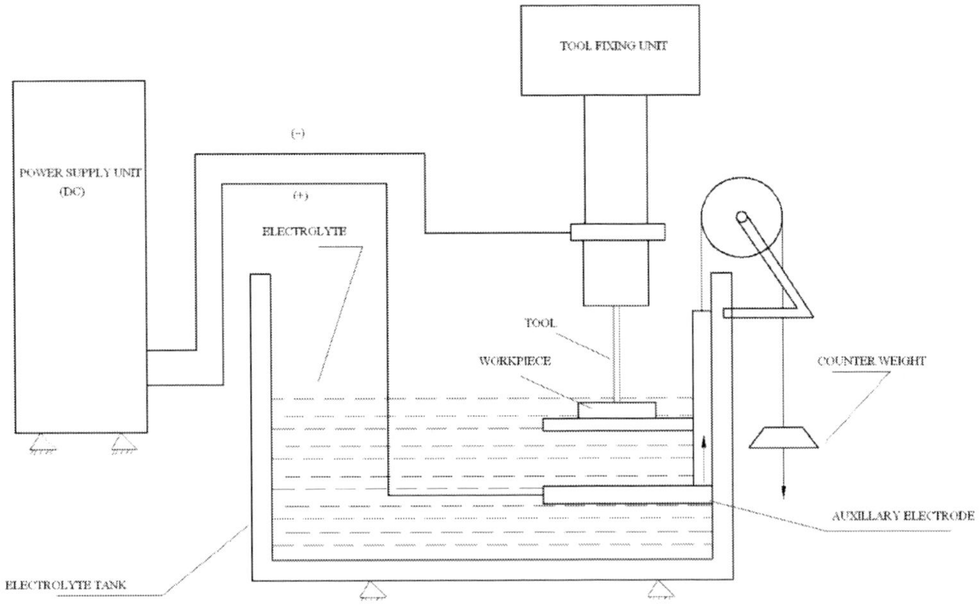

Figure 1. Configuration of DS–ECSM set-up.

2.1. Governing Equation, Initial and Boundary Conditions

The governing heat conduction equation within the workpiece domain (Figure 2) is given by:

$$\rho C_p \frac{\partial T}{\partial t} = \frac{1}{r}\frac{\partial}{\partial r}\left(k\,r\frac{\partial T}{\partial r}\right) + \frac{\partial}{\partial z}\left(k\frac{\partial T}{\partial z}\right) \tag{1}$$

in domain ABCD (Figure 2), where, T is temperature, ρ is density, C_p is specific heat, and k is thermal conductivity of the workpiece material. In the above equation thermal conductivity and specific heat of workpiece material are taken to be temperature dependent.

The initial condition for temperature of the entire domain at the start of machining is considered to be room temperature, T_0 .

When $t = 0$, $T(r,z) = T_0$ in domain ABCD (Figure 2) (2)

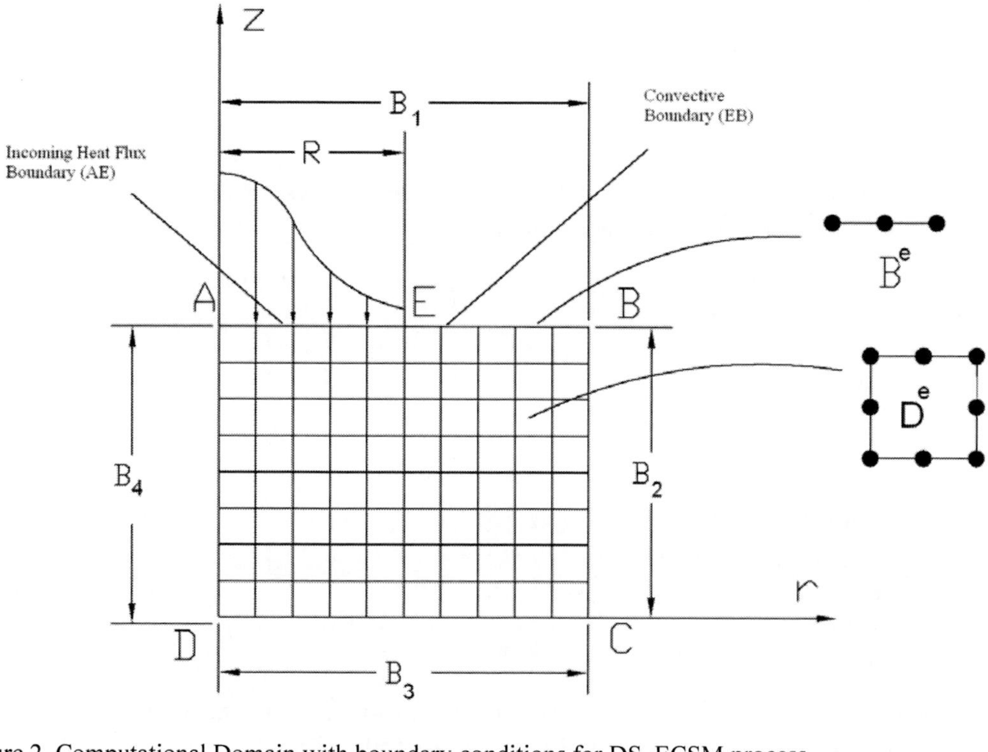

Figure 2. Computational Domain with boundary conditions for DS–ECSM process.

The boundary conditions of the heat affected zone (Figure 2) owing to single spark are as follows. On the top surface B_1, heat flux is considered only upto a radius equal to that of the spark radius (R). For the rest part of B_1, convective heat loss to electrolyte is considered. Further, the boundaries B_2 and B_3 are considered at such a large distance so that there is no heat transfer takes place across them. The boundary B_4 is an axis of symmetry. Thus the boundary conditions can be mathematically written as:

When t >0

$$-k\left(\frac{\partial T}{\partial z}\right)=\begin{cases} h(T-T_0) & if\ r>R \\ q_w(r) & if\ r\leq R \\ 0 & for\ off-time \end{cases} \qquad on\ B_1 \tag{3}$$

$$\frac{\partial T}{\partial n}=0 \quad on\ B_2, B_3, B_4 \tag{4}$$

Here, $q_w(r)$ is the quantity of heat flux owing to the spark, h is the convective heat transfer coefficient, T_0 is the room temperature of electrolyte and n is outward normal to the boundary.

In the present work a Gaussian heat flux distribution is assumed and the expression for heat flux at any radius r, $q_w(r)$ to the workpiece is derived as:

$$q_w(r)=\frac{4.45F_wVI}{\pi R^2}\exp\left\{-4.5\left(\frac{r}{R}\right)^2\right\} \tag{5}$$

where, R is spark radius, r is the radial distance from the axis of the spark, V is supply voltage, I is circuit current and F_W is energy partition.

The model proposed by Jain et al. (1999) for calculation of circuit current during DS–ECSM, is used in the present work:

$$I=0.1009V^{0.4815}k^{0.3420}\phi^{0.3420}d^{0.2881} \tag{6}$$

where, k is electrolyte conductivity, ϕ is tool diameter and d is depth of the tool inside electrolyte. Dependence of the electrolyte conductivity on the concentration (C) of the NaOH electrolyte is given by (Jain et al., 1999):

$$k=0.885958+4.62384C-0.193274C^2-0.00208907C^3 \tag{7}$$

The fraction of the generated heat entering the workpiece (F_W) depends on the thermal properties of workpiece, tool electrode material, and electrolyte used. No comprehensive method has so far been proposed to calculate the value of F_W during DS–ECSM. Earlier studies (Jain et al., 1999; Bhondwe et al., 2006) have used this value as 20%. Therefore in the present study F_W is taken as 20%.

For computational analysis of DS–ECSM process another important parameter required is the spark radius. However, quantification/measurement of the spark diameter is extremely difficult due to very high pulse frequencies. Basak and Ghosh (1997) have taken spark radius equal to 0.5×10^{-6} times circuit opening current in ampere for spark channel cylindrical in shape. Jain et al. (1999) assumed prismatic nature of spark with square cross section which is far from real life situation. Kulkarni et al. (2002) have given the crater diameter for different workpiece materials as 300μm based on their experiments. Therefore, in the present work, the spark radius is taken as 150μm.

In general, the material thermal properties such as specific heat capacity and thermal conductivity are temperature dependent. Earlier thermal models available in literature for DS–ECSM have assumed temperature independent thermal properties of the workpiece. In the present work, the temperature dependent thermal properties of silicon nitride workpiece (Figure 3) are taken into account. The best fit curves are obtained to find the nature of variation in polynomial equation form, which are used in the numerical calculations. The polynomial equations are given below:

$$\begin{aligned} k &= 37.1558 - 0.0314T + 1.26966\times10^{-5}T^2 - 9.242\times10^{-10}T^3 - 5.87147\times10^{-13}T^4 \\ C_p &= 413.21721 + 1.45487T - 0.00121T^2 + 5.29355\times10^{-7}T^3 - 9.30496\times10^{-11}T^4 \end{aligned} \quad (8)$$

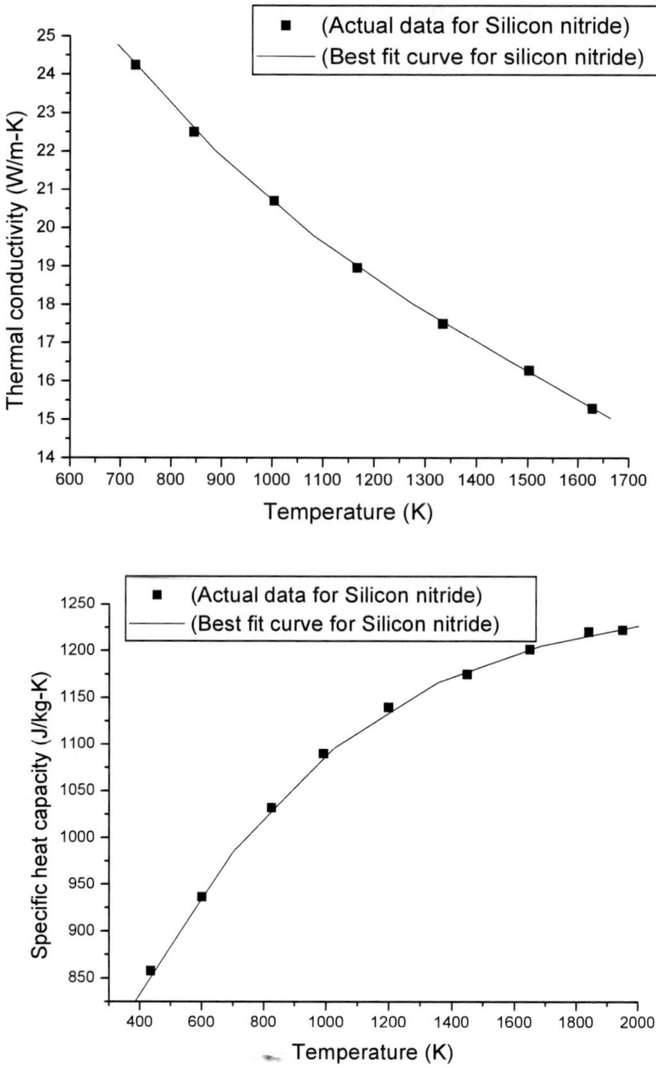

(a)

(b)

Figure 3. Variation of (a) thermal conductivity (b) specific heat with temperature.

2.2. Finite Element Formulation

In the present analysis the Galerkin weighted residual method (Reddy, 2005; Rao, 2005) has been applied to obtain the temperature distribution within the computational domain (Figure 2) due to heat flux of a single spark. The following elemental equations are obtained when Galerkin approach is applied to DS–ECSM process [Eqs.1-7].

$$[C]^e \left\{ \overset{\bullet}{T} \right\}^e + [K]^e \{T\}^e + [K]^b \{T\}^b = \{f_c\}^b + \{f_q\}^b \tag{9}$$

where,

$$\left.\begin{aligned} [C]^e &= \int_{D^e} \rho C_p \{N\}^e \{N\}^{eT} \, dD^e \\ [K]^e &= \int_{D^e} k [B]^{eT} [B]^e \, dD^e \\ [K]^b &= \int_{B_h} h \{N\}^b \{N\}^{bT} \, dB_h \\ \{f_c\}^b &= \int_{B_h} T_0 h \{N\}^b \, dB_h \\ \{f_q\}^b &= \int_{B_q} \{N\}^b q_w \, dB_q \end{aligned}\right\} \tag{10}$$

Here, $[C]^e$ is the elemental capacitance matrix, $[K]^e$ is the elemental conductivity matrix, $[K]^b$ is the boundary elemental conductivity matrix, $\{f_c\}^b$ is the boundary element convection vector, $\{f_q\}^b$ is the boundary element heat flux vector, $\{N\}^e$ is the interpolation functions vector for a typical area element, $\{N\}^b$ is the interpolation functions vector for a boundary element, $[B]^e$ is the matrix of derivatives of nodal interpolation functions for typical area element, B_h is the convective boundary and B_q is boundary of input heat flux. The Gauss quadrature technique is used to evaluate elemental matrices and vectors. When elemental quantities of Equation (10) are assembled, the following differential equations are obtained.

$$[GC] \left\{ \overset{\bullet}{T} \right\} + [GK] \{T\} = \{GF\} \tag{11}$$

where, $[GC]$ is the global capacitance matrix, $[GK]$ is the global stiffness matrix, $\{GF\}$ is the global heat flux vector, $\{T\}$ is global temperature vector and $\left\{ \overset{\bullet}{T} \right\}$ is time derivative of $\{T\}$.

Equation (11) is converted into algebraic equations after application of implicit Finite Difference Method (FDM) as follows.

$$[A]_{i+1}\{T\}_{i+1}=\{B\}_{i,i+1} \qquad i=0,1,2,3............n \tag{12}$$

where,

n is the number of time steps and $\{T\}_{i+1}$ is the temperature at $i+1^{th}$ step.

$$[A]_{i+1}=[GC]+\alpha\,\Delta t_{i+1}[GK]$$

$$\{B\}_{i,i+1}=\Delta t_{i+1}\left(\alpha\{GF\}_{i+1}+(1-\alpha)\{GF\}_{i}\right)+\left([GC]-(1-\alpha)\,\Delta t_{i+1}[GK]\right)\{T\}_i \tag{13}$$

where, Δt_{i+1} is the time interval of step ($i+1$). In the present work, value of α is chosen to be $\frac{2}{3}$ for Galerkin scheme. In each time interval Δt, the nodal temperatures at the new time level are calculated by using the set of simultaneous algebraic equations given by Equation (12). The solution thus marches in time, in steps of Δt until the desired final time is reached. In the present model, Δt is divided into two time steps Δt_1 and Δt_2. Here, Δt_1 and Δt_2 are pulse on-time and off-time of the spark respectively.

3. Modelling for MRR and ASR

3.1. Material Removal Rate (MRR)

Formulation of section 2.2 is used for the development of an FEM based code using Matlab 7.0.1. The developed code calculates the temperature field in the workpiece due to Gaussian distributed input heat flux of a single spark. The material removed by a single spark can be calculated by assuming that the material having temperature more than the melting temperature is completely removed. Hence, melting isotherm (Figure 4) is plotted using the result of temperature distribution in the zone of influence of single spark under the applied process conditions. The volume of the material removed by single spark is calculated as follows:

$$Vol = \int_{z_1}^{z_2} \pi r^2\, dz \tag{14}$$

The volume obtained by Equation (14) is for single spark. The volume of material removed per unit time is calculated by:

$$MRR = Vol \times Sparking\ frequency = Vol \times \frac{1}{t_p} \tag{15}$$

where, t_p is pulse time.

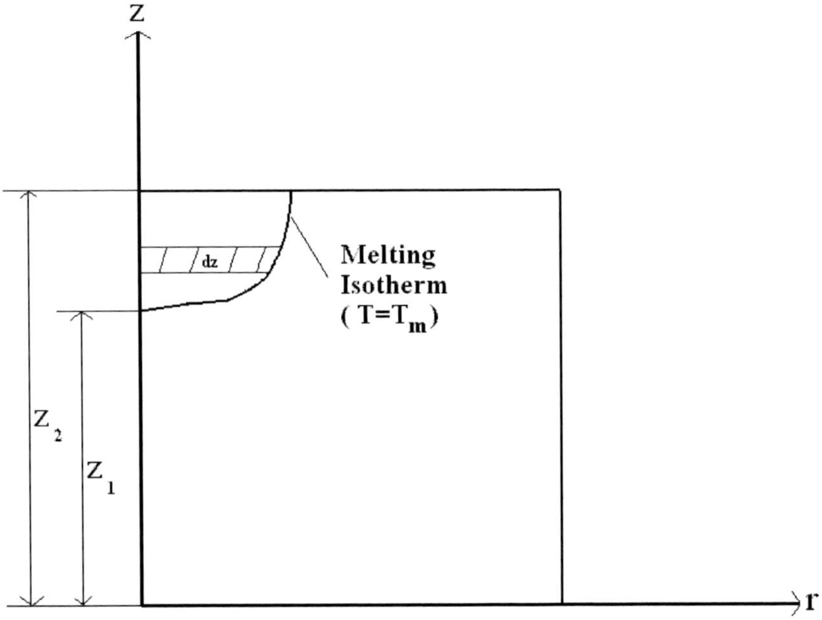

Figure 4. Crater above melting isotherm.

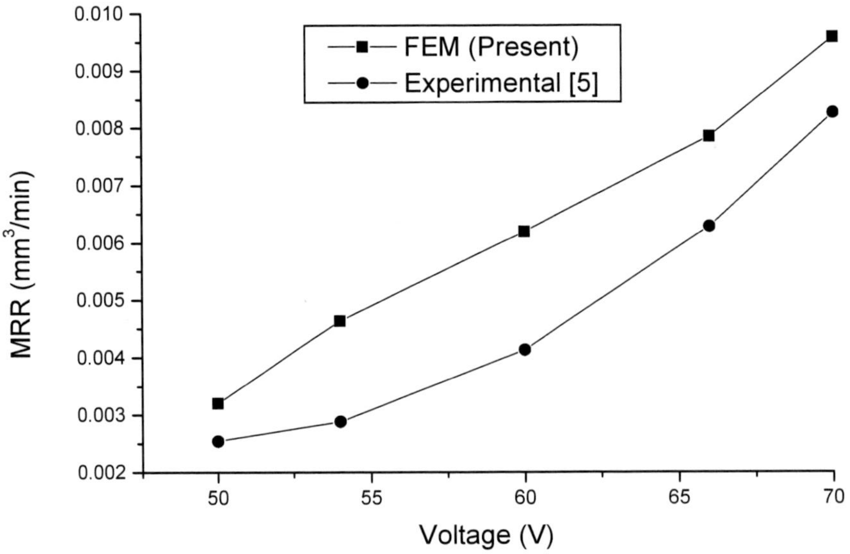

Figure 5. Variation of MRR with supply voltage.

3.2. Average Surface Roughness (ASR)

The present FE model developed can also be used to calculate the ASR. To calculate ASR, the crater formed due to single spark can be assumed to have circular paraboloid geometry (Figure 6) that is described by the following equation:

$$\frac{x^2+y^2}{r_c^2}=\frac{d_c-z}{d_c} \quad (16)$$

where r_c and d_c are the crater's radius and depth respectively. The maximum surface roughness can be assumed as being equal to the depth of the crater formed during an individual spark; however this is not the case for the ASR. The finished workpiece surface is the outcome of a series of overlapping craters as shown in figure 7a. It is therefore obvious that the maximum deviation of the workpiece surface will be equal to depth d_a, which thus defines the average surface roughness R_a.

The basic assumption for estimating R_a is that besides the formed crater C_1, another crater C_2 can be formed on condition that the occurrence of spark discharge can only be on the surface of the workpiece and not on the flanks of the crater. Therefore, the axis of the heat source radius of the second crater (C_2) will be at a distance equal to (r_c+R) from the axis of the first crater (Figure 7b). Based on Eq. (16), the equations of both craters can be derived from:

$$C_1:\ \frac{y^2}{r_c^2}=\frac{(d_c-z)}{d_c}$$
$$C_2:\ \frac{\left(y-(r_c+R)\right)^2}{r_c^2}=\frac{(d_c-z)}{d_c} \quad (17)$$

Setting parameters $y=y_p$ and $z=z_p$ in Eq. (17), the linear system can be solved for estimating depth d_a:

$$d_a=d_c-z_p=\frac{1}{4}\left[\frac{r_c+R}{r_c}\right]^2\times d_c \quad (18)$$

Therefore the maximum surface roughness and the ASR are given by equations (19) and (20) respectively:

$$R_{\max}=d_c \quad (19)$$

$$R_a=\frac{1}{4}\left[\frac{r_c+R}{r_c}\right]^2\times d_c \quad (20)$$

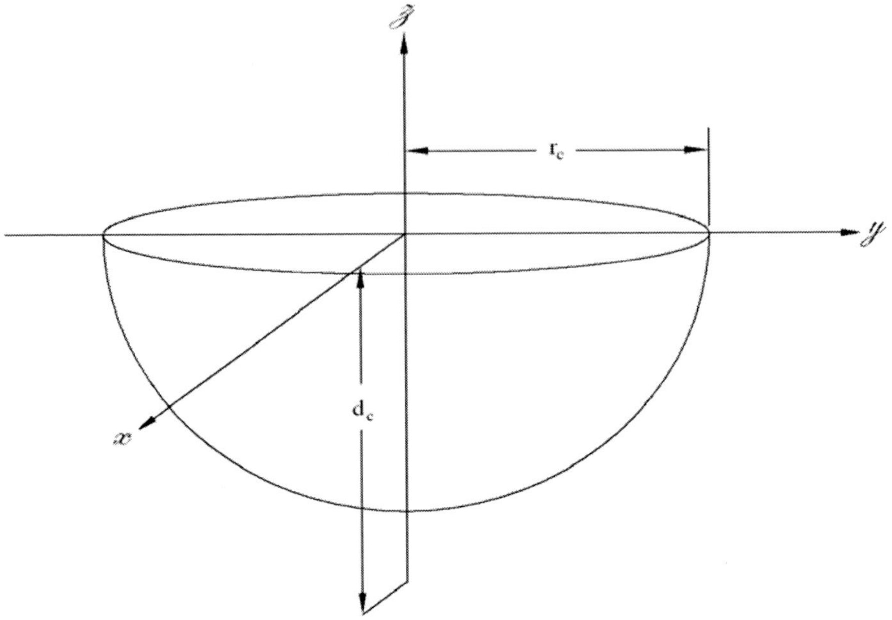

Figure 6. Crater geometry.

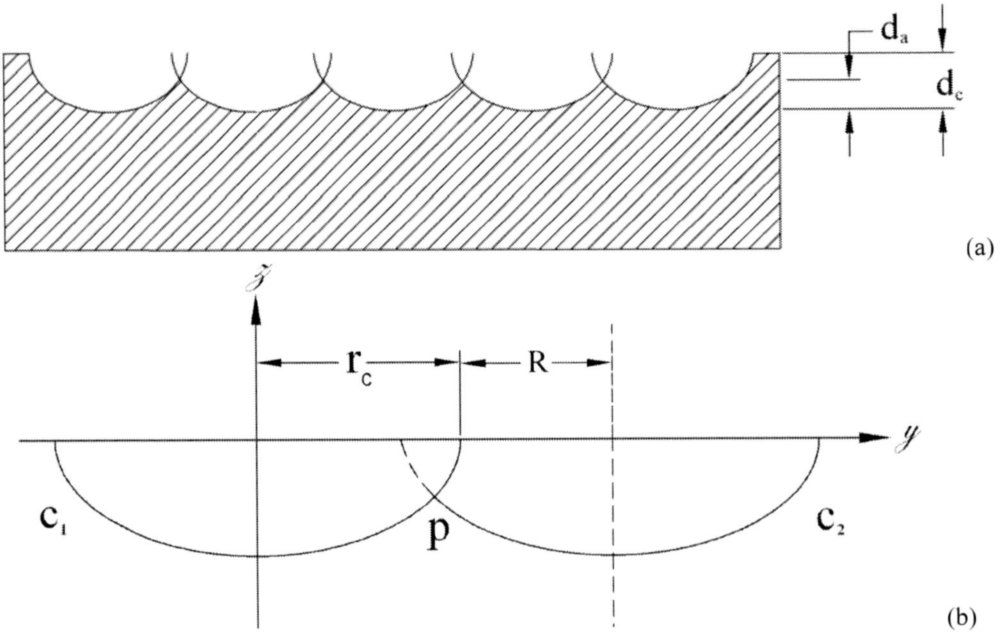

Figure 7. Schematic representation of (a) series of overlapping craters (b) two overlapping craters.

The effect of the supply voltage on the ASR is depicted in figure 8. Here it is observed that ASR value increases with increase in supply voltage, because at higher supply voltage higher energy discharge takes place during sparking, this causes deeper crater on the machined surface. Hence, ASR increases with increase in supply voltage. In the present case 7.4% increase in ASR is observed by increasing supply voltage from 50V to 70V.

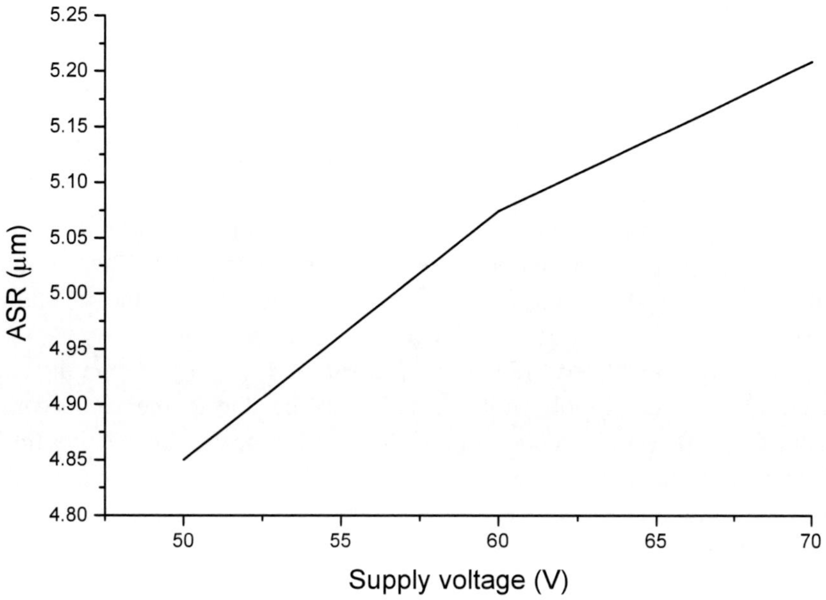

Figure 8. Effect of supply voltage on ASR.

4. Results and Discussion

4.1. Comparison of MRR

Sarkar et al. (2006) have experimentally measured MRR from silicon nitride workpiece of size 20 mm × 20mm ×5mm during DS–ECSM. In their experiment a stainless steel tool electrode with a diameter of 400 μm was used as cathode. NaOH electrolyte with 10 to 30 wt. % electrolyte concentration was used. The workpiece was clamped just below the tool tip, and the tool tip was immersed 2-3 mm below the upper level of the electrolyte. Experiments were carried out at five different voltage levels: 50 V, 54 V, 60 V, 66 V, and 70 V using pulsed DC power supply.

The above-mentioned problem is solved using present model for determination of MRR from silicon nitride workpiece during DS–ECSM. The wokpiece material and machining conditions used are taken same as used in the literature (Sarkar et al., 2006). The material properties of silicon nitride are taken from Guyer and Brownell (1989). Because of computational limitations of our computing system the domain size has been changed from 20 mm × 20mm ×5mm to 600 μm × 600 μm × 600 μm and since, the 3-D problem is approximated to a 2-D problem the workpiece domain is discretized into eight noded quadrilateral elements (Figure 2). Convergence test was carried out by increasing the number of elements in the mesh. The simulation showed that the nodal temperature of workpiece domain obtained were essentially unchanged, when the mesh size is in excess of 1024 elements. The mesh of 1024 elements is thus found to be adequate for convergence. Hence the mesh consisting of 1024 number of elements with total 3201 nodes are used for further calculations. The nodal temperature in the workpiece domain is found using computer with Pentium 4 processor.

MRR is calculated for different supply voltage. MRR calculated using present FEM based model is compared with experimental values obtained by Sarkar et al. (2006). The results obtained using present model show (Figure 5) similar pattern as given in literature (Sarkar et al., 2006). MRR calculated using present FEM based model increases with increase in supply voltage. Calculated MRR at supply voltage of 50V using present FEM based model is found as 0.0032mm^3/min whereas the experimental value of MRR for the same input condition is 0.0025mm^3/min. Further, when the supply voltage increases to 60V the MRR calculated using present FEM based model is found as 0.0062mm^3/min, whereas experimental value is 0.0041mm^3/min. At the supply voltage of 70 V the calculated MRR is found 0.0096mm^3/min against experimental MRR of 0.0083mm^3/min.

Figure 5 shows that calculated MRR using present model is higher than the experimental MRR for each value of the supply voltage. This may be due to the value considered for ejection efficiency in present model during DS–ECSM process. In the present model the ejection efficiency is assumed to be 100% (it is assumed that all the material, which is melted, is removed). In real situation, some part of the molten material is not completely removed but it adheres back (resolidify) to the parent material because of the quenching effect caused by liquid electrolyte. This plays a dominant role, as the ejection efficiency in DS–ECSM is very low. It may be as low as 10% or even lower than that (Jain et al., 1999). Further there is no exact data available regarding energy partition and spark radius for the combination of silicon nitride workpiece, NaOH electrolyte and stainless steel tool electrode. Above described reasons are responsible for getting different values of computational and experimental results.

4.2. Computational Experiments

There is no exact data available regarding the value of energy partition (F_W) and spark radius (R) in DS–ECSM process for a given tool electrode, workpiece, and electrolyte combination. Hence, computational experiments have been performed to determine the values of energy partition and spark radius. Silicon nitride is taken as workpiece material and present model discussed in section 2 is used for the computational experiments. The cutting conditions are given in Table 1 and material properties of workpiece are given in Table 2.

Table 1. Machining Conditions

Electrolyte	NaOH
h (W/m^2-K)	24870
C	6 % to 14 %
Ø (μm)	400
d (mm)	2
R (μm)	150
F_W	20%
t_{on} (μs)	1000
T_0 (K)	298
Supply voltage (V)	50V to 70V

Table 2. Material Properties of Silicon nitride (Si_3N_4)

C_p (J/kg-K)	752
k (W/m^2 K)	29
T_m (^{0}C)	1900
ρ (kg/m^3)	3260

Energy Partition (F_w)

Energy partition is the fraction of the total heat energy going to workpiece. First a computational experiment was performed for finding MRR for different values of F_w. For a given value of supply voltage, taking different values of energy partition (F_w) and keeping all other parameters in equation (4) constant, $q_w(r)$ is calculated. Taking this value of $q_w(r)$, MRR is computed by the procedure as discussed in section 3.1. Best fit curve is plotted to find the nature of variation of MRR with F_w in polynomial equation form. Putting the known value of MRR in the polynomial equation F_w is computed. Similar computational experiments were performed for different values of supply voltage.

Figure 9 shows the best fit curves of MRR vs. F_w for different values of supply voltage, which indicates that MRR increases with increase in F_w for a particular supply voltage.

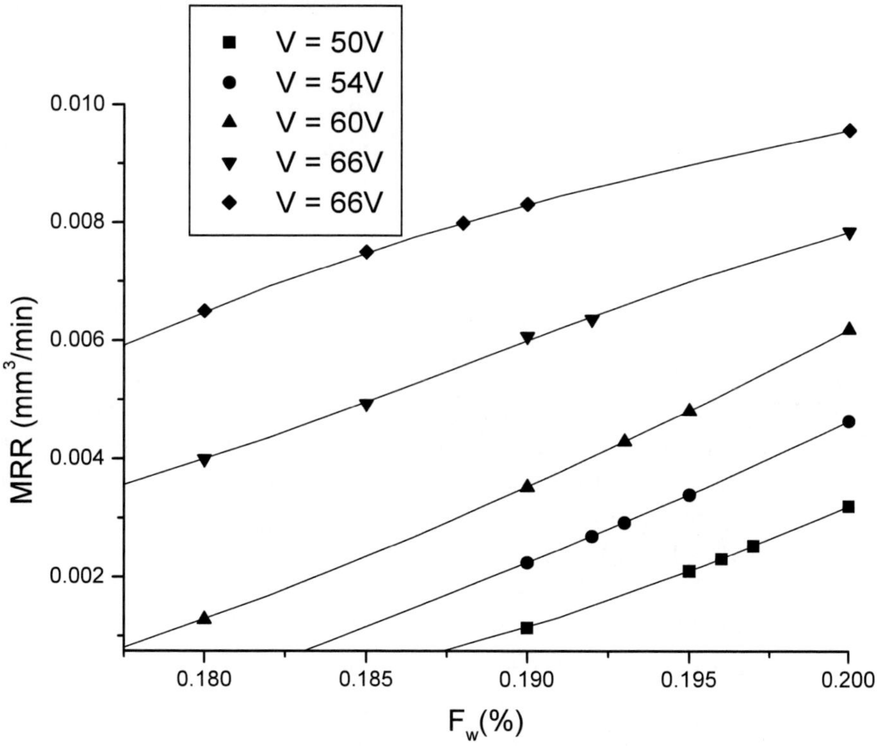

Figure 9. Variation of MRR with F_w for different values of supply voltage.

The polynomial equation of best fit curve for a supply voltage of 60V and 10 % electrolyte concentration is obtained as:

$$F_w = 0.1732 + 5.69292(MRR) - 317.00545(MRR)^2 + 15605.90636(MRR)^3 \tag{21}$$

Since the experimental value of MRR is 0.00413mm^3/min for a supply voltage of 60V and 10% electrolyte concentration (Sarkar et al., 2006), putting this value of MRR in Equation (21), it is found that F_w is 19.24%. Similarly, F_w is computed for different values of supply voltage and shown in Table 3. From Table 3 it is observed that as the supply voltage increases, F_w is decreasing.

Table 3. Computational Energy partition for different values of supply voltage

Supply Voltage (V)	Ejection Efficiency (%)	Spark Radius (μm)	Experimental MRR (mm^3/min)	Calculated Energy Partition (%)
50	100	150	0.00254	19.71
54	100	150	0.00288	19.28
60	100	150	0.00413	19.24
66	100	150	0.00628	19.13
70	100	150	0.00826	18.97

Table 4. Computational Spark radius for different values of supply voltage

SupplyVoltage (V)	Ejection Efficiency (%)	Energy Partition (%)	Experimental MRR (mm^3/min)	Calculated Spark Radius (μm)
50	100	20	0.00254	151.69
54	100	20	0.00288	154.30
60	100	20	0.00413	154.69
66	100	20	0.00628	154.84
70	100	20	0.00826	155.17

Spark Radius (R)

In case of DS–ECSM, material is mainly removed by thermal phenomena. So spark radius is an important parameter in this machining process. But no literature is available regarding the exact value of spark radius for DS–ECSM for given combination of tool, workpiece, and electrolyte. In this work computational experiments have been performed to determine the spark radius for different values of supply voltage.

First a computational experiment was performed for finding MRR for different values of R. For a particular supply voltage, taking different values of spark radius (R) and keeping all other parameters in equation (4) constant, $q_w(r)$ is calculated. Taking this value of $q_w(r)$, MRR is computed by the procedure as discussed in section 3.1. Best fit curve is plotted to find the nature of variation of MRR with R in polynomial equation form. Putting the known value of MRR in the polynomial equation R is computed. Similar computational experiments were performed for different values of supply voltage. Figure 10 shows the best fit curves of MRR vs. R for different values of supply voltage, which indicates that MRR decreases with an increase in R for a particular supply voltage. The equation of best fit curve for a supply voltage of 60V and 10% electrolyte concentration is obtained as:

$$R = 163.61702 - 2103.29664(MRR) - 14863.56274(MRR)^2 \tag{22}$$

Since the experimental value of MRR is 0.00413mm^3/min for a supply voltage of 60V and 10% electrolyte concentration (Sarkar et al., 2006) putting this value of MRR in Equation (22), it is found that R is 154.68μm. Similarly, for different values of supply voltage R is computed and shown in Table 4. From Table 4 it is observed that as the supply voltage increases the R is also increasing.

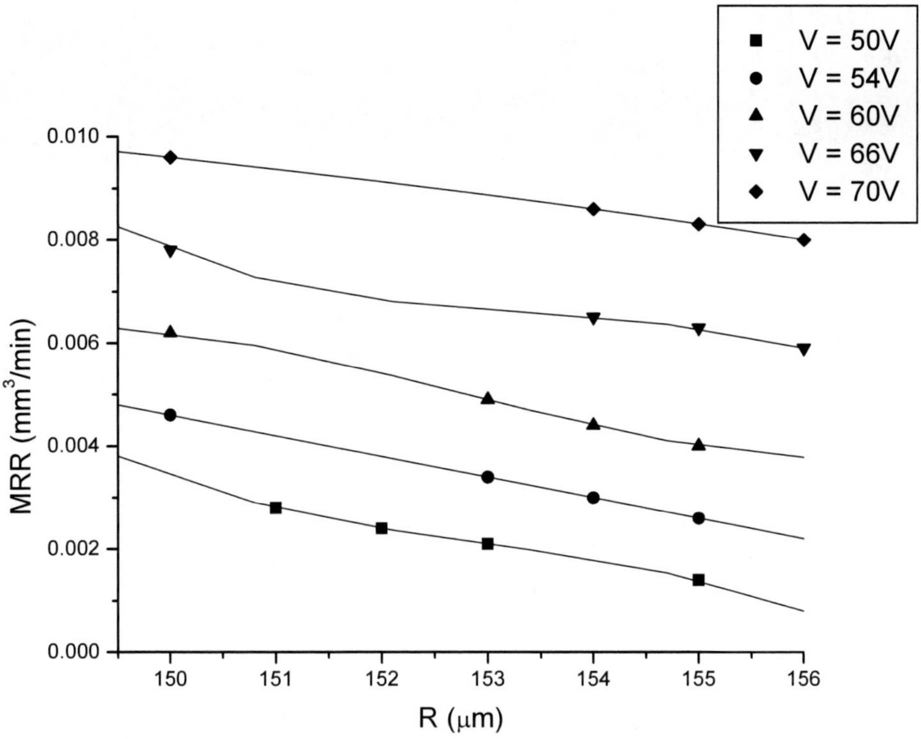

Figure 10. Variation of MRR with R for different values of supply voltage.

4.3. Parametric Studies

The influences of process parameters such as energy partition, duty factor, spark radius and electrolyte concentration on MRR and ASR during DS–ECSM have been studied. Silicon nitride is taken as workpiece material. The machining conditions are given in Table 1 and material properties of workpiece are given in Table 2. Governing equation, Initial and Boundary conditions as discussed in section 2.1 are taken same through out the parametric studies.

Effect of Energy Partition

Figure 11(a) shows the effect of energy partition (F_w) on MRR during DS–ECSM. Here it is observed that MRR increases with the increase in energy partition. 606% increase in MRR is observed by increasing the energy partition by 10%. Increase in the energy partition means the amount of heat going to the workpiece is more, which is responsible for larger volume of material melting and hence high MRR.

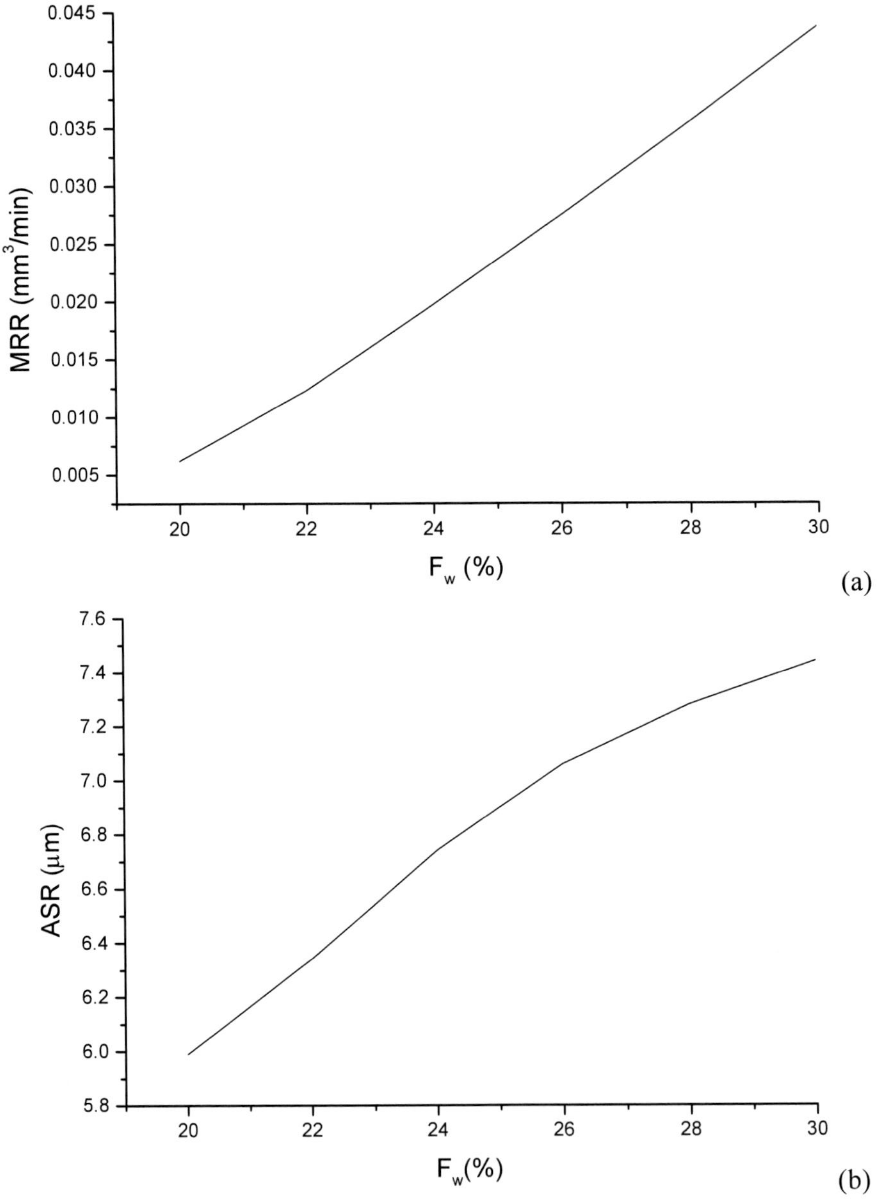

Figure 11. Effect of energy partition due to DS–ECSM of Silicon nitride workpiece with V =60V, C = 10 %, R=150 μm (a) on MRR (b) on ASR.

Effect of energy partition on ASR is shown in figure 11(b). ASR increases with increase in energy partition. A 24% increase in ASR is observed by increasing energy partition by 10%. Higher value of energy partition means more amount of heat energy penetrating into the workpiece, which results in larger material removal in depth direction causing deeper crater in machined surface and hence ASR increases.

Effect of Duty Factor

Duty factor is the ratio of spark on-time and pulse duration (on-time + off-time). Duty factor has been taken by considering off-time 500μs and varying the on-time. Figure 12(a) shows the effect of duty factor on MRR in DS–ECSM.

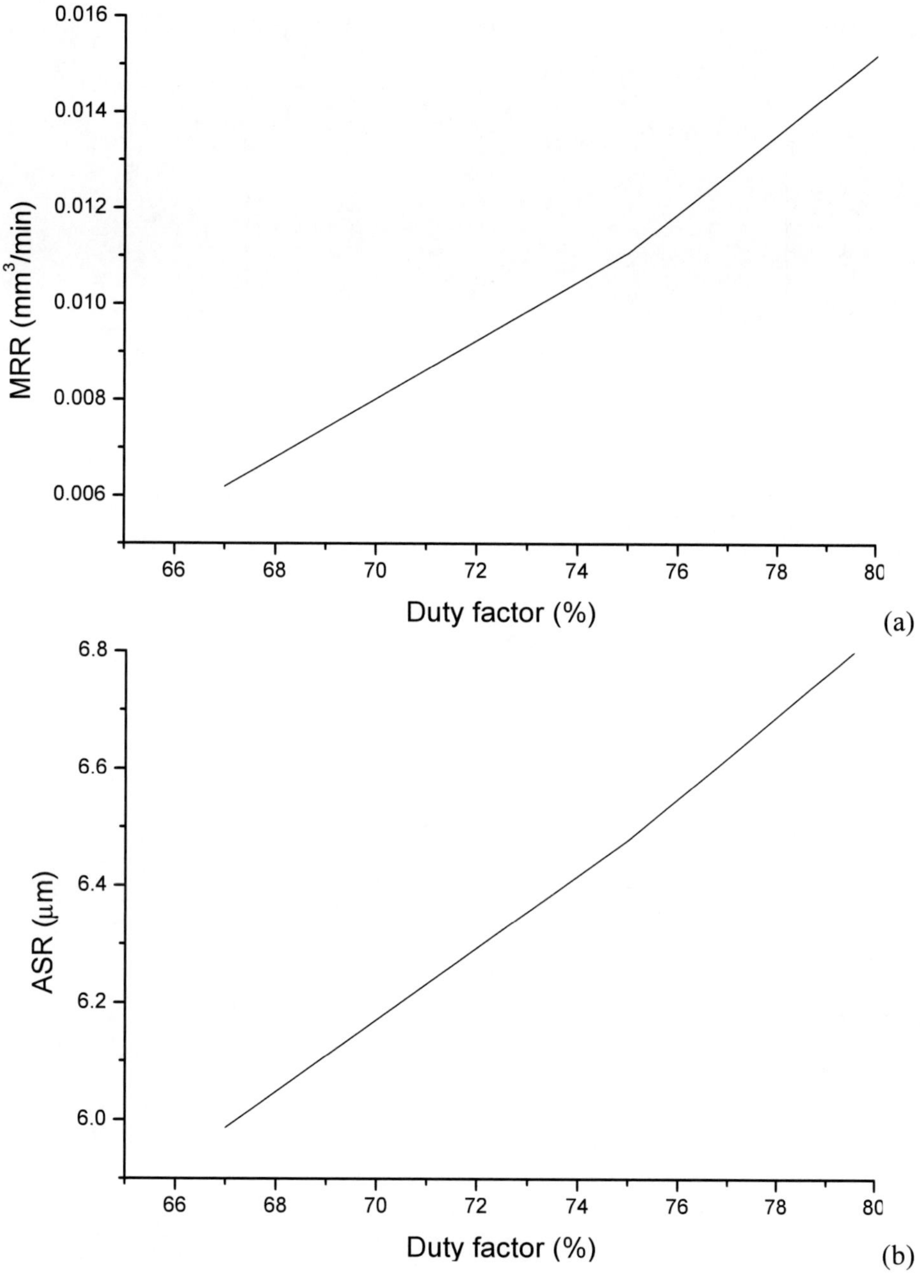

Figure 12. Effect of duty factor due to DS–ECSM of Silicon nitride workpiece with V = 60V, C = 10 %, F_w = 20% (a) on MRR (b) on ASR.

Here it is observed that MRR increases with the increase in duty factor. 145% increase in MRR is observed by increasing the duty factor by 13%. With increase in duty factor, the spark energy will be transferred to the workpiece for more time because of which volume of workpiece material melted is also more. Hence, the MRR is increased with increase in duty factor. Figure 12(b) shows the effect of duty factor on ASR. Increase in duty factor increases ASR. This is because of the fact that due to increase in duty factor the presence of the sparks prolonged and hence an increase in energy per spark on the workpiece surface. Because of which d_c increases which results in ASR increase. In the present case it is found that by increasing the duty factor by 13% ASR increases by 14%.

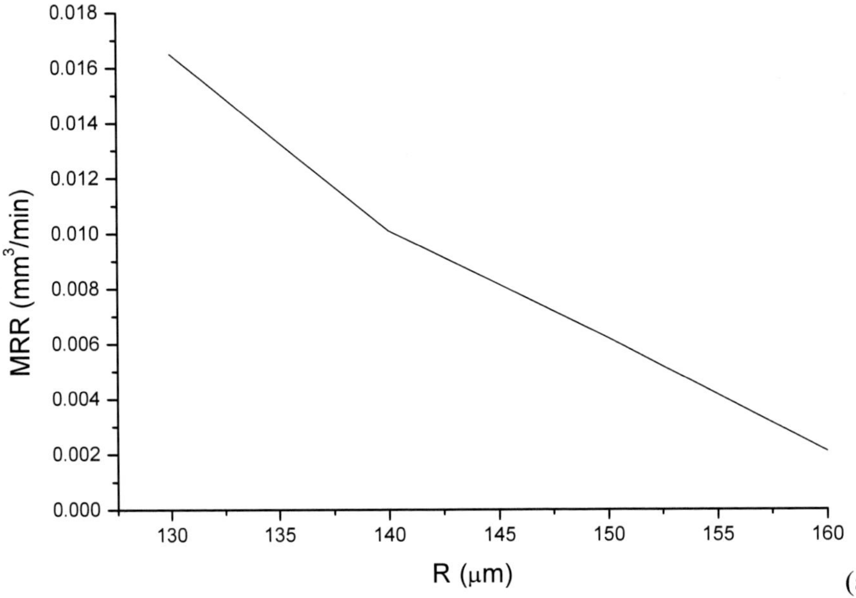

(a)

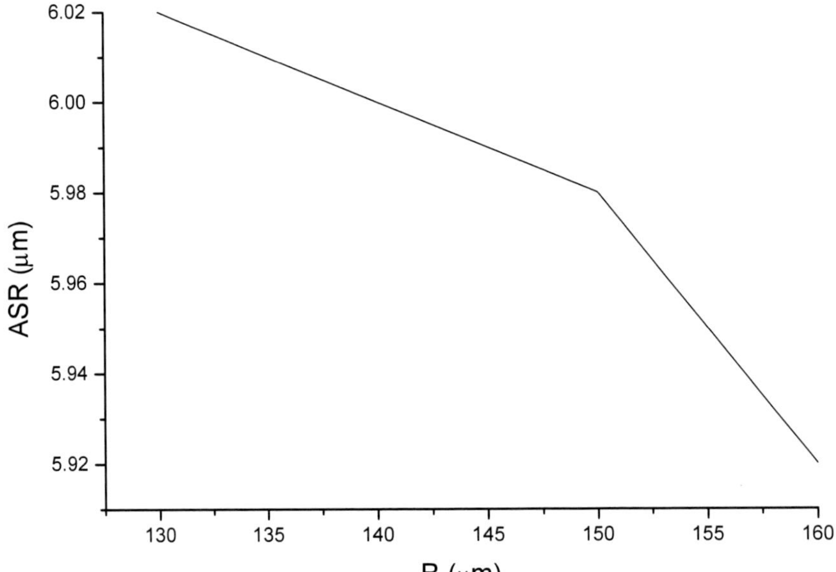

(b)

Figure 13. Effect of spark radius due to DS–ECSM of Silicon nitride workpiece with V = 60V, C = 10 %, F_w = 20% (a) on MRR (b) on ASR.

Effect of Spark Radius

The effect of spark radius on MRR is shown in figure 13(a). Here it is observed that MRR is decreasing from $0.0165mm^3/min$ to $0.0021mm^3/min$ if spark radius is changed from 130μm to 160μm. This shows a decreasing trend of MRR with respect to increase in spark radius. This is because, heat flux decreases with increase in spark radius. Due to lower heat flux less volume of material is melted and hence MRR decreases with increase in spark radius.

Effect of spark radius on ASR is shown in figure 13(b). Average surface roughness decreases with increase in spark radius. Here it is observed that 2% improvement in surface finish is achieved by 23% increase in spark radius. At lower value of spark radius more amount of heat energy penetrating into the workpiece, which results in larger value of depth of crater and lower value of crater radius. Since d_c is mainly responsible for deciding the ASR and hence surface roughness deteriorate.

Effect of Electrolyte Concentration

The effect of electrolyte concentration on MRR is shown in Figure 14(a). The MRR goes on increasing from 6% electrolyte concentration to 10% electrolyte concentration and then starts decreasing. This is due to the fact that the specific conductance of NaOH solution increases up to 10%, beyond which it starts decreasing (Jain et al., 1999). An increase in specific conductance means increased electrolyte conductivity and consequently more circuit current. An increase in circuit current would mean the accelerated electrolysis process. It would result in greater rate of evolution of hydrogen bubbles at the cathode. The increased rate of formation of gas bubbles at the cathode implies an enhanced rate of sparking and hence higher MRR. When the concentration is increased beyond 10%, rate of electrolysis decreases. Due to increased rate of reactions, mobility of the ions is decreased and in turn the specific conductance of the electrolyte is decreased. Therefore, beyond a certain value of electrolyte concentration (C=10%), change in the circuit current decreases. Since, the heat energy developed from the spark is proportional to the circuit current, the material removed will be less at higher values of concentration.

ASR also depends on electrolyte concentration. Figure14(b) shows the effect of electrolyte concentration on average surface roughness. The average surface roughness is less for 6% and 14% concentration as compared to 10% concentration. The reason for this variation is that as explained above the circuit current is maximum at 10% concentration of NaOH electrolyte. It is seen from equation (4) that heat energy developed from the spark is proportional to the circuit current. At higher value of circuit current more amount of heat energy penetrating into the workpiece which results in larger value of depth of crater. Because of which the ASR increases.

From the above study it is seen that, MRR during DS–ECSM process is influenced by four interacting process parameters viz. energy partition, duty factor, spark radius, and electrolyte concentration. It seems that a complex, non linear relationship exist between these process parameters and MRR. Hence, it is quite difficult to select the optimum process parameters for specific application of DS–ECSM process. To overcome this difficulty it is suggested by the authors that the results obtained from the present developed model can further be used to develop an intelligent model for simulation and multi-objective optimization to select the optimum process parameters for specific application of DS–ECSM process.

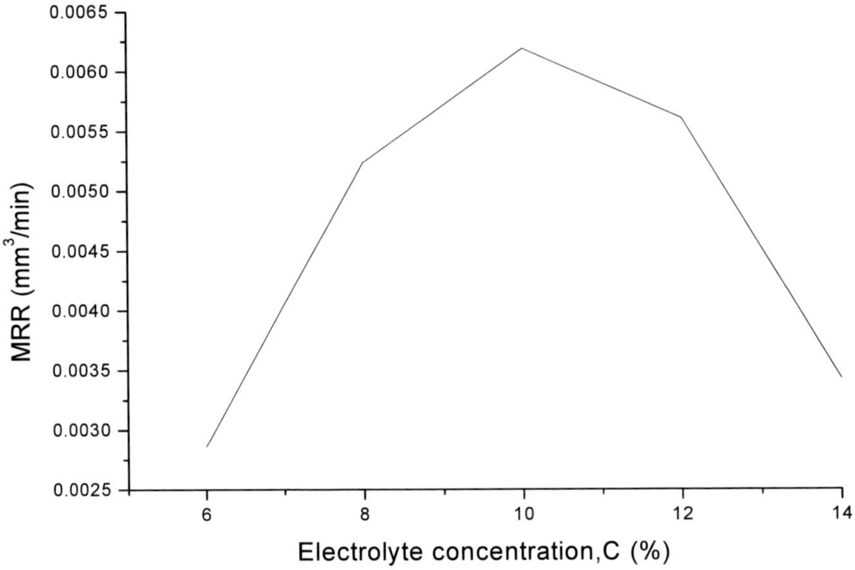

(a)

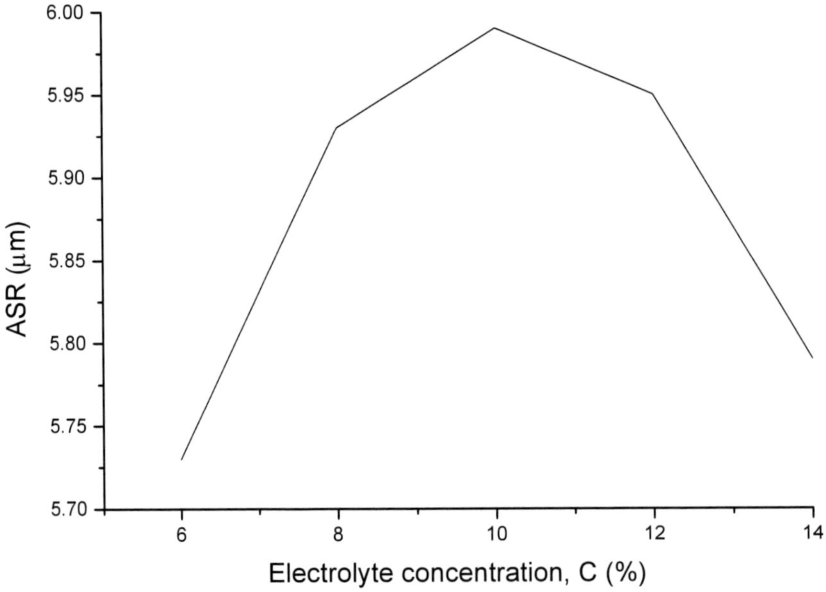

(b)

Figure 14. Effect of electrolyte concentration due to DS–ECSM of Silicon nitride workpiece with V =60V, R = 150 µm, F_w = 20% (a) on MRR (b) on ASR.

CONCLUSIONS

In the present work, an axisymmetric FEM based transient thermal model has been developed for the determination of MRR and ASR during DS–ECSM. Computational experiments have been performed to determine energy partition and spark radius. The present model is used to study the influence of different input parameters such as energy partition, duty factor, spark radius and electrolyte concentration on MRR and ASR of silicon nitride workpiece material. Following conclusions have been derived from the present analysis.

1. The trend of variation of computed MRR with supply voltage has been found similar to that observed experimentally. The computed MRR values are found higher than those obtained experimentally because of 100% ejection efficiency consideration.
2. Energy partition (F_w) and Spark radius (R) have been determined in the present study by computational experiments for given supply voltage and known experimental MRR. It has been found that F_w decreases with increase in supply voltage whereas R increases with increase in supply voltage.
3. With increase in energy partition (F_w), there is increase in MRR and ASR. Improvement in MRR by 606% and deterioration in surface finish by 24% have been found by increasing F_w by 10%.
4. Increase in MRR and ASR have been found with increase in duty factor. A 145% increase in MRR and 14% increase in ASR is observed by increasing duty factor by 13%.
5. Both MRR and ASR decreases with the increase in spark radius during DS–ECSM. Here it is observed that 2% improvement in surface finish is achieved by 23% increase in R but MRR decreases by 87%.
6. MRR and ASR are found to increase with increase in NaOH electrolyte concentration up to 10% but they start to decrease if concentration is further increased.
7. The results obtained from the present developed model can further be used for developing an intelligent model for simulation and multi-objective optimization of DS–ECSM process.

REFERENCES

Basak, I. and Ghosh, A. (1997) 'Mechanism of material removal in electrochemical discharge machining, a theoretical model and experimental verification', *Journal of Materials Processing Technology,* Vol. 71, pp. 350- 359.

Bhattacharyya, B., Doloi, B. N. and Sorkhel, S. K. (1999) 'Experimental investigations into electrochemical discharge machining (ECDM) of nonconductive ceramic materials', *Journal of Materials Processing Technology,* Vol. 95, pp.145-154.

Bhondwe, K. L., Yadava, V. and Kathiresan, G. (2006) 'Finite element prediction of material removal rate due to electrochemical spark machining', *International Journal of Machine Tools and Manufacture,* Vol. 46, pp. 1699-1706.

Cook, N. H., Foote, G. B., Jordan, P. and Kalyani, B. N. (1973) 'Experimental studies in electro-machining', *Transactions of ASME, Journal of Engineering for Industry*, Vol. 95, pp. 413-417.

Gautam, N. and Jain, V. K. (1998) 'Experimental investigations into ECSD process using various tool kinematics', *International Journal of Machine Tools and Manufacture,* Vol. 38, pp. 15-27.

Guyer, E. C. and Brownell, D. L. (1989) Hand book of applied thermal design, McGraw-Hill book Company, New York.

Jain, V. K., Dixit, P. M. and Pandey, P. M. (1999) 'On the analysis of the electro chemical spark machining process', *International Journal of Machine Tools and Manufacture,* Vol. 39, pp. 165-186.

Kulkarni, A., Sharan, R. and Lal, G. K. (2002) 'An experimental study of discharge mechanism in electrochemical discharge machining', *International Journal of Machine Tools and Manufacture,* Vol. 42, pp. 1121-1127.

Kurafuji, H. and Suda, K. (1968) 'Electrical discharge drilling of glass', *Annals of the CIRP*, Vol. 16, pp. 415-419.

Rao, S. S. (2005) *The Finite Element Method in Engineering*, fourth ed. Elsevier Publication, New Delhi.

Reddy, J.N. (2005) *An Introduction to Finite Element Method*, third ed. Tata McGraw-Hill Publishing Company Limited, New Delhi.

Sarkar, B. R., Doloi, B. N. and Bhattacharyya, B. (2006) 'Parametric analysis on electrochemical discharge machining of silicon nitride ceramics', *International Journal of Advanced Manufacturing Technology*, Vol. 28, pp. 873-881.

Tandon, S., Jain, V. K., Kumar, P. and Rajurkar, K. P. (1990) 'Investigations into machining of composites', *Precision Engineering*, Vol. 12, pp. 227-238.

Wuthrich, R. and Fascio, V. (2005) 'Machining of non-conducting materials using electrochemical discharge phenomenon-an overview', *International Journal of Machine Tools and Manufacture,* Vol. 45, pp. 1095-1108.

In: Machining and Forming Technologies, Volume 2
Editor: J. Paulo Davim
ISBN: 978-1-61122-820-5

Chapter 2

PREDICTION OF SURFACE ROUGHNESS IN ELECTRICAL DISCHARGE MACHINING OF D2 STEEL USING REGRESSION AND ARTIFICIAL NEURAL NETWORKS MODELING

***M. K. Pradhan**[*a]**, R. Das**[b] **and C. K. Biswas**[a]*
[a]National Institute of Technology, Rourkela,
Pin -769008, India
[b]Purushottam Institute of Engineering and Technology,
Rourkela, India

ABSTRACT

Surface roughness (Ra) is a significant upshot in the manufacturing process and it materializes a major part in the manufacturing system. It depends on different machining parameters and its prediction and control is a query to the researchers. In this study, a regression model and two artificial neural networks (ANNs) namely: Back propagation and radial basis function, were developed to predict surface roughness in electrical discharge machined surfaces.

In the development of predictive models, machining parameters of discharge current (Ip), pulse duration (Ton) and duty fraction (τ) were considered as model variables with a constant voltage 50 volt. For this reason, extensive experiments were carried out in order to collect surface roughness dataset. The developed models are validated with a new set of experimental data, and predictive behavior of models is compared, subsequently relative advantages of each model are analyzed.

Analysis of variance (ANOVA) and F-test were used to check the validity of regression model and to determine the significant parameter affecting the surface roughness. The statistical analysis exemplify that the Ip, Ton and τ were the factors in sequence have significant influence the on surface roughness. The microstructures of

* Corresponding author, E-mail: mohanrkl@gmail.com

machined surfaces were also studied by scanning electron microscopy (SEM). The SEM investigations revealed that EDMed produced were increased significantly with Ip.

Keywords: Surface roughness, Regression analysis, Back propagation, Radial basis function.

1. Introduction

With the continuous demand of components of new and advanced “difficult-to-machine” materials (tough super alloys, ceramic and composites) having stringent design necessities to fabricate complex shapes, tighter tolerance and very high machining cost, are incessantly challenging the modern manufacturing industry. These advance materials plays increasingly important role in the modern manufacturing industries, mainly in automobile, aircraft, tool, die, and mould making industries. The substantially increased properties such as strength, wear resistance, heat resistance, etc. yielded huge advantage to the manufacturing industries through superior product performance and product design. However, the traditional manufacturing processes are not capable to machine these materials inexpensively (Snoeys, 1986). Thus, to meet up such demands, the traditional machining processes are increasingly being replaced with advance manufacturing process, which make use of distinct class of energy for material removal using the material properties, such as electrical and thermal conductivity, melting temperature, electrochemical equivalent etc. Amongst all such non-traditional machining processes, electro discharge machining (EDM) is regarded as one of the most vital processes which is suitable for a wide range of applications in the said industries. It is a thermal process of eroding electrically conductive materials with a series of repetitive electric sparks. The mechanism of erosion is a complex phenomenon concerning several disciplines of science and branches of engineering. The theories revolving around the formation of plasma channel between the tool and the workpiece, thermodynamics of the repetitive spark causing melting and evaporating the electrodes, micro-structural changes, and metallurgical transformations of material, are still not clearly understood. However, it is widely accepted that the mechanism of material erosion is due to intense local heating of the workpiece causing melting and evaporation of workpiece. Therefore, for insight the process in a better way, extensive qualitative and quantitative analysis of the mechanism is vital and subsequently development of models, understanding the influence of various process parameters on the responses, confirmation of experimental results and enhancing the process performance by implementing/incorporating some of the theoretical findings are necessary (Jain and Jain , 2001)

Several research attempts has been made to model the EDM process, the interest is to recognize the effect of machining parameters on the responses. Tsai and Wang (2001) developed a semi-empirical model and testify that factors spark time, maximum current, polarity, input power, material density, conductivity of the material, specific heat capacity, heat conductivity, melting point, and boiling point of the material are affecting the surface roughness. Halkacı and Erden (2002) performed an experimental study and identified relations between Ra and spark time, in addition to Ra and power. Rebelo et el., (2000) looked into parameters for material removal rate and Ra when processing of hard copper-berilium alloys and the effect of discharge current and discharge time on surface integrity was presented. Petropoulos et el. (2004) presented a multi-parameter analysis of surface finish imparted to Ck60 steel plates by electro-discharge machining. The interrelationship between surface texture parameters and process parameters is highlighted.

In recent times, ANNs have materialized as an extremely flexible modeling tool for manufacturing sectors. ANNs are found to be effectual as computational processors for various associative recall, classification, data compression, combinational problem solving, adaptive control, modeling, forecasting, multisensor data fusion, and noise filtering. These competences of ANNs in capturing the mathematical mapping between input variables and responses are the driving force of implementing for modeling machining processes. The complexity and stochastic nature of EDM process has defied numerous attempts of ANN modeling it accurately.

Tsai and Wang (2001) demonstrated six different neural-networks models and a neuro-fuzzy network for the purpose of assessment the surface finish for different work material with the change of electrode polarity and concluded comparing the results and checking errors among these models that experimental results are in good agreement to the predictions. Pradhan et al., (2009), proposed two neural network models for the prediction of SR with input parameter Ip, Ton and τ, and compared with the experimental results. It is claimed that the said models could predict SR successfully within reasonable accuracy. Assarzadesh et el., (2007) offered a 3-6-4-2-size BPN model to ascertain a process model considering Ip, Ton and voltage as input to the network and MRR and Ra as output, which is able to predict process performance with reasonable accuracy, under varying machining conditions. Mandal et el. (2007) attempted to model the EDM process using ANN with back propagation as the learning algorithm. Ra, MRR and tool wear, with various input parameters and found suitable for predicting the response parameters. Markopoulos et el. (2008) proposed a ANN model for the prediction of Ra using Ip, Ton, and the processed material as input parameters and found satisfactorily prediction of Ra. Pradhan and Biswas (2008) presented a neuro-fuzzy model to predict MRR with different process parameters such as Ip, Ton and duty cycle, and the model predictions were found to be in good agreement with the experimental results Indurkhya et el.(1992) used 9-9-2 size back propagation network with input parameters such as Ip, Ton, Toff, machining depth, tool radius, orbital radius, radial step, vertical step, offset depth, etc. and material removal rate and Ra as output parameter. The results are compared with the multiple regression analysis and reported that ANN provides faster and more accurate results.

The literatures reveal that there are several attempts have been made to model the Ra with ANN, on the other hand, regression analysis is also attempted by many researchers, and however, very few attempts are on comparison between them. Apart from this, very few works are reported on AISI D2 tool steel, which has a growing range of application on the field of mould making industries.

In this work two different ANN model (BPN and RBFN) and regression have been employed, and comparison is made among them. The ANN models uses data for the training are obtained from the experiments with the input parameters Ip, Ton, and τ, which varied over a wide range, from roughing to near-finishing conditions and corresponding response Ra. The model is tested with the test data that are not earlier used to develop the model.

From the same set of data a second order regression model is also developed and ANOVA is used to check the sufficiency of the model. At the end of study, both the ANN and regression analysis results were compared with experimental data and prediction errors have been calculated for the testing data. The proposed models were found to be successful, for reliable predictions, which provides a possible way to avoid time and money-consuming experiments.

2. Experimental Details

Experiments were conducted on Electronica Electraplus PS 50ZNC die sinking machine shown in Figure 1.

Figure 1. Experimental set up.

A cylindrical pure copper was used as a tool electrode (of positive polarity) with a diameter of 30 mm and workpiece materials used were AISI D2 tool steel square plates of dimensions 15×15 mm^2 and thickness 4 mm. Commercial grade EDM oil (specific gravity = 0.763, freezing point= 94°C) was used as dielectric fluid. Lateral flushing with a pressure of 0.3 kg f /cm^2 was used. Keeping the voltage constant at 50 V, number of experiments was conducted to investigate the effects of Ip, Ton and τ on Ra, where τ is defined as:

$$\tau = \frac{Ton}{Toff} \tag{1}$$

The experimental conditions and the levels of the input parameters are shown in Table 1. Each treatment of the experiment was run for 15 minutes and the Ra was measured.

Table 1. Experimental Contortions

Sparking voltage in V	50
Current (Ip), in A	1 5 10 20 30 50
Pulse on Time (Ton), in μs	5 10 20 30 50 100 150 200 500 750
Duty fraction (τ)	1 12
Dielectric used	Commercial grade EDM oil
Dielectric flushing	Side flushing with pressure
Work material	AISI D2 tool steel
Electrode material	Electrolytic pure Copper
Electrode polarity	Positive
Work material polarity	Negative

3. Surface Roughness Measurement

The Ra is used to portray the technical surface quality of an engineering component. It has very significant influence on the manufacturing outlay of a product. A good quality surface enhances the fatigue strength, corrosion, and wear-resistance of the workpiece. There is a number of ways by which surface roughness of a component is described, such as roughness average (Ra), root-mean-square (rms) roughness (Rq) and maximum peak-to-valley roughness (Ry or Rmax), etc.

In this work, Ra is used, which is measured using Talysurf (Taylor Hobson, Surtronic 3^{+}) shown in Figure 2 , the experimental lay out plan of the inputs and the values of Ra are summerised in Table 2.

The profilometer was set to a cut-off length of 0.8 mm, filter 2CR, traverse speed 1 mm/second and 4 mm evaluation length. Roughness measurements, in the transverse direction, on the workpieces were repeated four times and average of four measurements of surface roughness parameter values was recorded. The measured profile was digitized and processed through the dedicated advanced surface finish analysis software Talyprofile for assessment of the roughness parameters. Ra can be defined as the arithmetic value of the profile from centerline along the sampling length. It can be express as

$$Ra = \frac{1}{L} \int \left| y(x) \right| dx \Big| \tag{2}$$

where L is the sampling length, y is the profile curve and x is the profile direction. The average Ra is measured within L = 0.8 mm. Centre-line average Ra measurements of electro-discharge machined surfaces were taken to provide quantitative evaluation of the effect of EDM parameters on surface finish.

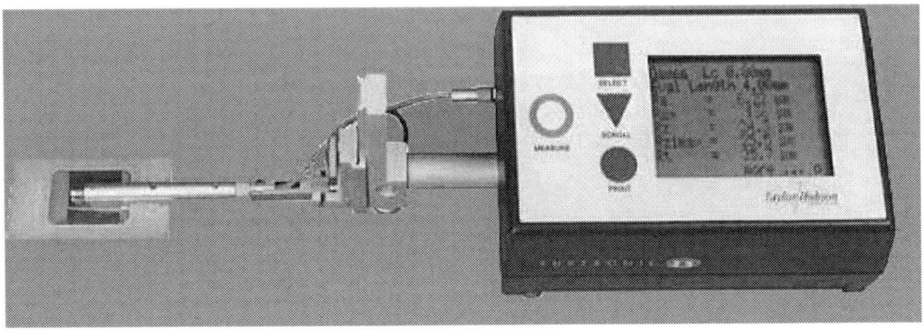

Figure 2. (a) Talysurf (Taylor Hobson, Surtronic 3+).

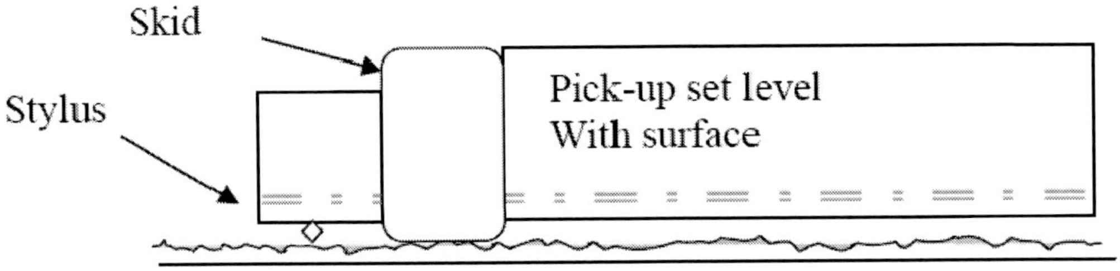

Figure 2. (b) the pick –up is supported on the workpiece by the skid.

Table 2. Input parameters and experimental values of Ra (training data)

Run	Ip (A)	Ton (μS)	Duty fraction	Ra (μm)	Run	Ip (A)	Ton (μS)	Duty fraction	Ra (μm)
1	1	5	12	2.81	18	5	20	1	3.62
2	1	5	1	1.77	19	5	50	12	4.33
3	1	10	12	2.3	20	5	50	1	5.69
4	1	10	1	2.95	21	10	100	12	7.5
5	1	20	12	2.47	22	10	150	12	8.13
6	1	20	1	3.75	23	10	200	1	9.31
7	1	30	12	2.15	24	10	50	12	6.34
8	1	50	12	2.06	25	10	50	1	4.26
9	1	50	1	4.3	26	10	75	12	7.05
10	1	30	1	4.45	27	10	75	1	6.65
11	5	5	12	2.48	28	10	100	1	7.52
12	5	100	12	5.25	29	10	150	1	8.38
13	5	100	1	5.79	30	10	200	12	8.79
14	5	5	1	3.23	31	20	100	12	8.01

Table 2. (Continued)

Run	Ip (A)	Ton (μS)	Duty fraction	Ra (μm)	Run	Ip (A)	Ton (μS)	Duty fraction	Ra (μm)
15	5	10	12	3.37	32	20	100	1	8.82
16	5	10	1	3.09	33	20	150	12	9.38
17	5	20	12	3.35	34	20	500	1	4.53
					35	20	200	12	10.75

4. Predictive Models for Surface Roughness

4.1. Regression Analysis

A second order polynomial regression model for three variables under consideration is proposed. The equation for Ra can be written as shown in Equation 3.

$$Ra = a_0+a_1I\ p+a_2\ Ton + a_3\ \tau + b_1I\ p^2 + b2\ (Ton)^2 + b_3\tau^2 + c_1I\ p\ Ton+ c_2\ Ton\ \tau +c_3Ip\tau \quad (3)$$

The co-efficient a_0 is the free term; the coefficients a1, a_2 and a_3 are the linear terms; the coefficients b_1, b_2 and b_3 are the quadratic terms; and the coefficients c_1, c_2, and c_3 are the interaction terms. The coefficients of regression model were estimated from the experimental results and the model is developed.

ANOVA is used to check the sufficiency of the second-order model, which includes test for significance of the regression model, and model coefficients. It is used for testing the null hypothesis of the experimental data with a confidence level of 95%. The p-value for the F-statistic, expressing the probability of observing a value of F at least, as large, if H_0 is true and treatments have no effect. If p-value less than 0.05, it is concluded that $H\alpha$ is true and the treatments have a statistically significant effect. Ra obtained from the experiments is compared with the predicted value calculated from the model.

Table 3 is an ANOVA summary of the terms in the model, corresponding coefficients (Coef.) of the terms, the standard error of a coefficient (SE Coef), t-statistic and p-value of the terms to help to decide whether to reject or fail to reject a null hypothesis.

It can be seen that the p-value of all the terms are below the α-value, hence they are significant in the model. The values of R^2 and R^2 adj are 94% and 93%, respectively. Where R^2 = 94% indicates that 94% of total variation in the response that is explained by predictors or factors in the model and R^2 adj is 93%, which accounts for the number of predictors in the model describes the significance of the relationship.

Table 4 illustrates the ANOVA analysis of the model, and the columns describing the degrees of freedom (DF), the sequential sums of squares (Seq SS), the adjusted sums of squares (Adj SS), the adjusted means squares (Adj MS), the F-statistic from the adjusted means squares, and its p-value. The sequential sum of squares is the added sums of squares given that prior terms are in the model, which depends on the model order.

The adjusted sums of squares are the sums of squares given that all other terms are in the model that does not depend upon the model order. It can be observed in this table, the p-value of regression model is less than 0.05, hence, the Ra fitting the regression model with the linear, and square terms are significant at the level of 95%. The square terms of τ along with all interaction terms are not included in the model as their p-values are exceeding the α-value. This way the simplified truncated model is shown in the following equation.

$$Ra = 1.91846 + 0.23683\ Ip + 0.04354\ Ton - 0.03750\ \tau - 0.00295\ Ip^2 - 0.00009\ Ton^2 \quad (3)$$

where Ra is in μm, Ip in Amps and Ton is in μs.

Table 3. Results of the ANOVA for R_a

Term	Coef	SE Coef	T	P
Constant	1.91846	0.287141	6.681	0.000
Ip	0.23683	0.075480	3.138	0.004
Ton	0.04354	0.004220	10.318	0.000
Duty fraction	-0.03750	0.021485	-1.746	0.091
Ip*Ip	-0.00295	0.003231	-0.914	0.368
Ton*Ton	-0.00009	0.000008	-11.194	0.000
S=0.6802 R-Sq = 94.0% R-Sq(adj) = 93.0%				

Table 4. The ANOVA table for the fitted models

Source	DF	Seq SS	Adj SS	Adj MS	F	P
Regression	5	210.354	210.3541	42.0708	90.93	0.000
Linear	3	137.629	99.7015	33.2338	71.83	0.000
Square	2	72.725	72.7249	36.3624	78.60	0.000
Residual Error	29	13.417	13.4170	0.4627		
Total	34	223.771				

4.2. Neural Network Modeling

An ANN is a new computing tool that process information using neurocomputing approach. This is different from the classical computation. In past, several studies have been reported on the development of neural networks based on different architectures. Neural networks are characterized by their architecture, activation function and learning algorithms. Each type of neural networks would have its own input-output characteristics, and therefore it could be applied only on some specific process. A neural network is represented by weighted interconnections between processing elements. These weights are the parameters that actually define the non-linear function performed by the neural network.

The process of determining such parameters is called training or learning, relying on the presentation of many training patterns. The architecture of the neural network having *n* inputs and *m* outputs is given in the Figure 1.

In this study, two neural networks are used for modeling the Ra in the EDM process. Two networks are discussed as follows.

(1) Back-propagation Neural Network (BPNN)
(2) Radial basis function neural network (RBFNN)

4.2.1. Back-Propagation Neural Network

Back-propagation networks are composed of layers of neurons. The input layer of neurons is connected to the output layer of neurons. The network has the general architecture as shown in Figure 3. Input layer neurons are linear, whereas neurons in the hidden and output layers have sigmoidal functions.

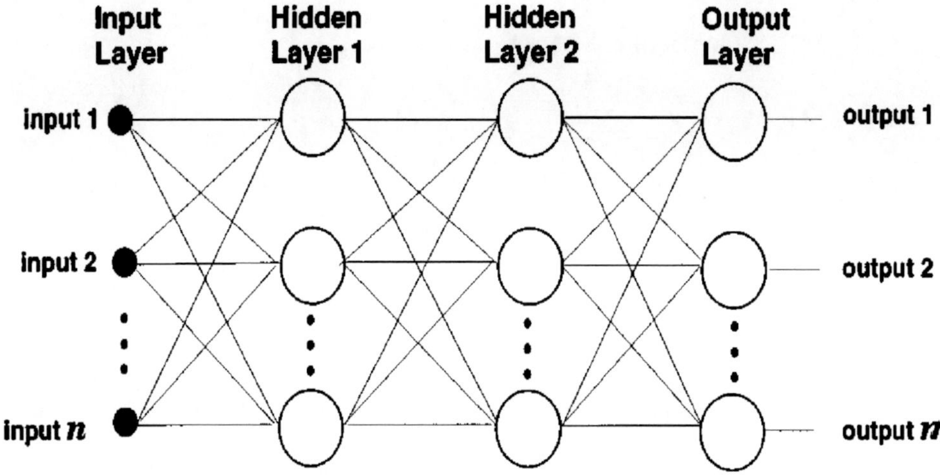

Figure 3. Architecture of Neural Network.

Table 5. Description of BPNN

	Input	Hidden	Output
Number neurons	$n+1$	$m+1$	r
Function	linear	sigmoidal	Sigmoidal
Neuron index range	$i=0,\ldots,n$	$h=0,\ldots,m$	$j=0,\ldots,r$
Activation	x_i	z_h	y_j

Vector or scalar variables will be subscripted or superscripted by the iteration index, *k*. All neurons use similar functions. For linear neurons in the input layer,

$$S(x) = x$$

and for sigmoidal neurons in the hidden and output layers,

$$S(x) = \frac{1}{1+e^{-x}}$$

We assume a set of Q training vector pairs:

$$T = \{(X_k, D_k)\}_{k=1}^{Q}$$

$X_k \in R^n$, $D_k \in R^r$, where D_k is a vector response desired when input X_k is presented as input to the network.

The vector pairs in T are assumed to be samples representative of some unknown function $f : R^n \rightarrow R^r$ which we wish the neural network to approximate. The error gradient for a pattern are computed and used to change the weights in the network. For *k*th training pair (X_k, D_k) then define the error:

$$E_k = D_k - S(Y_k)$$

and

$$\varepsilon_k = \frac{1}{2} E_k^T E_k$$

where ε_k is the sum of the squares of the each individual output error. The mean square error ε, is computed over the entire training set T on a specific neural network:

$$\varepsilon = \frac{1}{Q} \sum_{k=1}^{Q} \varepsilon_k$$

The error calculated is used to compute the change in the hidden to output layer weights and the change in input to hidden layer weights such that a global error measure gets reduced. Then all weights of the network are updated. It will be repeated until the global error falls below a predefined threshold.

4.2.2. Radial Basis Function Neural Network

Through a defined learning algorithm, the network performs the adjustments of its weights so that error between the actual and desired responses is minimized. Once trained, the network performs the interpolation in the output vector space. In the previous section we discussed that back propagation network can be trained to perform a nonlinear mapping between the input and output vector spaces. In this section we present an alternative network that is capable of accomplishing the same task. This is the radial basis function neural network.

RBFNN is based on supervised learning. It consists of three layers: an input layer, a single layer of nonlinear processing neurons, and output layer. Input layer contains n neurons and output layer contains r neurons. It has a feed forward structure consisting of a single hidden layer of m locally tuned units which are fully interconnected to an output layer of r linear units. All hidden units simultaneously receive the n-dimensional real valued input vector X. The main difference from that absence of hidden layer weights.

RBFNNs are best suited for approximating continuous or piecewise continuous real valued mapping $f : R^n \rightarrow R^r$, when n is sufficiently small. RBFNN can be viewed an approximating a desired function $f(x)$ by superposition of non orthogonal, bell-shaped basis functions. The degree of accuracy of these RBFNN can be controlled by three parameters: the number of basis functions used, their location and their width.

In the present work we have assumed a Gaussian basis function for the hidden units given as Z_j for $j = 1,2,\cdots,m$

$$Z_j = \exp\left(\frac{\left\|X - \mu_j\right\|^2}{2\sigma_j^2}\right)$$

and μ_j and σ_j are mean and the standard deviation respectively of the *j*th unit receptive field and the norm is the Euclidean.

Given an input vector X, the output value $O_k(X)$ of the *k*th output node is equal to the sum of the weighted outputs of the hidden nodes and the bias of the kth output node and is described by

$$O_k(X) = \sum W_{kj} Z_j + b_k$$

where W_{kj} is the weight between the *j*th hidden node and *k*th output node.

The sum of the squared error function can be considered as an error function E to be minimized over the given training set. The training of the RBFNN is radically different from the classical training of standard feedforward neural network. In this case, there is no changing of weights with the use of the gradient method. In RBFNN with the chosen type of radial basis function training resolves itself into selecting the center and width and calculating the weights of the output neurons. Heuristic operation on a given defined training set starts from an empty subset of the basis function. Then the empty subset is filled with succeeding basis functions with their center marked by the location of elements of the training set. It decreases the sum-squared error or the cost function. In this way, a model of the network constructed each time is being completed by the best element. Construction of the network is continued till the criterion demonstrating the quality of the model.

Initially, the architecture and the topology of the networks i.e. the number of hidden layers and the number of neurons in each layer in the networks are decided. The process parameters the discharge current (Ip), pulse on time (Ton) and duty fraction (τ) are taken as the inputs and Ra is taken as output. Thus, there are three input nodes and one output node.

The variation of process parameters for different experimental set (Run) are as presented in the Table 1.

The size of the network becomes very large for large number of training patterns. As such, the data for training is selected judiciously. 35 training data sets are considered for both the networks to compare the performances. Besides, 9 testing sets outside the training data set are selected for testing the neural networks. The training data sets and the test data sets are taken from Experiment. Both the ANNs were trained with the above datasets to reach the error goal. The performance of two neural network models is studied with the special attention to their generalization ability and the training time.

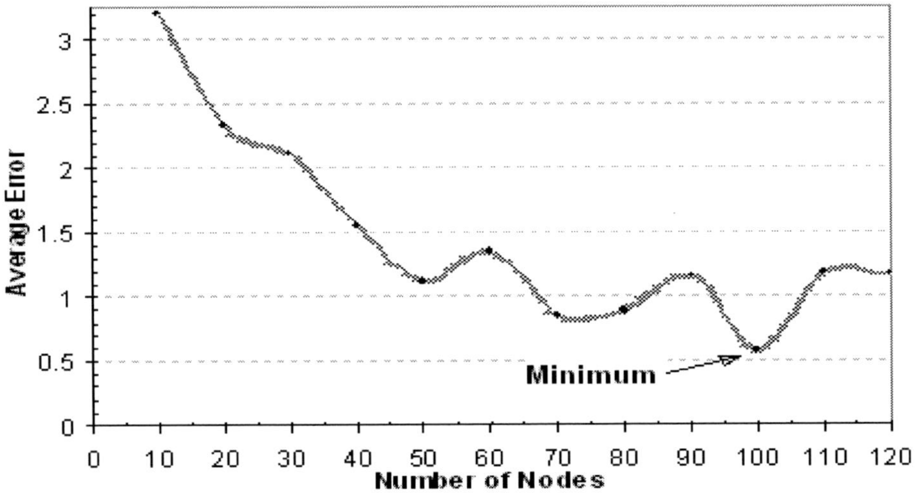

Figure 4. Comparison of average errors for various nodes on BPNN.

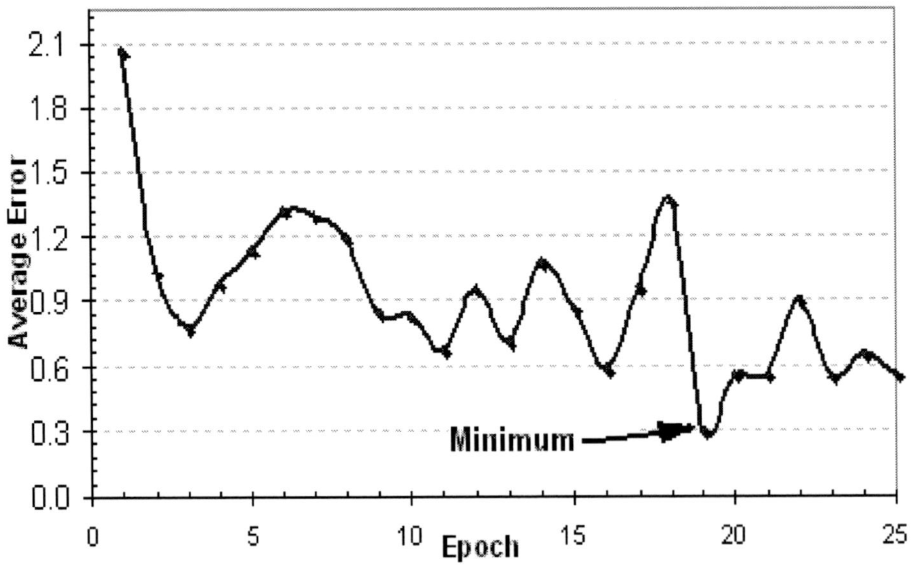

Figure 5. Comparison of average errors for various epochs on BPNN.

It is a difficult task to find an optimal configuration of BPNN. There are no exact rules for setting the proper number of neurons in the hidden layer to avoid over fitting or under fitting to make the learning phase convergent. For the best performance of the BPNN, the proper number of nodes in the hidden layer is selected through a trial and error method based on the number of epochs needed to train the network. It was observed that the network performed well with 100 nodes and 500 epochs. The learning behavior of BPNN model for Ra is shown in Figure 7. The RBFN is auto configuring in the sense that it has only one hidden layer with a growing number of neurons during learning to achieve an optimal configuration. The only parameter to be varied to obtain the best generation ability is the spread factor (SF). Computations are carried out for different values of spread factor. It was observed that the best generalization ability of the network achieved with a spread factor of value 45(Figure 7). The training performances of the same training data sets are shown in Figure 8. The error goal is reached in only 19 epochs in RBFN, while 500 epochs are required by the BPNN.

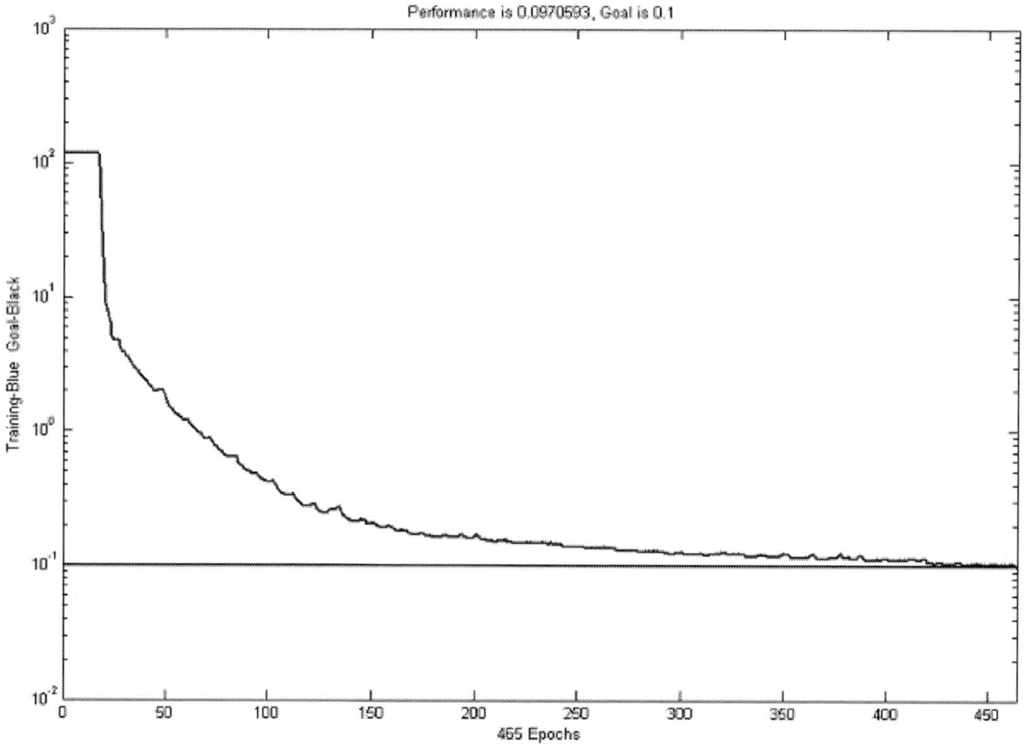

Figure 6. Learning behavior of BPNN model for surface roughness.

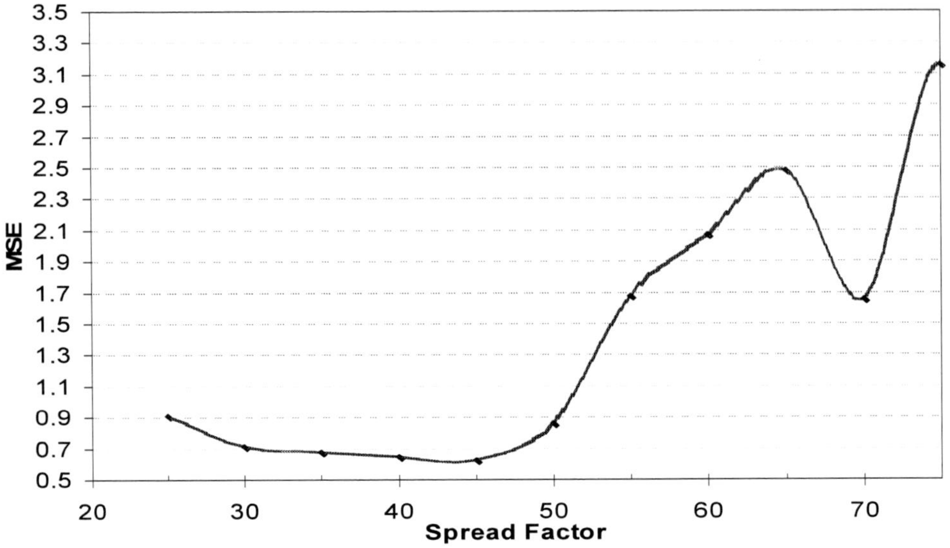

Figure 7. Comparison of average error for various nodes and spread factor.

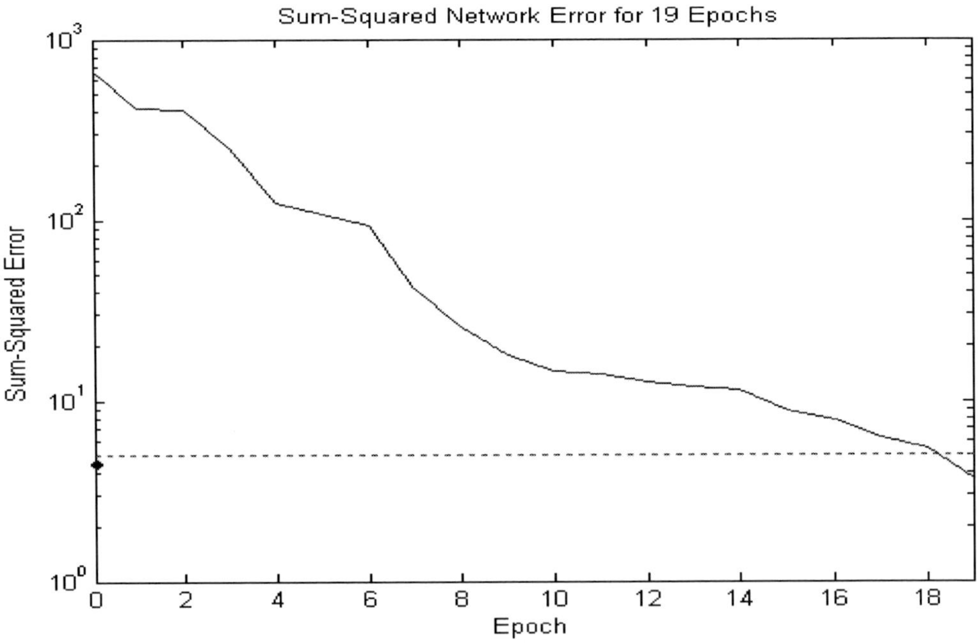

Figure 8. Learning behavior of RBFN model for surface roughness.

5. Result and Discussion

The experiments were conducted and the impact of the machining parameters such as Ip, Ton and τ on Ra is analyzed. Ra is convincingly dependent of spark energy, and the size of the bauble, which is the prime cause of material removal. The spark input energy and the size of crater is dependent on discharge voltage, Ip and Ton (Dibitonto, 1989). The response Ra is predicted by a regression model, and two NN model. The effects of each of parameter on Ra are compared with different models and discussed.

Figure 9 displays the normal probability plot of the residuals, which is used to test the normal distribution of the errors. The distribution of independent random errors of observation is expected to take a normal distribution. It can be seen that the residuals are almost falling on a straight line, which indicates that the errors are normally distribute and it appears to be a good model for these data (NIST/SEMATECH).

The plot of the residues verse experiment number for the model under consideration are illustrates that the errors are distributed randomly as there is no noticeable pattern or unusual structure present in the data as depicted in Figure 10, which suggests that the models fit the data well. Also the residuals which are very large or very small than the rest are typically called outliner and a few such outliners may distort the analysis. The residues, which are calculated as the difference between the predicted and observed value lies in the range of –2.2 to 1.86, –1.77 to 2.65 and –1.04 to 1.69 for regression, BPN and RBFN respectively.

Experimental observation is also compared with the predicted values of each model, regression, BPN and RBFN in the Figure 11 to demonstrate the accuracy of the predictive model. The represented data refer to both training and validation data sets. These plots are also presented straight lines to make them easier to infer. It is seen that the predicted and experimental values in the models could predict very accurately and except for one or two outliner, almost all the values are close to the line. It could be noted that closer the value to the line, more is the accuracy.

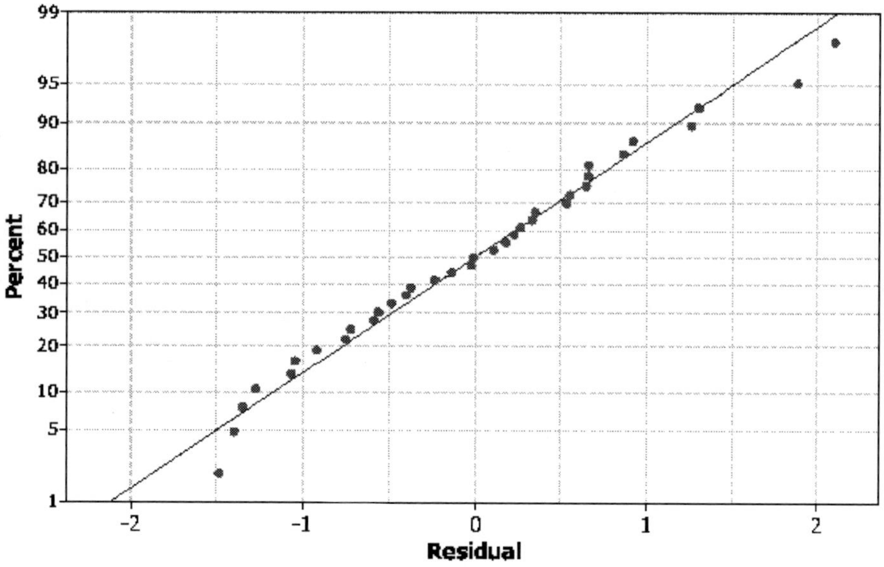

Figure 9. Normal plot of residuals.

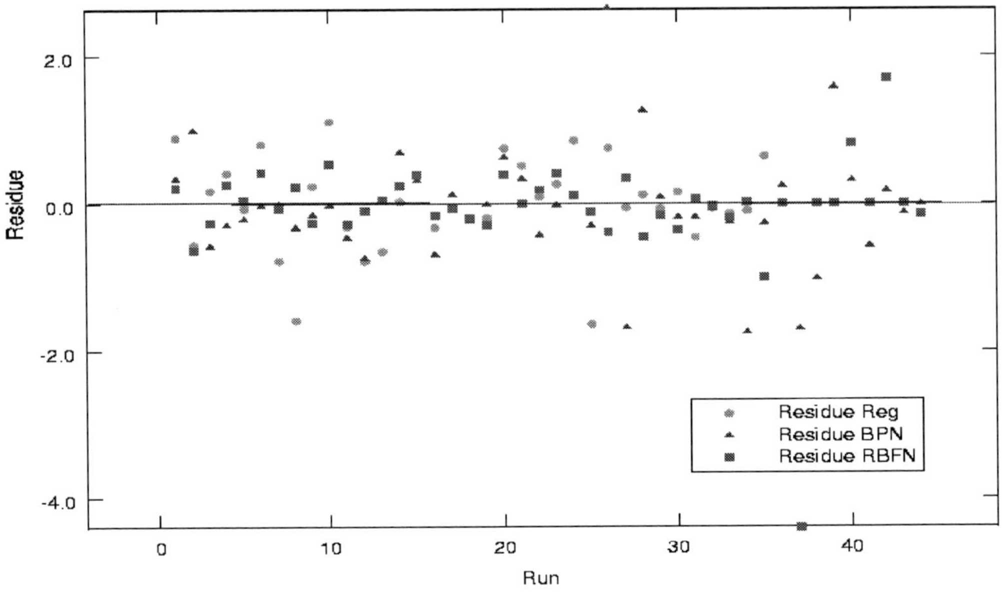

Figure 10. Plot of residuals vs. Expt No.

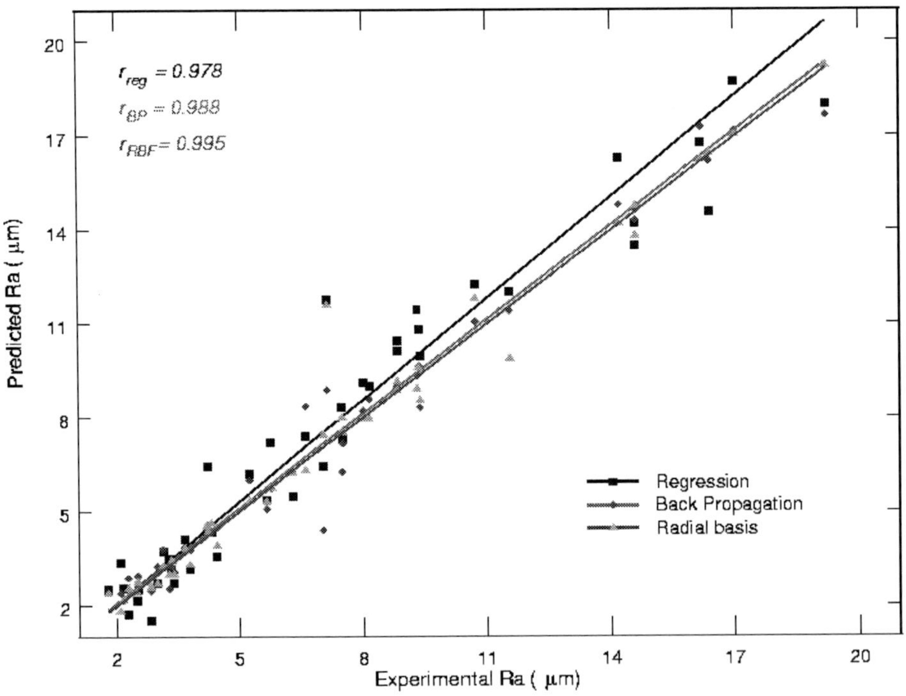

Figure 11. Predicted vs. experimental Ra.

The three lines match fairly closely for lower values of Ra. However, for large values of Ra, the Regression line tends to generate higher values for the predicted values than does the BPN and RBFN. BPN and RBFN are similar at all the ranges, nevertheless the RBFN model is comparatively more accurate than the BPN and regression model, having correlation coefficient 0.998, 0.988 and 0.978, respectively, which conforms the effectiveness of the models and that all the three models are moderately well fitted with the experimental value. However, the models can be sequenced as regression, BPN and RBFN, respectively, in terms of accuracy. This sequence is also conformed by the correlations coefficients.

5.1. Parametric Influences on Surface Roughness

5.1.1. Effect of Ip

To establish the effect of EDM on the Ra of the AISI D2 steel, the surface profiles of the topography of the EDM surfaces were measured. Figure 13 shows the measurement results. The Ra on the machined surface varied from 8 μm to 17μm. The higher pulsed current caused more frequent melt expulsions, leading to the formation of a deeper and larger crater on the surface of the workpiece and resulted in a poorer surface finish. This is an agreement with that of (Salonitis et al, 2009), and (Bhattacharyya et al, 2007).

5.1.2. Effect of Ton

The effect of pulse duration (Ton) on Ra is depicted in Figure 12 and 13 for Ip = 20, and 30 A respectively. It can be observed that the Ra increases steadily with the increase in Ton. This increase in Ra is sharper for Ip =30 than that of Ip=20, where it shows a peak after which Ra decreases. It is expected that when the pulse time increases, the Ra usually increases up to a maximum value after which it starts to decrease. The higher pulsed duration caused more frequent melt expulsions, since the discharge energy and the impulsive force increase with the peak current; more melted material will be removed to produce deeper and larger discharge craters, a rougher surface is thus obtained. Large amount of deep and large crater, are found on the work piece. With the further increase of Ton the bigger and sallow craters are formed and therefore the Ra decreases. This is an agreement with that of (Wu, et al., 2005).

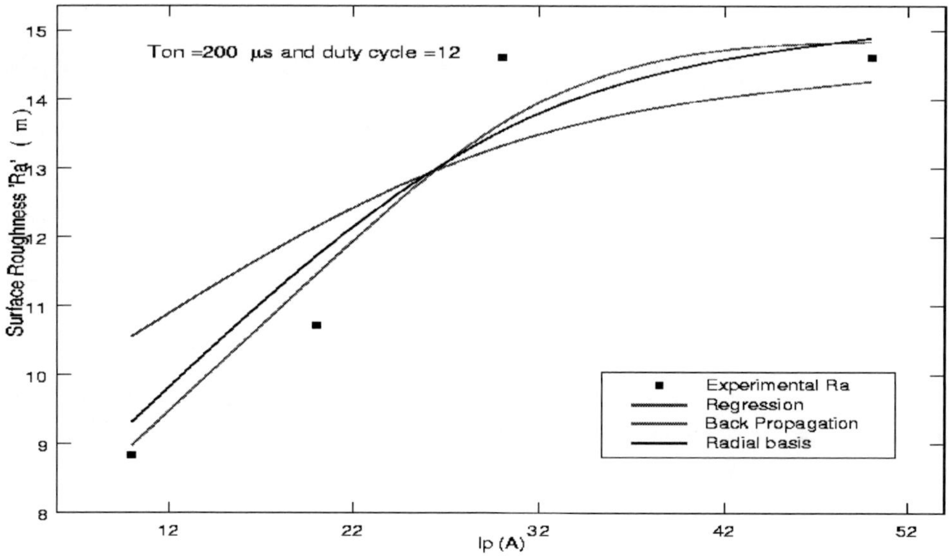

Figure 12. Effect of Ip on Ra.

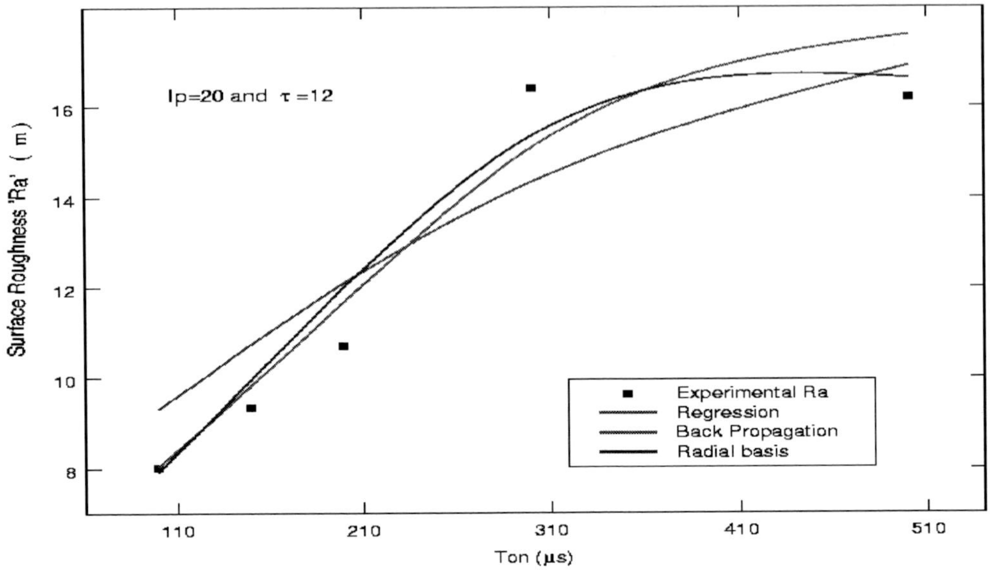

Figure 13. Effect of Ton on Ra at Ip= 20 and τ =12.

Moreover, the similarity in nature of the curves, as observed in Figure 13, and 14 obtained through developed mathematical model and that obtained using experimental data confirms the suitability of the pattern of the developed l models for the surface roughness.

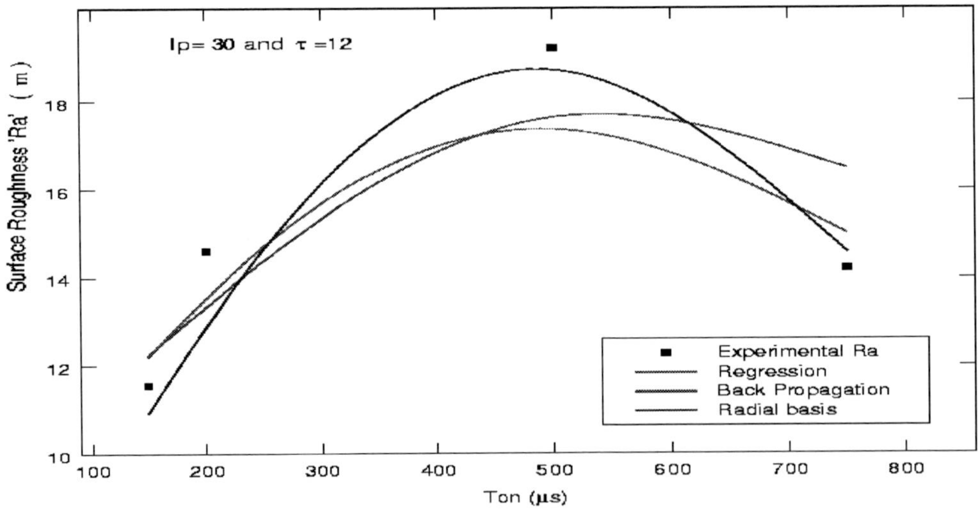

Figure 14. Effect of Ton on Ra at Ip= 30 and τ =12.

Comparing with the results of Figure 13, figure 14 and Figure 15 we find that good machined finish can be obtained by setting the machine parameters at low pulsed current and small pulse-on duration, but that will be more time consuming.

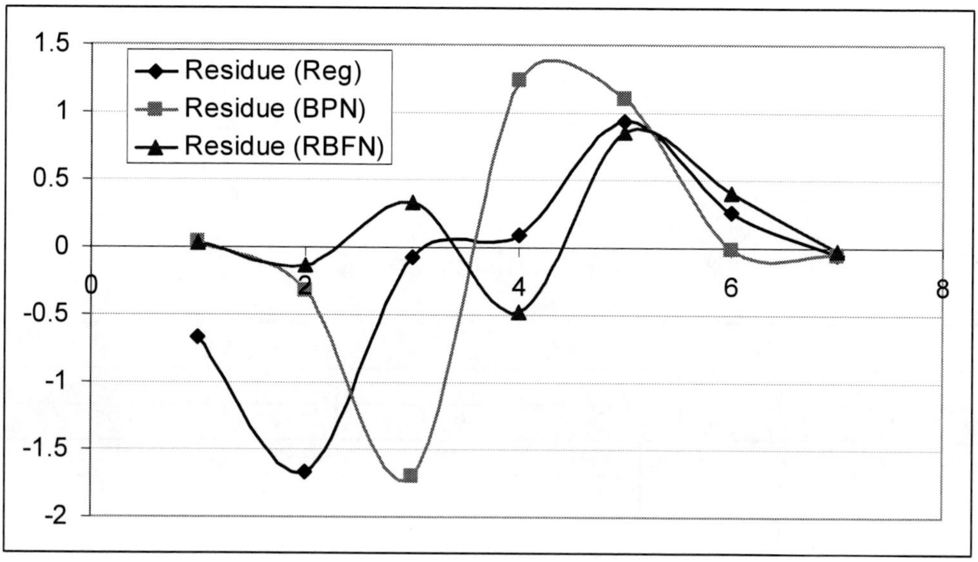

Figure 15. Residuals calculated as the difference between experimental and predicted values for the validation data set.

5.2. Prediction Error

The validation of all the models is carried out with the testing data set that is not previously used to develop the model. In order to assess the accuracy of the prediction models, percentage error and average percentage error are used. Prediction error has been defined as follows:

$$\text{Prediction error} = \frac{|\,\text{Expt. MRR - Fitted MRR}\,|}{\text{Expt. MRR}} \times 100 \qquad (4)$$

In Table 6, the process parameters of testing data, their corresponding experimental Ra, percentage error and the average percentage error are shown. The average percentage error of these model validations are about 4.11%, 9.01%, and 20.21% for RBF, BPN, and regression model, respectively. Consequently, the models can be sequenced as RBFN, BPN, and regression model, in terms of prediction accuracy.

ANNs are compared separately with results obtained by experiments and the average error obtained for both the networks. RBFN model is poorer at two input values but for the rest of the input range, within which the normal operating range lies, both the models have almost identical generalization ability. The test results accuracy measured in terms of mean absolute error (MAE) for 7 test data are found to be 0.297188 for the BPNN and 0.574888 for RBFN. In the case of RBFN, the number of epochs is equal to the number of neurons in the single hidden layer of the network. The error goal is reached in only 19 epochs, while 465 epochs are required by the BPNN. The amount of the work done in each epoch is not equivalent for both ANN's.

Figure 15 shows the error for each model, calculated as the difference between the experimental findings and predicted values. It was found that except at two places, both the models predict the roughness.

The test data and prediction error is shown in Table 6. The process variables and the experimental Ra are. In order to evaluate the accuracy of the prediction model, percentage errors are used. Percentage of prediction errors is shown in the in the Table.

Table 6. Test data and Prediction errors

Sl No	Ip (A)	Ton (μs)	τ	Ra (μm)	%Error Reg	% error BPN	% error RBFN
1	5	100	1	5.78	-11.5114	0.73	0.55
2	10	50	1	4.24	-39.29	-7.68	-3.21
3	10	75	1	6.64	-1.107	-25.61	4.97
4	10	100	1	7.5	1.22	16.56	-6.36
5	10	150	1	9.4	9.99	11.77	9.06
6	10	200	1	9.32	2.76	-0.13	4.38
7	20	100	1	8.84	-0.58	-0.59	-0.22
Average prediction error					9.49	9.01	4.11

It can be noted that the RBFN is the best while predicting the Ra with 9.06 % prediction error. However, the other two models are also predicting the Ra except at one or two outliners. The experimental results and predicted results of 'Ra' by the BPN and RBFN were plotted on the same scale, as shown in figure-9.

6. Analysis Surface Roughness Using SEM

The EDMed surfaces of AISI D2 tool steel were examined using the SEM and depicted in the Figures 16 (a)-(c). Figure 16(a) represents the surface generated when EDMed with Ip=10, however Figure 14(b), with Ip=20 and figure 14(c), with Ip=30. With the increase in Ip the spark energy increases, and hence forms bigger craters. The diameter and the depth of craters of the EDMed surface increase with an increase in the discharge current, and hence the Ra consequently increases. Significantly different Ra value with the change in Ip is represented by these three samples. The size of the crater can be clearly visualized with the increasing discharge current. In the SEM micrographs bigger craters due to larger amount of molten work-material with the spark erosion can be identified on EDM surfaces.

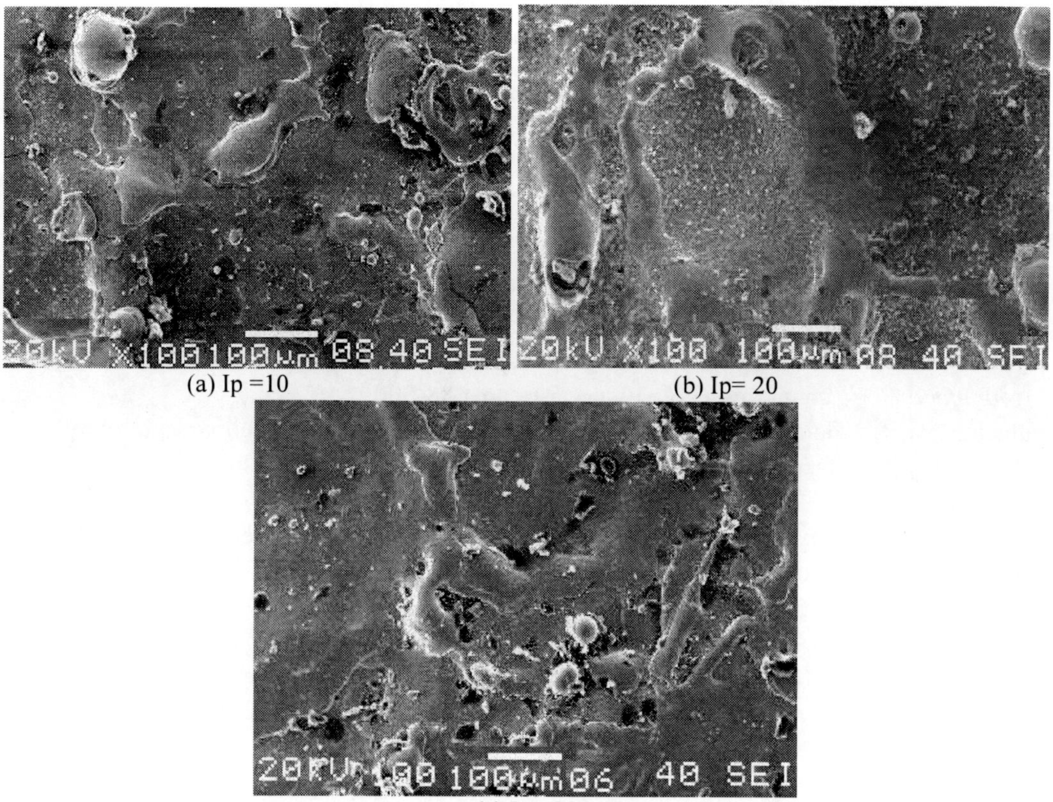

(a) Ip =10 (b) Ip= 20

(c) Ip= 30

Figure 17. SEM of EDMed surfaces of AISI D2 Steel.

Conclusion

In the present investigation, the prediction and comparative analysis was carried out on EDMed AISI D2 tool steel surfaces. The study was to analyze relationship between machining parameters and surface roughness in EDM process by a multiple regression analysis and two ANN models. Comparisons were made of the said models, after testing their performances on 7 randomly selected test cases. The Regression model and two ANN models have performed satisfactorily in the prediction of Ra. However, the ANN models are comparatively more accurate. BPN is slightly better performer than RBFN, but RBFN is faster. . It is important to note that for BP networks the required number of nodes in the hidden layer was found by trial and error method, where as, the RBF networks have only one hidden layer with a growing number of neurons. Based on the ANOVA and *F*-test, the most dominant parameter on the surface roughness was found as pulse on time, while the second ranking factor was discharge current. Duty fraction was comparatively less effective on surface roughness.

ACKNOWLEDGMENT

The authors are extremely thankful to the reviewers for their valuable comments and suggestions which contributed significantly to improve the quality of the paper.

REFERENCES

Assarzadeh, S. and Ghoreishi, M. (2008), 'Neural-network-based modeling and optimization of the electro-discharge machining process', *International Journal of Advanced Manufacturing Technology*, vol. 39, no. 5-6, pp. 488-500.

Bhattacharyya, B., Gangopadhyay, S. and Sarkar, B.R. (2007), "Modelling and analysis of EDMED job surface integrity", Journal of Materials Processing Technology, vol. 189, no. 1-3, pp. 169-177.

Dibitono, D.D., Eubank, P.T., Patel, M.R., Barrufet, (1989.) M.A.: Theoretical model of the electrical discharge machining process I: A simple cathode erosion model. Journal of Applied Physics, 66, 4095–4103.

Debabrata Mandal Surjya K. Pal Partha Saha, (2007) 'Back propagation neural network based modeling of multi-responses of an electrical discharge machining process *International Journal of Knowledge-based and Intelligent Engineering Systems,* Vol. 11, No- 6, pp 381-390.

Halkaci H S and Erden A. (2002) 'Experimental investigation of surface roughness in electric discharge machining (EDM)'. *6th Biennial Conference on Engineering Systems Design and Ana*lysis, Y′ stanbul, Turkey, pp 8–11.

Indurkhya, Gopal, Rajurkar, K. P.(1992), 'Artificial Neural Network approach in modeling of EDM process', *Intelligent Engineering Systems Through Artificial Neural Networks*, Vol. 2, pp-845-850.

Jain, N.K. and Jain, V.K. (2001), 'Modeling of material removal in mechanical type advanced machining processes: A state-of-art review', *International Journal of Machine Tools and Manufacture*, vol. 41, no. 11, pp. 1573-1635.

Markopoulos, A.P., Manolakos, D.E. and Vaxevanidis, N.M. 2008, 'Artificial neural network models for the prediction of surface roughness in electrical discharge machining', *Journal of Intelligent Manufacturing,* vol. 19, no. 3, pp. 283-292.

NIST/SEMATECH e-Handbook of Statistical Methods (2006) [http://www.itl.nist.gov/div898/handbook/]

Petropoulos, G., Vaxevanidis, N.M. and Pandazaras, C. (2004), 'Modeling of surface finish in electro-discharge machining based upon statistical multi-parameter analysis', Journal of Materials Processing Technology, vol. 155-156, no. 1-3, pp. 1247-1251.

Pradhan, M. K. and Biswas, C. K. (2008). 'Neuro-fuzzy model on material removal rate in electrical discharge machining in AISI D2 steel'. *Proceedings of the 2nd International and 23rd All India Manufacturing Technology, Design and Research Conference*, 1: 469–474.

Pradhan, M. K., Das, R. and Biswas, C. K. (2009), Comparisons of neural network models on surface roughness in electrical discharge machining', *Proc. IMechE Part B: J. Engineering Manufacture., Vol. 223, Number 7, pp- 801-808/ 2009.*

Rebelo, J.C., Morão Dias, A., Mesquita, R., Vassalo, P. and Santos, M. (2000), "Experimental study on electro-discharge machining and polishing of high strength copper-beryllium alloys", *Journal of Materials Processing Technology,* vol. 103, no. 3, pp. 389-397.

Shatla, M., Kerk, C. and Altan, T. 2001, 'Predictions on surface finish in electrical discharge machining based upon neural network models', *International Journal of Machine Tools and Manufacture*, vol. 41, no. 10, pp. 1385-1403.

Snoeys, R., Staelens, F. and Dekeyser, W. (1986) 'Current Trends in Non-Conventional Material Removal Processes', *Annals of the CIRP* 35, No-2, pp. 467–479.

Salonitis, K., Stournaras, A., Stavropoulos, P. and Chryssolouris, G. 2009, "Thermal modeling of the material removal rate and surface roughness for die-sinking EDM", *International Journal of Advanced Manufacturing Technology*, vol. 40, no. 3-4, pp. 316-323.

Wang, P. and Tsai, K. (2001), 'Semi-empirical model on work removal and tool wear in electrical discharge machining', *Journal of Materials Processing Technology*, Vol. 114, No. 1, pp. 1-17. *Handbook of Statistical Methods.*

Wu, K.L., Yan, B.H., Huang, F.Y. and Chen, S.C. (2005), "Improvement of surface finish on SKD steel using electro-discharge machining with aluminum and surfactant added dielectric", International Journal of Machine Tools and Manufacture, vol. 45, no. 10, pp. 1195-1201.

In: Machining and Forming Technologies, Volume 2
Editor: J. Paulo Davim
ISBN: 978-1-61122-820-5

Chapter 3

ON THE RESPONSE SURFACE MODELLING OF WIRE ELECTRICAL DISCHARGE MACHINING OF AL/SiC_P METAL MATRIX COMPOSITES (MMCS)

N. G. Patil* and P. K. Brahmankar
Department of Mechanical Engineering
Dr. Babasaheb Ambedkar Technological University,
Lonere – 402 103, Maharashtra, India

ABSTRACT

In this study, wire electrical discharge machining of aluminium matrix composites has been experimentally investigated. Three work materials viz. A359/SiC_p/10%, A359/SiC_p/20% and A359/SiC_p/30% have been machined at various combinations of machining parameters such as pulse on-time, off-time, wire speed and wire tension. Empirical models have been developed to predict the cutting rate, surface roughness and kerf width of the machined composites. The models were developed by using the technique of response surface methodology. The constants and coefficients of these models were determined by employing the results of the experiments. The experiments were conducted by using carefully planned response surface design matrix (central composite design). The model predictions are in good agreement with the experimental results. The R^2 values of the proposed cutting rate, surface finish and kerf width models are above 0.99, 0.99 and 0.98 respectively. Volume fraction of ceramic particle and pulse on-time were found to influence cutting rate and surface finish significantly. Kerf width was significantly affected by volume fraction, pulse on-time, off-time, wire speed, and wire tension.

Keywords: WEDM, MMCs, response surface modelling (RSM), surface finish, composite machining.

* Corresponding author, E-mail: nileshgpatil@rediffmail.com

1. Introduction

Wire electrical discharge machining (WEDM), Figure 1, is a widely accepted non-traditional materials removal process used to manufacture components with intricate shapes and profiles. Ho and Newman (2004) discussed about the wide acceptance to WEDM in the machining of various materials used in modern tooling applications. Altpeter and Perez (2004) mentioned that compared to die sinking electrical discharge machining, WEDM is a technology with increasing economic importance in terms of units sold per year. The ultimate goal of the WEDM process is to achieve an accurate and efficient machining operation without compromise in the machining performance. The settings for the various process parameters required in the WEDM process play crucial role in producing an optimal performance. Thus, the goal can be met mainly by understanding the inter-relationship between the large numbers of factors affecting the process and identifying the optimal machining conditions.

Many attempts have been made to obtain the relationship between pulse conditions and material removal by solving time dependent heat transfer equations assuming various heat source models. Patel et al. (1989) developed anode erosion model considering Gaussian heat flux on the anode surface and shown that the plasma flushing efficiency is same for both cathode and anode. Erden et al. (1995) compared the performance characteristics and accuracy obtained from many of the earlier models. Recently, Yeo et al. (2008) compared five EDM models from Snooyes, Van Dijck, Beck, Jilani, and DiBitonto in terms of the temperature distribution, crater geometry, and material removal at the cathode. Comparative analyses on the material removal rate (MRR) ratio of the predicted result to experimental data for discharge energy range from 0.33 to 952 mJ showed that DiBitonto's model yielded the closest proximity of 1.2–46.1 MRR ratio. They have suggested that the disk heat source models can be enhanced by improving the approximation of the heat flux and energy fraction.

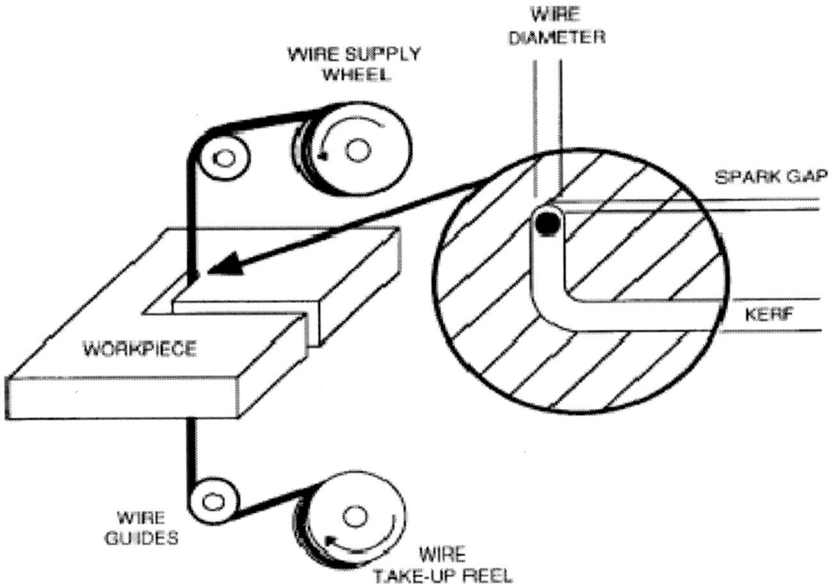

Figure 1. Details of WEDM spark gap.

All of these and most of the many other, theoretical models are concerned with die-sinking, in which cathodic workpiece polarity and rather long discharge times are employed.

For WEDM, in which the work piece polarity is anodic and the pulses are of short duration with a highly transient discharge current, very few such models exist; Spur and Schonbeck (1993) developed a theoretical model studying the influence of the workpiece material and the pulse type properties on the WEDM. They had shown the significance of thermal conductivity and melting temperature in WEDM.

In addition, many different types of problem-solving quality tools have been used to investigate the significant factors and its inter-relationships with the other variables in obtaining an optimal WEDM performance. Scot et al. (1991) used a factorial design to determine most favourable combination of WEDM parameters. They found that pulse current, pulse time and discharge voltage are the most significant control factors. Tarng et al. (1995) employed neural network to model the WEDM process to assess the optimal cutting parameters using an adjustable objective function. Liao et al. (1997) proposed an approach of determining the parameter settings using by taguchi methods. Material removal rate and surface finish were found to be affected by table feed rate and pulse on-time. The modelling of the WEDM process by means of empirical modelling techniques has also been applied to effectively relate the process parameters to the different performance measures of the WEDM process. Spedding and Wang (1997) developed response surface model using response surface methodology (RSM). They also used artificial neural network technology for the development of cutting rate and surface finish model. The predictions by artificial neural network model were better than that of the RSM model.

Although, large number of efforts to model this process empirically has been made, these studies were done on monolithic materials such as steels. Therefore, the aim of this study was to develop models for cutting rate, surface finish and kerf width in machining of ceramic reinforced aluminium matrix composites.

2. Machining of Metal Matrix Composites

In recent years, the critical need for less expensive structural materials that can provide an optimum level of performance has generated considerable research interest in the development and application of metal matrix composites (MMCs). Clyne and Withers (1992) discussed that use of MMC provides significant benefits including performance such as component's life, and improved productivity. Kannan and Kishawy (2006) mentioned that MMCs provide economic advantage through energy savings or lower maintenance cost and environmental benefits of lower noise levels and fewer air-borne emissions. Compared to the monolithic metals, MMCs have higher strength-to-weight ratios, better fatigue resistance, better elevated temperature properties, lower coefficients of thermal expansion, improved thermal conductivity, good damping characteristics, and excellent wear resistance. Further, it provides high flexibility in design. However, the utilization of the MMC in different industries is not as generalized as expected due to difficulties in the machining of MMCs. Cost effective machining has not been, yet, proven. Although primary fabrication processes such as casting, forging and extrusion have been well developed by a number of companies, cutting and finishing operations are still less well understood. Hung et al. (1999) did extensive studies in machining of MMCs. They found that the ceramic particles that enhance the mechanical properties of MMCs, and make them desirable also significantly reduce cutting tool life due to rapid abrasive wear. Chadwick and Heath (1990) studied into machining of MMCs and shown that PCD tools give longer service lives than carbide tools by 2-200 times and cause less subsurface damage. Further, P.J. Heath (2001) shown that tool life of PCD tools is much longer as compared to cemented carbide tools. J.P. Davim (2002) compared the performance of the diamond tools with CVD coated inserts and brazed PCD inserts in machining $Al/SiC_p/20\%$ composite. The life of PCD inserts were found to be almost 10 times longer, with a cutting speed 5 times higher than that of CVD coated tool. Further, the associated VB wear was about 1/3 of that of CVD tool. Over the years, it has been found that the tool wear is rapid for cutting tools other than polycrystalline diamond (PCD). The literature review on conventional machining of MMCs reveal that only PCD is a suitable tool to machine these materials. However, it is too costly and sub-surface damages, work hardening of matrix, delamination between matrix and particles, crack of particle or matrix are associated. Hence, poor machinability makes it one of the most challenging types of materials to be processed especially when intricate parts and precision parts are required.

The non-conventional machining processes such as abrasive water jet machining (AWJM) and laser machining can be applied (Muller and Monaghan, 2000). However, the equipment is too expensive and problems such as poor surface quality and limitation of work piece height are associated. Thus, electrical discharge machining (EDM) has evolved as one of the promising choices to machine these materials.

Many researchers have worked in electrical discharge machining of MMCs. Their findings include surface softening, effect of ceramic reinforcement on surface finish, effect of process parameters on surface finish and material removal rates. The common conclusions of all these studies suggest that the ceramic particles do not melt and may only dislodge after melting of the surrounding matrix. Extensive studies have been made on die-sinking EDM of these materials. Comparatively, very few studies have been undertaken in wire electrical discharge machining (WEDM) of MMCs. Further, most of these studies have been done by using one-factor-at-a-time approach which may not explain the effects of interaction among the various factors.

Some of the crucial findings are mentioned here. Roux et al. (1993) studied on the influence of EDM on surface integrity of Al/SiCp/15% MMC. Their findings suggest negligible influence of the reinforcements on surface roughness. The surface roughness was found to be related to pulse energy alone. Ramulu and Taya (1989) reported that the surface profiles were irregularly spaced and random on EDM machined Al/SiCp/15% MMC. Ramulu et al. (2001) shown that the surfaces were dominated by the matrix globules and not the SiC particles. Lau et al. (1995) suggested that machining of MMCs may not be possible using WEDM due to very frequent wire breakages. Yan et al. (2005) did extensive one-factor-at-a-time experimental study in WEDM of Al/Al_2O_{3p} composites. Their work concluded that the ceramic reinforcement was more significant than the energy level on surface finish of the MMCs. The surface roughness values were found to increase with increased volume fraction of the reinforcement. Almost similar findings were reported earlier by Yue and Dai (1996). Rozenek et al. (2001) worked on the wire electrical discharge machining characteristics of metal matrix composites. In this study, SiC and Al_2O_3 particulate reinforced aluminium matrix composites were machined by WEDM using one-factor-at-a-time methodology. They found current, voltage and on-time as significant parameters on cutting rate and surface finish. Patil and Brahmankar (2006) found that surface roughness of WEDM machined Al/SiCp composites was significantly influenced by the volume fraction of the SiC reinforcements. Further, the surface finish was found to improve with increased volume fraction. Thus, the reported literature shows that capable machining studies established in the past have not been optimized, expect a few papers reporting optimization of single type of MMC using taguchi analysis and grey relational analysis. Chiang and Chang (2006) have used grey-relational methods along with Taguchi's methodology to optimize Al/Al2O3P/20% MMC. They found that arc on and off-times were most significant on the machining performance.

Literature shows that the understanding of the WEDM process behaviour in machining of these materials as well as their machined surfaces is not complete. The above studies show promise for machinability of these materials using WEDM, however, they have not been incorporated into any modelling activity. Therefore, this work was an attempt towards better understanding of WEDM of these materials using response surface methodology.

3. Experiments

Ceramic particulate reinforced aluminium matrix composites were used as work material. The matrix alloy is A359 cast aluminium alloy. The reinforcements are silicon carbide (SiC) particles. The volume fraction of ceramic reinforcements was 10%, 20% and 30%.

The experiments were conducted by using Robofil 290, a wire electrical discharge machine (WEDM) manufactured by Charmilles technology. Figure 2 shows the photograph of the WEDM machine.

Cutting rates were taken from the machine display as well as average values in terms of mm/min. The surface roughness was measured by using Mitutoyo surface roughness tester (S J 301). The cut-off length was selected as 0.8 mm. During roughness measurements, total five measurements were taken on each specimen at different locations. The final value was taken as average of these measurements. The kerf widths were measured by using Nikon microscope with an optical magnification of 100X.

Figure 2. WEDM machine.

Figure 3 shows the photographs of roughness tester and microscope employed in this study. The SEM investigations were conducted by JEOL JSM–6380A analytical SEM. The details of experimental condition are shown in Table 1.

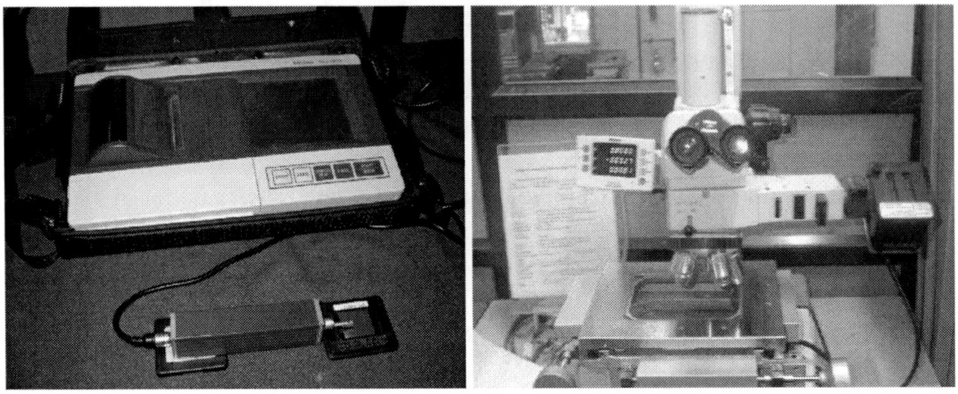

Figure 3. a) Surface roughness tester and b) Microscope for kerf width measurement.

Table 1. Details of experiments (fixed factors)

Machining parameters	Fixed level
Workpiece height	20 mm
Length of cut	20 mm
Angle of cut	Vertical
Location of the workpiece	Centre of the table
Work material[a]	A359/SiC_p/10%, 20% and 30%
Wire electrode	Coated brass wire, 0.25mm diameter, CuZn50
Temporary reduction in frequency (FF)	100
Short pulse time (T_{ac})	0.4μs
Ignition pulse current (IAL)	8 A
Servo reference voltage, (Aj)	45V
Open circuit voltage, (V)	-80 V
Injection pressure, (Inj)	11.5 MPa
Strategy (ST)	1
Dielectric	Deionized water, 15 μS/cm
Dielectric temperature	22 ^{0}C

4. DESIGN OF EXPERIMENTS

In the present study, experiments were designed on the basis of experimental design technique proposed by Box and Hunter (1957). The uniform-precision rotatable central composite experimental design (CCD) as demonstrated by Myers and Montgomery (1995) was used to improve the reliability of the results and to reduce the size of experiments without loss of accuracy. This design consists of 32 factorial (coded level ± 1) runs, 10 axial runs as well as 10 central trials. The strategy of complete randomization was adopted for conducting the experiments. On the basis of extensive preliminary investigations, literature and the experience of machine operators input parameters (control factors) were selected to develop empirical models for WEDM in machining of Al/SiC$_p$ composites. Pulse on-time (μs), pulse of-time (μs), wire feed speed (m/min) and wire mechanical tension (N) were selected as the input parameters. Table 2 shows the levels and range of these parameters for the experiments as well as their codes. The experimental matrix employed in this study is shown in Table 3. The coded numbers of the input parameters in Table 2 and Table 3 are obtained from the following transformation:

Pulse on-time, $$X_1 = \frac{A - A_m}{\Delta A}, \tag{1}$$

Pulse off-time, $$X_2 = \frac{B - B_m}{\Delta B}, \tag{2}$$

Wire speed, $$X_3 = \frac{WS_m - WS_m}{\Delta WS}, \tag{3}$$

Wire tension, $$X_4 = \frac{WB - WB_m}{\Delta WB}, \tag{4}$$

Volume fraction, $$X_5 = \frac{Vp - Vp_m}{\Delta Vp} \tag{5}$$

where, X_1, X_2, X_3, X_4 and X_5 are the coded values of the variables A, B, WS, WB and volume fraction respectively. A_m, B_m, WS_m and WB_m are the values of pulse on-time, off-time, wire speed and wire tension at zero level respectively. The intervals of variation in A, B, WS and WB are ΔA, ΔB, ΔWS and ΔWB respectively. The volume fraction was the only factor which was varied at three levels 10%, 20% and 30% of the SiC particulate reinforcements. Therefore, the axial runs for volume fraction were conducted at the factorial levels of ceramic volume fraction.

Table 2. Experimental parameters settings

Level	Pulse on-time, A (μs)	Pulse off-time, B (μs)	Wire speed, WS (m/min.)	Wire tension, WB (N)	Volume fraction, V (%)
-2	0.2	10	8	8	10
-1	0.4	12	9	9	10
0	0.6	14	10	10	20
+1	0.8	16	11	11	30
+2	1	18	12	12	30

5. Results and Discussion

The RSM is a dynamic and foremost important tool of design of experiment (DOE) wherein the relationship between response(s) of a process with its input decision variables is mapped.

5.1. Response Surface Modelling (RSM)

The modelling of the WEDM process in the machining of Al/SiC$_p$ composites can be regarded as a problem of correlating the input parameters which include the ceramic particulate volume fraction in the MMCs with its output parameters like cutting rate, surface roughness and kerf width. This is dealt with by means of response surface methodology (RSM). The most important performance measures in WEDM are material removal rate (or cutting speed), workpiece surface finish and kerf (cutting width). Pulse on-time, pulse off-time, wire speed, and wire tension, are the machining parameters which affect the performance measures. In the present study, RSM models were developed for cutting rate, surface finish, and kerf width.

In RSM, the quantitative form of relationship between the desired response and the independent input variables (Montgomery, 2001) could be represented as:

$$Y_i = f(x_1, x_2, x_3, x_4, x_5) \pm \varepsilon \tag{6}$$

where Y_i is the desired response, **f** is the response function, x_1, x_2, x_3, . . ., x_n are the, independent input variables and ε is the fitting error. The appearance of response function is a surface as plotting the expected response of **f**. The identification of suitable approximation of **f** will determine whether the application of RSM is successful or not. In this study, the approximation of **f** will be proposed using the fitted second order polynomial regression model, which is called the quadratic model. The quadratic models of y_i for 5 input parameters can be written as following:

$$Y_{CR} = \alpha_0 + \sum_{i=1}^{5} \alpha_i x_i + \sum_{i=1}^{5} \alpha_{ii} x_i^2 + \sum_{j>i}^{5} \alpha_{ij} x_i x_j \tag{7}$$

$$Y_{SF} = \beta_0 + \sum_{i=1}^{5} \beta_i x_i + \sum_{i=1}^{5} \beta_{ii} x_i^2 + \sum_{j>i}^{5} \beta_{ij} x_i x_j \tag{8}$$

$$Y_{KERF} = \gamma_0 + \sum_{i=1}^{5} \gamma_i x_i + \sum_{i=1}^{5} \gamma_{ii} x_i^2 + \sum_{j>i}^{5} \gamma_{ij} x_i x_j \tag{9}$$

where α_0, β_0, and γ_0 are constants for CR, R_a, and kerf respectively. Further, α_i, β_i and γ_j represent the coefficients of linear terms. The quadratic and interaction terms are denoted by α_{ii} and α_{ij} respectively for the cutting rate model. Similarly, β and γ denotes these corresponding coefficients for surface finish and kerf width models respectively. Here, x_i is the variable corresponding to process parameters such as on-time, off-time, wire speed, wire tension and ceramic volume fraction. The response surface may contain linear terms, squared terms and interaction terms. The proposed quadratic model of Y_i in this study investigates over the entire process parameter space. The cutting rate, surface roughness (R_a) and kerf width as indicated byY_{CR},Y_{SF} and Y_{KERF} respectively were analyzed. The necessary data for building the response models were collected by the design of experiments. These experimental results are shown in the CCD design matrix (Table 3).

Table 3. Design matrix central composite design (CCD)

Exp. No.	A	B	WS	WB	V	CR, mm/min.	Ra, µm	Kerf, mm
1	0.4	12	9	9	10	3.3	2.76	0.330
2	0.8	12	9	9	10	7.1	3.48	0.370
3	0.4	16	9	9	10	2.6	2.73	0.326
4	0.8	16	9	9	10	5.6	3.44	0.370
5	0.4	12	11	9	10	3.3	2.74	0.330
6	0.8	12	11	9	10	7.1	3.42	0.380
7	0.4	16	11	9	10	2.4	2.71	0.330
8	0.8	16	11	9	10	5.5	3.44	0.370
9	0.4	12	9	11	10	3.2	2.83	0.330
10	0.8	12	9	11	10	7.1	3.54	0.362
11	0.4	16	9	11	10	2.5	2.75	0.330
12	0.8	16	9	11	10	5.6	3.40	0.367
13	0.4	12	11	11	10	3.2	2.75	0.330
14	0.8	12	11	11	10	7.1	3.48	0.370
15	0.4	16	11	11	10	2.5	2.72	0.330
16	0.8	16	11	11	10	5.4	3.46	0.370
17	0.4	12	9	9	30	2.6	2.04	0.315
18	0.8	12	9	9	30	5.5	2.48	0.350
19	0.4	16	9	9	30	2.0	2.05	0.320
20	0.8	16	9	9	30	4.2	2.44	0.355
21	0.4	12	11	9	30	2.6	2.03	0.320
22	0.8	12	11	9	30	5.4	2.42	0.355
23	0.4	16	11	9	30	1.9	2.10	0.320
24	0.8	16	11	9	30	4.2	2.5	0.356
25	0.4	12	9	11	30	2.5	2.08	0.320
26	0.8	12	9	11	30	5.5	2.48	0.340
27	0.4	16	9	11	30	1.9	2.04	0.320
28	0.8	16	9	11	30	4.3	2.40	0.350
29	0.4	12	11	11	30	2.5	1.98	0.320
30	0.8	12	11	11	30	5.4	2.39	0.350
31	0.4	16	11	11	30	1.9	2.02	0.320
32	0.8	16	11	11	30	4.2	2.45	0.350
33	0.2	14	10	10	20	1.0	2.00	0.290

Table 3. (Continued)

Exp. No.	A	B	WS	WB	V	CR, mm/min.	Ra, μm	Kerf, mm
34	1.0	14	10	10	20	7	3.10	0.360
35	0.6	10	10	10	20	4.8	2.54	0.340
36	0.6	18	10	10	20	2.8	2.53	0.340
37	0.6	14	8	10	20	3.7	2.54	0.330
38	0.6	14	12	10	20	3.6	2.53	0.340
39	0.6	14	10	8	20	3.7	2.56	0.340
40	0.6	14	10	12	20	3.7	2.55	0.330
41	0.6	14	10	10	10	4.3	3.14	0.360
42	0.6	14	10	10	30	3.4	2.24	0.344
43	0.6	14	10	10	20	3.8	2.50	0.344
44	0.6	14	10	10	20	3.7	2.48	0.345
45	0.6	14	10	10	20	3.8	2.50	0.346
46	0.6	14	10	10	20	3.7	2.56	0.348
47	0.6	14	10	10	20	3.8	2.52	0.348
48	0.6	14	10	10	20	3.6	2.52	0.352
49	0.6	14	10	10	20	3.6	2.54	0.354
50	0.6	14	10	10	20	3.7	2.50	0.350
51	0.6	14	10	10	20	3.8	2.53	0.350
52	0.6	14	10	10	20	3.6	2.49	0.350

The constants and coefficients for these models were determined by using the CCD experimental results. The final forms of these models are shown as below:

$$Y_{CR} = 5.6898 + 1.93078A - 0.6438B - 0.025WS - 0.01WB - 0.0149V + 0.0808A^2 + 0.0308B^2 - 0.175A*B - 0.0021A*V + 0.0006B*V \quad (10)$$

$$Y_{KERF} = 0.3920 + 0.0223A - 0.00156B + 0.00165WS - 0.00145WB - 0.000372V - 0.005529A^2 - 0.00178B^2 - 0.00303WS^2 - 0.00303WB^2 + 0.000001V^2 + 0.000875A*WS - 0.00175A*WB - 0.000023\,A*V \quad (11)$$

$$Y_{SF} = 3.9574 + 0.4288A - 0.03459B - 0.00875WS - 0.02586WB - 0.01014V - 0.00001V^2 - 0.000023\,A*V + + 0.01969B*WS - 0.01031B*WB + 0.00014B*V - 0.00013WB*V \quad (12)$$

It is to be noted from the final equations that some of the coefficients are not considered. Only significant parameters and their coefficients are included in the final equation. The remaining insignificant parameters are omitted. The Student's *t*- test was employed to determine the significance and non-significance of these parameters and their coefficients. Further, the adequacy of the proposed models was also checked by the variance analysis (*F*-test). The R value for the cutting rate, surface finish and kerf width models are 0.99, 0.99 and 0.98 respectively. It indicates very good fit and agreement between experiments and the model predictions. Further, the lack-of-fit is insignificant in cutting rate, surface finish and kerf width models. Table 4, Table 5 and Table 6 show the analysis of variance of the presented response surface regression equations.

Table 4. Analysis of Variance for cutting rate

Source	DF	SS	MS	F	P	
Model	11	115.091	10.463	2000	0.000	Significant
Residual error	40	0.178	0.00445			
Lack-of-Fit	31	0.122	0.00393	0.63	0.836	Not significant
Pure error	9	0.056	0.00622			
Total	51	115.269				
R^2 = 99.7%, R^2(adj) = 99.7%						

Table 5. Analysis of Variance for surface roughness, R_a

Source	DF	SS	MS	F	P	
Model	11	9.8625	0.8965	2000	0.000	Significant
Residual error	40	0.02187	0.000547			
Lack-of-Fit	31	0.01643	0.00053	0.88	0.634	Not significant
Pure error	9	0.00544	0.000604			
Total	51	9.88438				
R^2 = 99.8%, R^2(adj) = 99.7%						

Table 6. Analysis of Variance for kerf width

Source	DF	SS	MS	F	P	
Model	15	0.0165	0.0011	163.2	0.000	Significant
Residual error	36	0.000243	0.000007			
Lack-of-Fit	27	0.000155	0.0000006	0.58	0.865	Not significant
Pure error	9	0.000088	0.00001			
Total	51	0.1674				
R^2 = 98.6%, R^2(adj) = 97.7%						

5.2. Influence of Process Parameters on Cutting Rate

Cutting rate determines the productivity of WEDM. It is the forward feed rate of the wire electrode in to the work material during machining. It is found to be influenced by the energy supplied in the gap. The greater energy supplied in the gap, the more material will be removed and consequently the gap size will increase. This increased gap size will be compensated by the increased rate of wire forward feed. Thus, the supplied power in the gap directly influences the cutting rate. Generally, the supplied power is related to the gap voltage, discharge current and pulse on-time. In the WEDM machine, the voltage remains almost constant and the values of peak current is determined by the slope of the rise of the current. Therefore, for a fixed slope of current rise, the value of peak current depends on the pulse on-time. Thus, the energy supplied in the gap is directly dependent on the pulse on-time. In the present study also the cutting rate was found to increase with increased pulse on-time. Other significant machining parameter on cutting rate was pulse off-time. Pulse off-time determines the pulse frequency at fixed value of on-time. Increased off-time reduces the pulse frequency and consequently reduces the energy supplied per unit time. Therefore, an increase in off-time reduces cutting speed. Similar trend was found during the machining experiments. Other machining parameters, in the present study, wire speed was also found to have significant effect on the cutting rate. However, the level of significance was relatively lower than pulse on-time and volume fraction. This result also agrees with many other earlier findings on WEDM.

The composite material specific parameter, i.e. ceramic reinforcement volume fraction was also chosen as input parameter in this study. This input parameter was found to be the most significant parameter on cutting rate. The volume fraction of ceramic reinforcements affects the material properties of the composite materials. The variation in volume fraction causes significant changes in the thermo-physical properties of the composite materials. In addition, the presence of ceramic particles in the machining electrode gap adversely affects the gap conditions. Thus, in both the cases the volume fraction influences the performance of the WEDM process. In the present study, cutting rate was found to decrease when composite with higher ceramic volume fraction was machined. These results agree with earlier studies by Yan et al. (2005) and Rozenek et al. (2000). Thus, the ceramic particles were found to adversely alter the gap conditions as well as material properties for WEDM process. Figure 4 shows the effect of input parameters on cutting rate.

Figure 5 and Figure 6 show the response surface for the combined effects of significant parameters on cutting rate. Some more trials were conducted other than the CCD design matrix to check the values of cutting rate at -2 and +2 levels of on-time, off-time, wire speed and wire tension for Al/SiC$_p$/10% and 30% reinforced MMCs (Table 7).

Experiments show that cutting rate varied from 1.3 to 8.4 mm/min. for Al/SiC$_p$/10% MMC, and ranged between 0.9 and 6.4 mm/min. for Al/SiC$_p$/30% MMC. The cutting rate was found to be ranged between 1 mm/min. and 7.1 mm/min. for Al/SiC$_p$/20% MMC. Further, the Al/SiC$_p$/20% composite was also machined at -1 and +1 levels of on-time, and off-time to compare between the predictions and experimental results of cutting rate. Figure 7 shows the comparison between predicted and the actual values.

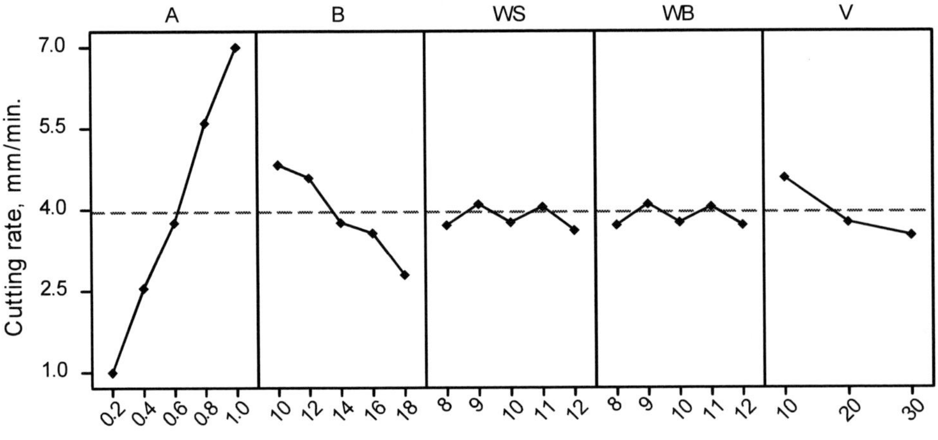

Figure 4. Influence of input parameters on cutting rate.

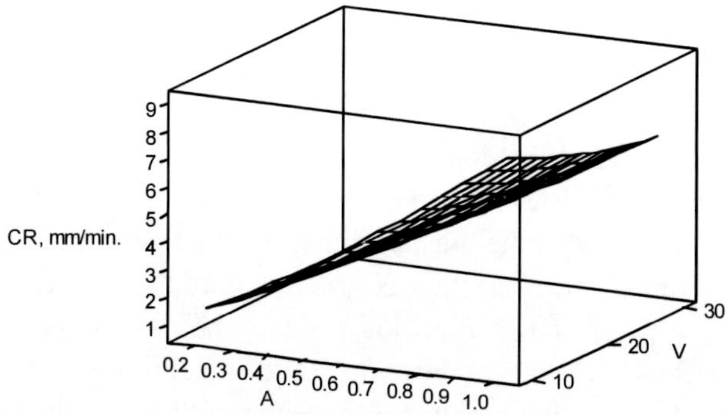

Figure 5. Response surfaces for combined effects of on-time and volume fraction.

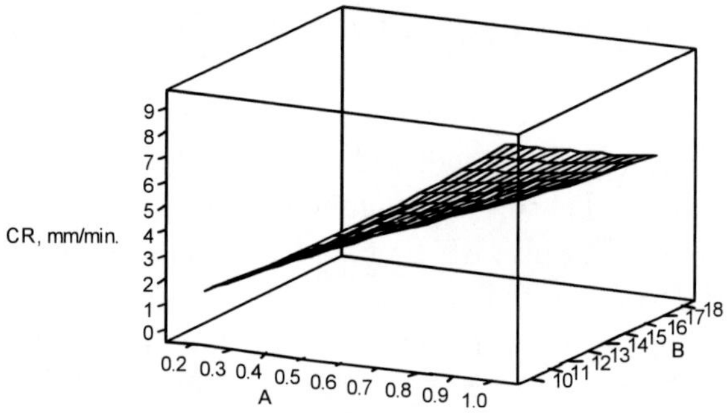

Figure 6. Response surfaces for combined effects of on-time and off-time.

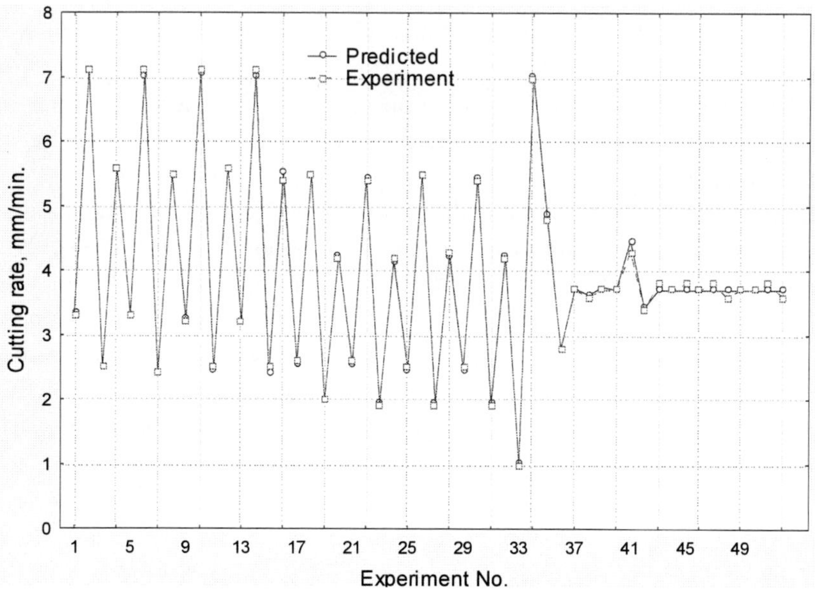

Figure 7. Comparisons between the actual and predicted values.

5.3. Influence of Process Parameters on Surface Roughness

In the present study, the quality of surface was measured in terms of the arithmetic average value of the surface profile denoted by R_a. This parameter gives a good general description of height variations. Based on the literature and our findings, it is believed that there is no significant difference in the surface finish in different directions in machining of particulate reinforced composites due to their isotropic nature. Therefore, R_a value was chosen to represent the topography of WED machined metal matrix composites. The distinctive morphology of a surface which has undergone EDM machining is due to the enormous amount of heat generated by discharges, which causes melting and vaporization. Further, in machining of ceramic particulate reinforced composites, the presence of ceramic particles in the material makes it more distinct from other metallic surfaces machined with EDM.

Pulse on-time was found to significantly influence the R_a value of the machined surface. At higher on-time the supplied energy in the gap is greater. Therefore, at high level of supplied energy, the volume of molten metal is also high. Part of this large molten metal flushes away from the machined surface by the high pressure dielectric fluid. The surface is then form by the tiny craters. The depth of these craters determines the roughness of the machined surface. Therefore, with increased on-time, the depth of craters also increases which consequently increases the surface roughness.

Other machine parameters pulse off-time and wire tension also affects the roughness of the machined surfaces. However, the significance of wire tension was found be low and the trend is not clear. The roughness was found to reduce with increased off-time (Figure 11).

It is well known that ceramic particles do not melt and/or evaporate during EDM machining. Rather, they either dislodge from the molten metal or remain protruding on the machined surfaces. In both the cases, thus, these particles affect the characteristic of the machined surfaces. Therefore, the volume fraction of the ceramic reinforcement significantly affects the roughness. The roughness was found to decrease with increase in the ceramic volume faction. The observed difference between surface roughnesses of MMCs with different volume fraction of particle reinforcements can be attributed to the following reasons:

- The difference in thermo-physical properties such as thermal diffusivity, thermal coefficient of expansion and specific heat with different volume fraction of reinforcing particles.
- If the particles are completely dislodged or pulled out of the machined surface, the depth of these craters contributes toward the roughness profile. However, in WEDM the molten metal is not completely flushed away, rather part of the metal remains as resolidified layer. Thus, while flowing, the melt might smear the craters created by the discharge as well as due to particle pull-out. Thus, resulting into shallow craters.
- The particles may impede the flow of the molten metal resulting into shallow craters.
- In addition to above causes, the presence of insulating ceramic particles, may increase the ignition delay time which may reduce the available energy per unit time in the electrode gap. Therefore, the craters are shallow with an increased percentage of insulating ceramic particles.
- The reaction between aluminium alloy and the SiC particles at higher temperatures such as WEDM conditions may cause diffusion of carbon from SiC. This might

affect the roughness as most of the particles are not present on the machined surfaces.

Thus, the increased percentage of the ceramic reinforcement contributes to reduce the resulting surface roughness either by particle shielding and/or due to the smearing of the pull-out craters. Figure 8 shows micrographs of the machined surfaces. The topography of the machined surfaces reveals that, most of the ceramic particles were released and the surfaces were dominated by the resolidified matrix alloy. The dominating matrix globules are depicted clearly at higher magnification as shown in Figure 8 (a), same results were reported by Gatto et al. (1997) during machining of Al/SiC composites. Figure 8 (b) shows the presence of few dislodged particles on the machined surface of Al/SiC_p/30% composite. The presence of SiC particles was confirmed by EDX analysis, Figure 8 (c).

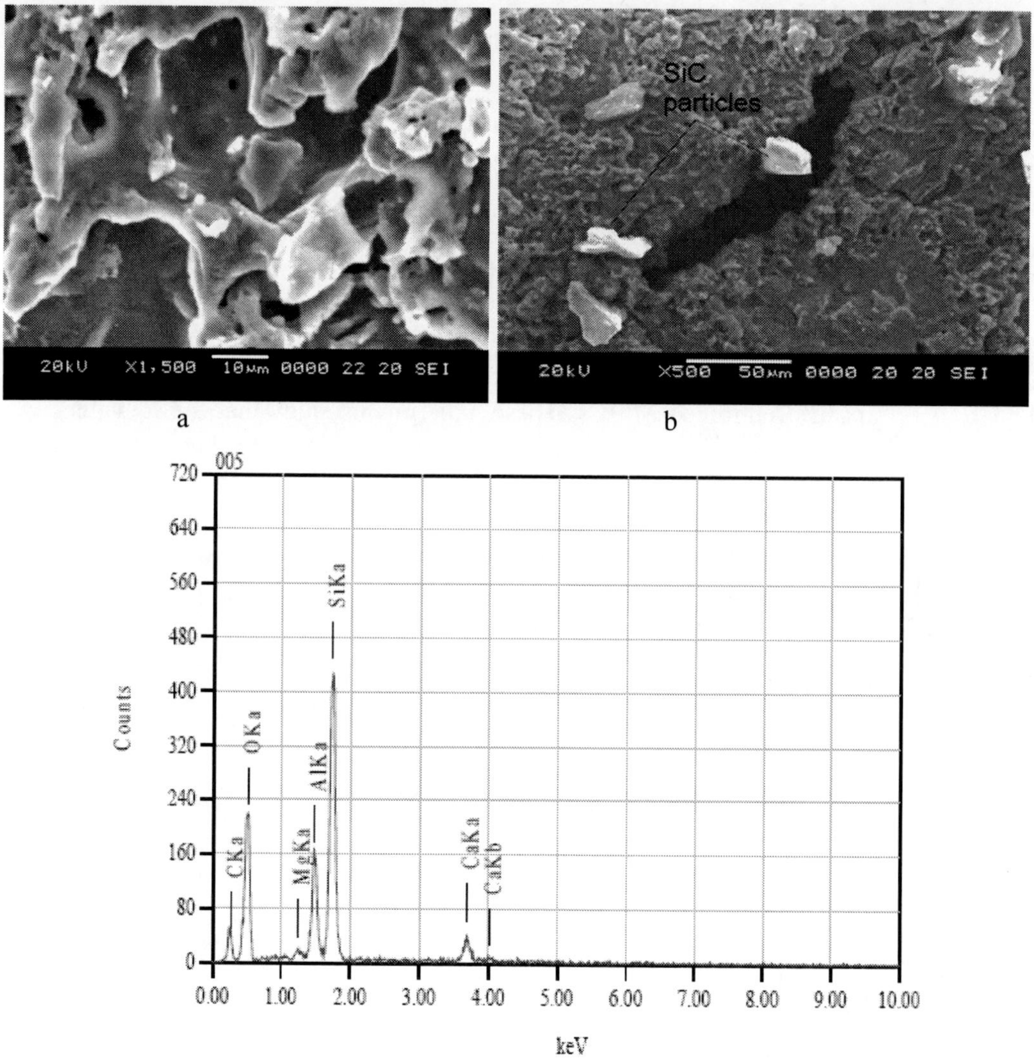

Figure 8. WEDM surfaces (a) machined sample of Al/SiC_p/20% MMC, (b) Al/SiC_p/30% MMC, (c) EDX of particles in figure (b).

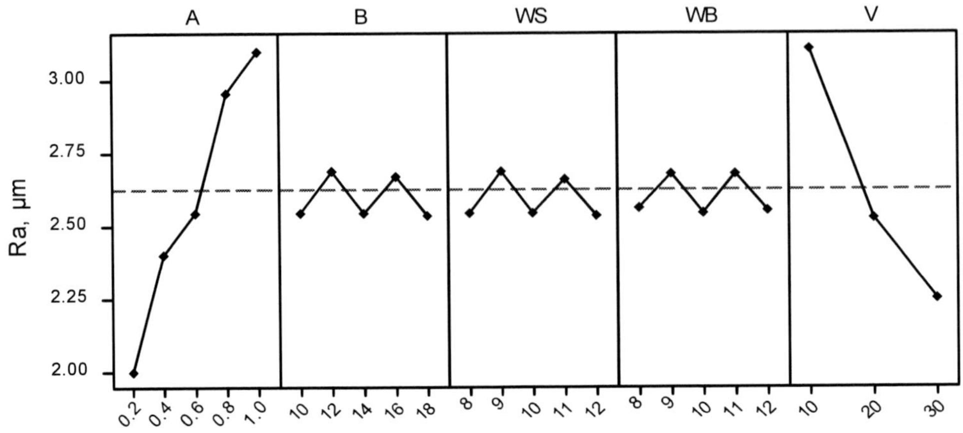

Figure 9. Influence of input parameters on surface roughness, R_a.

Therefore, the roughness decreases with increased volume fraction of the ceramic reinforcement. This positive effect of the particle reinforcements on the WEDM machined surfaces are reported by Muller and Monaghan (2000) in their studies on die-sinking EDM of Al/SiC composites.

However, these results of improved surface finish with increasing ceramic particle volume fraction do not agree with the reported studies by Yan et al. (2005) where it has been shown that the roughness increases with increased volume fraction of the reinforcements. Some more trials were conducted other than the CCD design matrix to check the values of surface roughness at -2 and +2 levels of on-time, off-time, wire speed and wire tension for Al/SiC$_p$/10% and 30% reinforced MMCs. These experiments show that the R_a values were found to vary between 1.84 µm to 2.74 µm in Al/SiC$_p$/30% composites and ranged between 2.4 and 3.9 in Al/SiC$_p$/10% MMC (Table 7).

In addition, the Al/SiC$_p$/20% composite was also machined at -1 and +1 levels of on-time, and off-time to compare between the predictions and experimental results of surface roughness. The surface roughness was found to be ranged between 2 µm and 2.9 µm. Figure 8 shows the effects of input parameters on surface roughness.

Further, Figure 10 shows the combined effect on-time and volume fraction on surface roughness. Figure 12 shows the comparison between measured and predicted roughness values.

Table 7. Additional experiments

Exp. No.	A	B	WS	WB	V	CR, mm/min.	Ra, µm	Kerf, mm
1	0.2	14	10	10	20	1	2	0.290
2	0.4	16	9	9	20	2	2.44	0.32
3	0.4	12	9	9	20	2.6	2.43	0.324
4	0.8	12	9	9	20	5.6	2.76	0.354
5	0.8	16	9	9	20	4.5	2.74	0.354
6	0.6	14	10	10	20	3.7	2.54	0.345
7	1	14	10	10	20	7.1	3.1	0.360

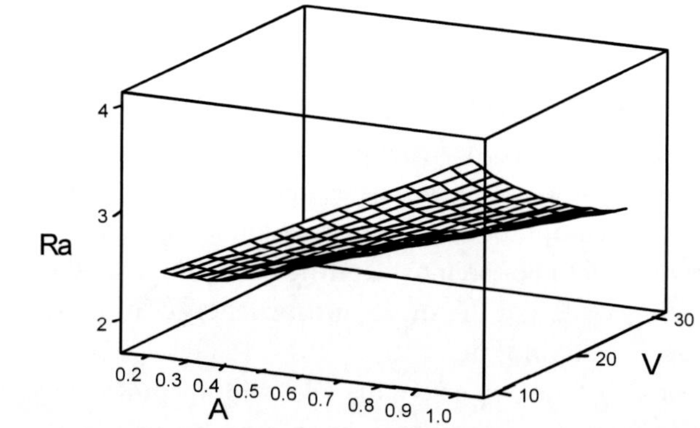

Figure 10. Response surfaces for combined effects of on-time and volume fraction on R_a

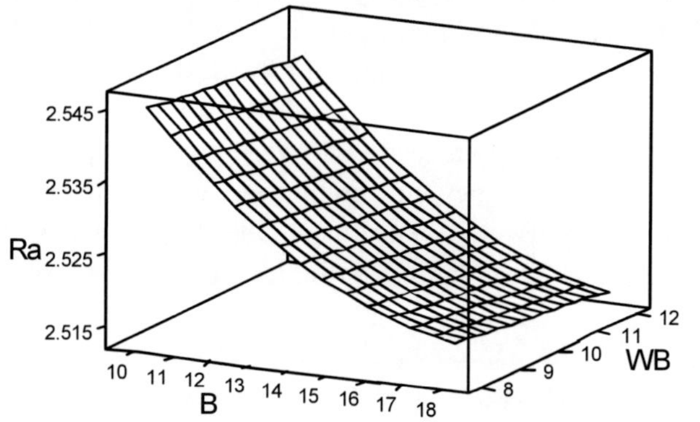

Figure 11. Response surfaces for combined effects of off-time and wire tension on R_a

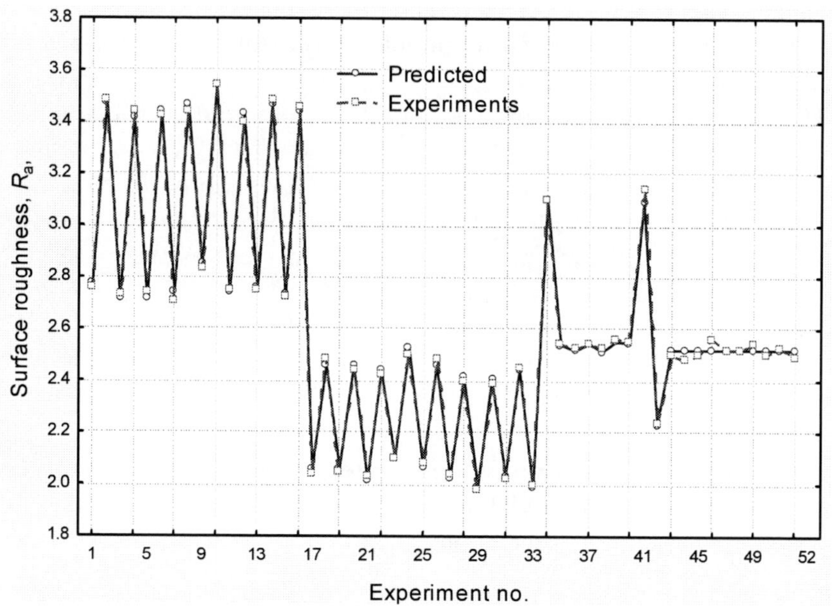

Figure 12. Comparisons between the actual and predicted values of surface roughness, R_a

5.4. Influence of Input Process Parameters on Kerf Width

The kerf (Figure 1), which determines the dimensional accuracy of the finished part, is of utmost importance. The internal corner radius to be produced in WEDM operations is also limited by the kerf. The wire–workpiece gap usually ranges from 0.025 to 0.075mm and is constantly maintained by a computer controlled positioning system.

Pulse on-time was found to be the most significant input parameter on kerf width. Longer on-times supply larger current, and therefore, greater energy in the spark gap. Thus, the gap width increases with increased on-time.

Pulse off-time was also found to affect the kerf width significantly. Kerf width increases with increased off-time. However, it was found to reduce beyond 14μs off-time.

Wire speed was also found to be significant on the kerf width. The spark gap and kerf was found to increase with increased wire speed. This is attributed to the lower wire wear at higher wire speed. The high wire speed subjects the wire to low discharge energy per unit time due to less time spent in the workpiece part. In case of low wire speed, the wire wear is larger which reduces the wire diameter compared to wire that has spent less time in the gap. This reduced wire diameter reduces the kerf width due to constant spark gap. However, at very low cutting speed the rate of wire wear was very high to cause severe wire breakages. Further, at higher wire speeds, i.e. beyond the highest selected level in this study, the wire stability of gap is adversely affected. This instability in the gap causes significant wire vibration and instability. Thus, these large amplitudes of wire vibration may also be cause of wire breakages at very high wire speed. This effect was found to be more significant in machining of MMCs where the gap contamination is more than unreinforced metal such as steel. The reported studies on the wire vibration and wire speed have demonstrated the correlation between wire speed and wire vibration (Lambert, Murphy, 2002 and Shahruz, 2003).

Wire mechanical tension was also found to be significant on kerf width. The kerf was found to be almost constant at -1 and +1 levels of wire tension. However, it increased from the lowest to highest levels of wire tension in this study. This may be attributed to the wire vibration and slackness at lower wire tension. At higher wire tension the kerf was found to be minimum. This might be due to more stability of wire and in turn efficient utilization of the discharge energy.

The volume fraction of ceramic reinforcements was found to be very significant on kerf width. The kerf width decreases with increased volume fraction of the SiC particles. This was attributed to the change in thermo-physical properties as well as the presence of ceramic particles in the spark gap.

Kerf width was found to vary from 0.3 to 0.38m in machining of Al/SiC_p/10% composites. It was in the range of 0.29 and 0.36 mm in the machining of Al/SiC_p/30% MMC. In the machining of Al/SiC_p/20% composites, the kerf width was found to be ranged between 0.294 to 0.363 mm.

Figure 13 shows the effect of process parameters on kerf width. Figure 14, 15 and 16 show the response surfaces for kerf width. Figure 17 shows the comparisons between model predictions and the experimental observations.

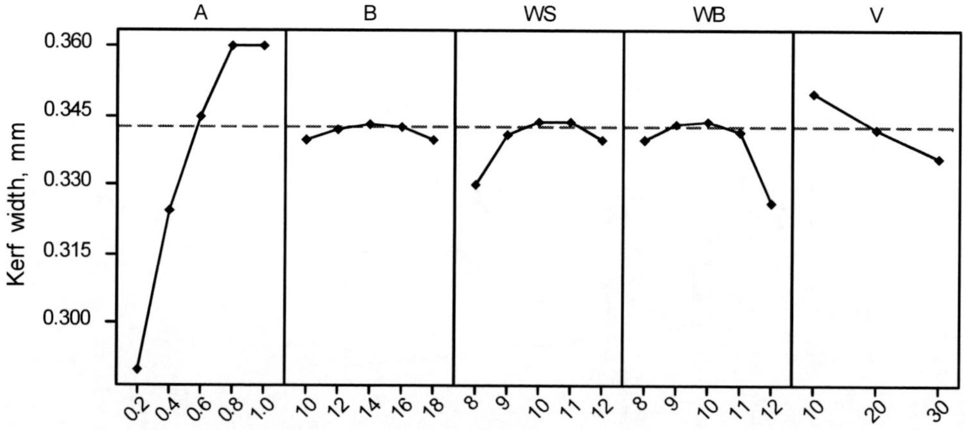

Figure 13. Influence of input parameters on kerf width.

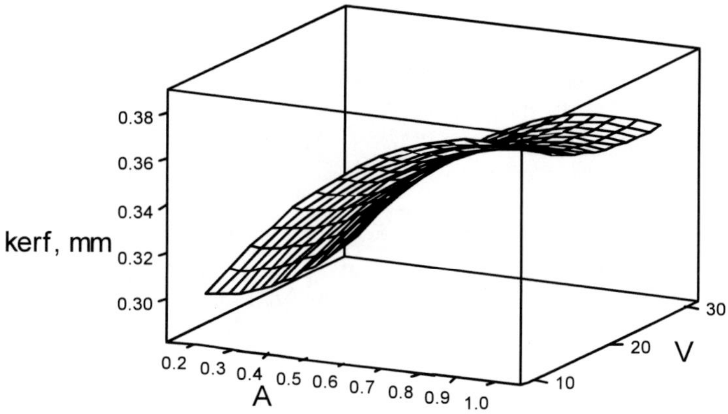

Figure 14. Response surfaces for combined effects of on-time and volume fraction on kerf width.

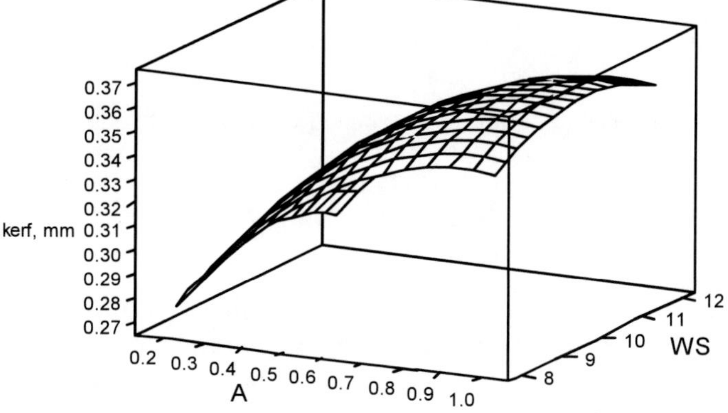

Figure 15. Response surfaces for combined effects of on-time and wire speed on kerf width.

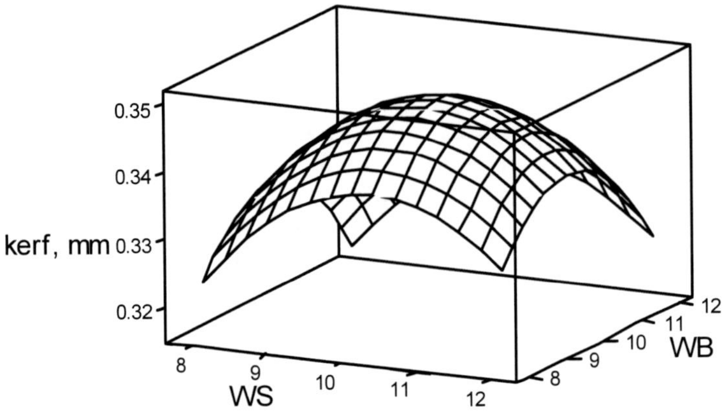

Figure 16. Response surfaces for combined effects of wire speed and wire tension on kerf width.

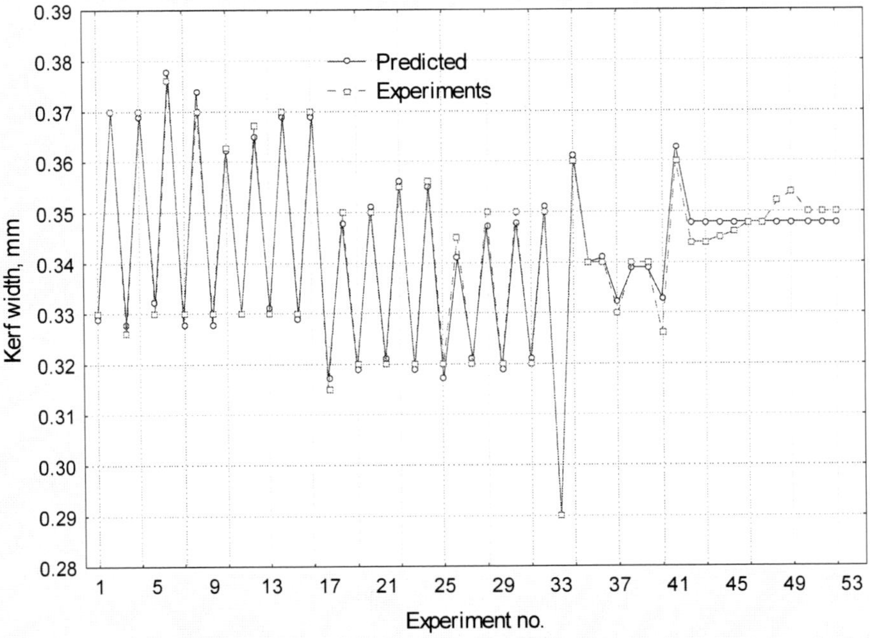

Figure 17. Comparisons between the actual and predicted values of kerf width.

CONCLUSIONS

The wire electrical discharge machining of A359/SiC_p composite with varying volume fraction of 10%, 20% and 30% were investigated. Empirical models were developed for cutting rate, surface roughness and kerf width in WEDM of these materials. The models were developed by using the technique of response surface methodology. The machining parameters selected for the models were pulse on-time, off-time, wire speed and wire tension.

The constant and coefficients of these models were determined by using the results of designed experiments.

Finally it can be concluded that:

1. WEDM is an adequate process to machine metal matrix composites with good surface finish and dimensional precision.
2. Volume fraction of ceramic reinforcement, pulse on-time and off-time were found to be significant on cutting rate.
3. Volume fraction of ceramic reinforcement and pulse on-time were found to be significant on surface finish
4. Volume fraction of ceramic reinforcement, pulse on-time, off-time, wire speed and wire tension were found to be significant on kerf width.
5. Higher volume fraction of the ceramic reinforcements resulted in to reduced cutting rates.
6. The surface finish of composites was found to improve with increased volume fraction of SiC reinforcements.
7. The machined surfaces of A359/SiC_p composites were dominated by the globules of matrix material.
8. The cutting rates varied between 1.3 mm/min. and 8.4 mm/min. in Al/SiC_p/10% composite and ranged within 0.9 mm/min. and 6.4 mm/min for Al/SiC_p/30%.MMC.it was found to vary from 1mm/min. to 7.1 mm/min. for Al/SiC_p/20%.
9. The surface roughness varies between 2.54 μm and 3.9 μm in Al/SiC_p/10% composite and ranged within 1.7 μm and 2.74 μm for Al/SiC_p/30%.MMC. It was found to vary from 2 μm to 3.1 μm for Al/SiC_p/20% composite.
10. Kerf width was found to vary from 0.3 to 0.38 mm in machining of Al/SiC_p/10% composites. It was in the range of 0.29 and 0.36 mm in the machining of Al/SiC_p/30% MMC. In the machining of Al/SiC_p/20% composites, the kerf width was found to be ranged between 0.294 to 0.363 mm.
11. The R values of the proposed cutting rate and surface finish model are above 0.99 and for the kerf width the R value is 0.98.
12. In future, further studies can address the performance measures such as surface quality including surface alterations in these materials. Similarly, other performance measures such as pulse frequency during machining of MMCs can be investigated. The future research may be in the direction to avoid the wire breakage and wire shifting in machining of these materials.

ACKNOWLEDGMENTS

The authors are thankful for to Mr. Kapase H.D., Mr. Patil R.D., Mr. Shivaiah, Mr. Bagul J.D. and Mr. Nere of Indo–German Tool Room, Aurangabad, India for the facility of WEDM machine. We express our deep appreciation of DURALKAN USA for supplying us with the composite materials for this investigation.

REFERENCES

Altpeter, F., Perez, R., 2004. Relevant topics in wire electrical discharge control. *Journal of Materials Processing Technology*. 149, 147-151.

Box, G.E., Hunter, J.S., 1957. Multifactor experimental designs, *Ann. Math. Stat.* (1957) 28–195.

Chadwick, G.A., Heath, P.J., 1990. Machining metal matrix composites. *Metals and Materials,* 73-76.

Chiang, Ko-Ta, Chang, Fu-Ping., 2006. Optimization of the wedm process of particle-reinforced material with multiple performance characteristics using grey relational analysis. *Journal of Materials Processing Technology*. 180, 96-101.

Clyne, T.W., Withers, P.J., 1992. *An introduction to Metal matrix composites*. Cambridge University Press, London.

Davim, J.P., 2002. Diamond tool performance in machining metal matrix composites. *Journal of Materials Processing Technology*. 128, 100-105.

Erden, A., Arinc, F., Kogman, M, 1995. Comparison of mathematical models for electric discharge machining. *J. Mater. Process. Manuf.* Sci. 4, 163-176.

Heath, P.J., 2001. Developments in applications of PCD tooling. *Journal of Materials Processing Technology.* 116, 31-38.

Ho, K.H., Newman, S.T., 2004. State of the art in wire electrical discharge machining (WEDM). *International Journal of Machine Tools and Manufacture*. 44, 1247-1259.

Hung, N.P., Loh, N.N., Venkatesh, V.C., 1999. Machining of metal matrix composites. Machining of Ceramics and Composites, Marcel Dekker, Inc., New York. 295-298.

Kannan, S., Kishawy, H.A., 2006. Surface characteristics of machined aluminium metal matrix composites. *International Journal of Machine Tools and Manufacture*. 46/15, 2017-2025.

Langhaar, H.L., 1957. Dimensional analysis and theory of models. John Wiley and Sons, Inc., New York.

Lau, W.S., Yue, T.M., Lee, T.C., Lee, W.B., 1995. Un-conventional machining of composite materials. *Journal of Materials Processing Technology* 48. 199-205.

Liao , Y.S., Huang, J.T., Su, H.C., 1997. A study on the machining parameters optimization of wire electrical discharge machining. *Journal of materials processing technology*. 71(3), 487-493.

Muller, F., Monaghan, J., 2000. Non-conventional machining of particle reinforced metal matrix composite. *International Journal of Machine Tools and Manufacture*. 40, 1351-1366.

Myers, R.H., Montgomery, D.C., 1995. *Response surface methodology: process and product optimization using designed experiments.* John Wiley and Sons. Inc., New York.

Patel, M.R., Barrufet, M.A., Eubank, Ph.T, Dibitonto, D.D., 1989. Theoretical models of the electrical discharge machining process-II: The anode erosion model. *Journal of Applied physics.* 66/9, 4104-4111.

Patil, N.G., Brahmankar, P.K., 2006. Some investigations into wire electro-discharge machining performance of Al/SiC_p composites. *Int. J. Machining and Machinability of Materials.* 1, 412-431.

Patil, N.G., Brahmankar, P.K., 2009. Some investigations into combined effect of ceramic reinforcement and process parameters into electro-discharge machining of Al/SiC_p composites. *Journal of Machining and Forming Technology 1*, 113-128..

Ramulu, M. Taya, M., 1989. EDM machinability of SiCw/Al composites. *Journal of Materials Science* 24, 1103-1108.

Ramulu, M., Paul G., Patel J., 2001. EDM surface effects on the fatigue strength of a 15 vol% SiCp/Al metal matrix composite material, Composite Structures, 4, 79-86.

Roux T. L., Wise L.H., Aspinwall, D.K., 1993. The effect of electrical discharge machining on the surface integrity of an aluminium-silicon carbide metal matrix composite. *Processing of Advanced Materials*. 3, 233-241.

Rozenek, M., Kozak, J., Dabrowski, L., Lubkowski, K., 2001. Electric discharge characteristics of metal matrix composites. *Journal of Materials processing Technology* 109, 367-370.

Scot, D., Boyina, S., Rajurkar, K.P., 1991. Analysis and optimization of parameter combinations in wire electrical discharge machining. *International Journal of Production Research* 29/ 11, 2189-2207.

Spedding, T. A., Wang, Z.Q., 1997. Study on modelling of wire EDM process. *Journal of materials processing technology* 69, 18-28.

Spur, G., Schonbeck, J., 1993. Anode erosion in wire-EDM- a theoretical model. *Annals of the CIRP*. 42(1), 253-256.

Tarng, Y.S., Ma, S.C., Chung, L.K., 1995. Determination of optimal cutting parameters in wire electrical discharge machining. *Int. J. Mach. Tools Manufact*. 35(12), 1693-1701.

Yan, B.H., Tsai, H.C., Huang, F.Y., Lee, L.C., 2005. Examination of wire electrical discharge machining of Al2O3p/6061Al composites. *International Journal of Machine Tools and Manufacture*. 45(3), 251-259.

Yeo, S.H., Kurnia, W., Tan, P.C., 2008. Critical assessment and numerical comparison of electro-thermal models in EDM. *Journal of Materials Processing Technology*. 203, 241-251.

Yue, T.M., Dai, Y., 1996. Wire electrical discharge machining of Al_2O_3 particle and short fiber reinforced aluminium based composites. *Materials Science and Technology* 12, 831-835.

In: Machining and Forming Technologies, Volume 2
Editor: J. Paulo Davim
ISBN: 978-1-61122-820-5

Chapter 4

EXPERIMENTAL INVESTIGATION ON cBN FINISH-TURNING OF HARDENED AISI 52100 STEEL

*C. Aslam and M. S. Shunmugam**
Department of Mechanical Engineering
Indian Institute of Technology Madras
Chennai-600036, India

ABSTRACT

Finish-turning of hardened steel to grinding-like accuracies requires proper selection of cutting tools and process parameters. In the present work, the bearing steel AISI 52100 hardened to 55 HRC is finish-turned using cBN inserts, and performance measures such as tool wear, surface finish, cutting tool temperature and vibration are investigated. For the first part of the study, experiments have been carried out employing orthogonal array in the design of experiments.

Significance of process parameters and their influence on performance measures are brought out by analysis of variance and response equations. Optimal conditions are identified for each of the performance measures. The second part of the investigation deals with performance of cBN under the optimal conditions. In such optimal conditions, the cutting tool may take long time to wear out and reach a failure stage. To simulate the condition of tool wear, cutting tools with different flank wear are obtained using aggressive cutting conditions.

Results obtained using such cutting tools with a pre-measured flank wear are presented and discussed in this paper.

Keywords: Finish hard-turning, Design of Experiments, Flank wear, Surface roughness, Tool temperature, Tool Vibration, Analysis of variance, Multiple regression, Optimization.

---

* Corresponding Author, E-mail: shun@iitm.ac.in

1. INTRODUCTION

Finish hard-turning is a single point machining process carried out on hard materials (where hardness is taken to be greater than 45 HRC - Rockwell C hardness) with appropriate feed rate and depth of cut conditions. Recent improvements in machine tool rigidity and the development of ceramic/cBN cutting tools allow machining of special materials including super alloys and hardended steel (Ghani et al., 2002 and Costes et al., 2007). Even though hard turning consists of small depths of cut and feed rates, material removal rates can be much higher than those achieved by grinding, depending on the application. The use of polycrystalline cubic boron nitride cutting tools in hard turning applications is continuously growing with the number of commercially available grades increasing, allowing new application areas to be explored (Lahiff et al., 2007). Turning hardened materials to grinding-like accuracies requires the right combination of part, machine and process parameters.

Aslan et al. (2007) optimized the cutting parameters when turning hardened AISI 4140 steel (63 HRC) with a ceramic tool. Combined effects of three cutting parameters, namely cutting speed, feed rate and depth of cut on two performance measures, namely flank wear (V_B) and surface roughness (R_a), were investigated employing an orthogonal array and analysis of variance. Also the relationship between the parameters and the performance measures were determined using multiple linear regressions. Luo et al. (1999) conducted experiments to study variations of tool wear with the cutting speed and the hardness of the work material. Pavel et al. (2005) studied experimentally the effect of tool wear on surface finish in case of continuous as well as interrupted hard turning. They found a good correlation between flank wear and roughness of the machined surface, and as the cutting time progresses surface roughness increases due to the increase in tool wear. Penalva et al. (2007) investigated experimentally the effect of tool wear on surface roughness in hard turning. They found that there is a good replication of tool on the roughness profile. Their strategy allows tool wear estimation by simple roughness measurements using shop floor instrument.

Sullivan and Cotterall (2001) and Muller and Renz (2003) studied temperature variations in single point turning using infrared pyrometer and brought out the influence of cutting speed and tool wear on the temperatures in a turning process. Abukashim et al. (2005) investigated heat partition in high speed turning of high strength alloy steel. They carried out temperature measurements on the tool rake face during orthogonal cutting at various cutting speeds. Liu et al. (2002) studied experimentally the cutting temperature and tool wear in turning of hardened bearing steel. They investigated and compared tool life and cutting temperature.

Dimla and Lister (2000) described an experimental and analytical method for detecting and monitoring the wear on a cutting tool. They developed on-line tool condition monitoring system involving cutting forces and vibration signature measurements. Analysis of vibration signal in time and frequency domain was carried out. Scheffer and Heyns (2001) implemented a monitoring system utilizing simultaneous vibration and strain measurements on the tool for the wear of a synthetic diamond tools which are specifically used for the manufacturing of aluminum pistons. During their investigation, a large number of features indicative of tool wear were automatically extracted from different parts of the original signals.

Cutting tool wear is most often discussed in terms of flank and crater wear. Traditionally, tool life is defined by flank wear due to the significant influence of this parameter on the surface finish and dimensional accuracy of the machined part. In case of cBN inserts, resistance to tool wear is a function of many variables, including cBN content, grain size, binder material, tool geometry, cutting edge geometry (sharp vs chamfered), workpiece properties, and cutting conditions. Cutting conditions have different degrees of influence on surface roughness, tool temperature and vibration of the cutting tool. This necessitates process optimization to determine optimal values of cutting parameters, such as cutting speed, feed rate and depth of cut, and machining studies to fully evaluate the performance of these tools in machining hardened materials. Tool wear not only reduces tool life, but also affects surface finish and increases cutting forces, tensile residual stress, tool temperature and vibration of the cutting tool.

Since cutting force is not a major consideration in finish hard-turning, performance measures such as surface roughness, tool wear, cutting temperature and vibration on the tool are considered in the present investigation. First part of the present work brings out the effect of cutting conditions in finish hard-turning, namely cutting speed, feed rate and depth of cut on different performance measures. The optimum cutting conditions are identified for each performance measures. In the second part of the work, finish hard-turning was carried out with the optimum conditions identified in the first part and variations in surface roughness, tool wear, cutting temperature and vibration of the tool are studied. Since such studies would take very long time, attempt has been made to simulate the flank wear using heavy cutting conditions. The flank wear of the tools was measured and the cutting tools with pre-measured flank wear were taken for turning experiments in optimum conditions.

2. Experimental Details

Work material was chosen as AISI 52100 (commercially known as EN 31 steel) alloy steel in the form of round bar with 60 mm diameter and 170 mm length. The chemical composition of the material is indicated in Table.1. The work material is a bearing steel which is used with a hardness of 55 HRC. Therefore, the round bar was hardened and tempered to 55 HRC. Required hardness was obtained through heat treatment after skin turning. The bar was heated uniformly to 800^0C to 820^0C. After keeping the bar at high temperature and allowing 30 minutes per 25 mm of ruling section, the bar was quenched immediately in oil. Tempering was carried out at 300^0C for achieving a hardness of 55 HRC by heating uniformly and thoroughly at the selected tempering temperature with a holding time of at least one hour per 25 mm of thickness.

Table 1. Chemical composition of AISI 52100 (% wt)

C	Si	Mn	Cr	Co
0.95-1.20	0.10-3.35	0.30-0.75	1.0 – 1.6	0.025

Finish hard-turning experiments were conducted on a standard precision lathe with main motor capacity of 15 kW and spindle speeds ranging from 140 rpm to 5600 rpm. cBN cutting tool insert from M/s Sandvik (S-TNGA110308S01031A CB7015) with an emissivity value of 0.79 at the tool and a tool holder PTGNR1616H11 with an approach angle of 91^0 were used for the experiments. The following is the cutting tool geometry: Rake angle -6^0, Cutting edge length 11 mm, Nose radius 0.8 mm and Insert thickness 3.18 mm.

The cutting parameters were selected keeping in mind that the finish hard-turning operation is generally used as an alternative to grinding. The parameter levels were chosen within the ranges recommended by the cutting tool manufacturer. Three levels selected for each cutting conditions are given in Table 2. Each test was carried out for a fixed machining time.

Table 2. Allocation of parameters in L9 OA and experimental results

Exp. No.	Depth of Cut (mm)	Feed Rate (mm/rev)	Cutting Speed (m/min)	Roughness Ra (μm)	Tool Temperature (^{0}C)	Flank wear V_B(15 min) (mm)	Vibration Acceleration (m/s^2)
1	0.05	0.05	145	0.253	668	0.19	426.60
2	0.05	0.10	185	0.886	793	0.03	324.30
3	0.05	0.16	225	1.422	814	0.06	270.53
4	0.10	0.05	185	0.224	704	0.15	416.52
5	0.10	0.10	225	0.350	837	0.08	324.30
6	0.10	0.16	145	1.044	709	0.20	478.22
7	0.15	0.05	225	0.291	890	0.18	522.50
8	0.15	0.10	145	0.200	795	0.08	643.17
9	0.15	0.16	185	1.132	831	0.17	499.86

Flank wear was measured with a tool-makers microscope of 0.5 μm accuracy. Average wear-land width V_B is taken as a measure of flank wear on the cutting tool. Surface roughness measurements were performed by using a Perthometer S5P with a cut off length of 0.8 mm and a tracing length of 5.6 mm and the three measurements were averaged to get the surface roughness parameters. Tool temperature of cBN tool was measured by using infrared pyrometer with a temperature measurement range of -50^0C to 1000^0C and with an adjustable emissivity range of 0.3 to 0.95. The accuracy of instrument at 23^0C and emissivity of 0.95 is +/- 2 % of reading. Vibration was measured by using an accelerometer (Bruel and Kjer Type 4332) attached to the tool holder. This accelerometer measured the vibration in the direction of main cutting force. This accelerometer was connected to a CRO through a charge amplifier (Bruel and Kjer 2626). The vibration signals were measured just before the machining was stopped for the measurements of flank wear and surface roughness. The acceleration signals were sampled at a frequency of 20 kHz while recording 2000 samples per channel. The processing of the measured signals in both time and frequency domains was achieved through Mathwork's Matlab coding programme on a PC.

3. Orthogonal Array Experiments

First set of experiment was carried out to find the significance of the cutting parameters, namely cutting speed, feed and depth of cut for each performance measures such as surface roughness, tool wear, cutting temperature and vibration on the tool. The results are useful in developing the relationship as well as in arriving at the optimal conditions for each of the performance measures.

Three levels were chosen for each of the factors. Using L9 orthogonal array, 9 experimental runs corresponding to a fraction of parameter combinations are given in Table 2 (Montgomery, 1991). The first column was assigned to the depth of cut, the second column to the feed rate, the third column to the cutting speed. One test was performed for each combination resulting in a total of 9 tests (no replications) and machining time for each test was maintained as 15 min. Also a random order was used for running the tests.

3.1. Analysis of Variance

The experimental results were analyzed using analysis of variance (ANOVA) to identify the factors significantly affecting the performance measures. The results of the ANOVA with the surface roughness, tool wear, tool temperature and vibration of the cutting tool are shown in Tables 3(a), 3(b), 3(c) and 3(d) respectively. These tables show the S^1-values, that is, the realized significance levels, associated with the F-tests for each source of variation. Also, the last column of the tables shows the percent contribution (ρ %) of each source to the total variation indicating the degree of influence on the result.

Table 3. ANOVA Results
DF: Degree of freedom, E: Error, F: F – Ratio,
P: Pooled, S: Sum of square, V: Variance, ρ: Percentage contribution

a) Surface Roughness (Ra)

Source	P	DF	S	V	F	S'	ρ , %
V (Cutting speed, m/min)		2	0.1966	0.0983	4.5462	0.1533	8.52
a (Depth of cut, mm)		2	1.4588	0.7294	33.7393	1.4156	78.67
f (Feed rate, mm/rev)		2	0.1008	0.0504	2.3319	0.0576	3.20
ERROR	(e)	2	0.0432	0.0216			
(e)	Unpool	2	0.0432	0.0216		0.1730	9.61
Total	Auto	8	1.7995	0.2249			

b) Tool Temperature

Source	P	DF	S	V	F	S'	ρ %
Cutting speed (m/min)		2	14,384	7,192	13.07	13,284	31.03
Depth of cut (mm)		2	4,452	2,226	4.05	3,352	7.83
Feed rate (mm/rev)		2	22,874	11,437	20.78	21,773	50.86
ERROR	(e)	2	1,100	550			
(e)	Unpool	2	1,100	550		4,402	10.28
Total	Auto	8	42,812	5,351			

Table 3. (Continued)

c) Flank wear (VB)

Source	P	DF	S	V	F	S'	ρ %
Cutting speed (m/min)		2	0.0050	0.0025	1.1905	0.0008	2.44
Depth of cut (mm)		2	0.0194	0.0097	4.6190	0.0152	46.34
Feed rate (mm/rev)		2	0.0042	0.0021	1.0000	0.0000	0.00
ERROR	(e)	2	0.0042	0.0021			
(e)	Unpool	2	0.0042	0.0021		0.0168	51.22
Total	Auto	8	0.0328	0.0041			

d) Vibration acceleration

Source	P	DF	S	V	F	S'	ρ ,%
Cutting speed (m/min)		2	72,585	36,292	55.11	71,268	65.37
Depth of cut (mm)		2	2,334	1,167	1.77	1,017	0.93
Feed rate (mm/rev)		2	32,790	16,395	24.89	31,472	28.87
ERROR	(e)	2	1,317	658.			
(e)	Unpool	2	1,317	658		5,268	4.83
Total	Auto	8	109,026	13,628			

ANOVA result in Table 3(a) for surface roughness indicates that depth of cut is significant at 95% confidence level. Its contribution is the highest among other parameters. From the confidence level of 95%, it can be understood that such an effect of the depth of cut on surface roughness is expected on more than 95 out of 100 occasions when the experiments are repeated many times. It can be seen that cutting speed and feed rate are not significant for surface roughness within the selected range of cutting conditions. Table 3(b) indicates that feed rate is significant at 95 % confidence level for tool temperature. Cutting speed also has a significant effect on tool temperature at 90% confidence level. On comparing the percentage of contribution (ρ %) of the different factors for tool temperature, it can be seen that feed rate has the highest contribution of about 50.86%. From Table 3(c) for tool flank wear, it can be seen that depth of cut has the highest contribution of about 46.34%. Table 3(d) indicates that cutting speed and feed rate are significant at 95% confidence level for vibration acceleration of the cutting tool. The percentage contribution (ρ %) of cutting speed is the highest having a value of about 65.37%. Feed rate also has a percentage contribution of 28.87%. The least significant factor is depth of cut.

3.2. Average Responses and Optimal Conditions

Figure 1 gives the main factor plots for different performance measures. Surface roughness (Figure 1(a)) decreases with the increase in cutting speed when cutting speed is increased from 145 m/min to 185 m/min. Surface roughness does not change when the cutting speed increases further to 225 m/min. It can be seen that surface roughness increases with the increase in depth of cut. Usually the influence of feed is higher on the surface roughness than the depth of cut in turning annealed steel. In the present work on hard-turning, however, depth of cut has highest influence for the range of parameters used. When feed rate is increased from 0.05 to 0.10 mm/rev, surface roughness increases drastically. But further increase in feed rate results in a slight decrease in surface roughness. The results of analysis of variance show that the best surface finish is achieved with a low feed rate, low depth of cut and a high cutting speed. In order to minimize the surface roughness, optimum cutting conditions are cutting speed 185 m/min, depth of cut 0.05 mm and feed rate 0.05 mm/rev.

Variation of cutting temperature with respect to cutting speed is shown in Figure 1(b). It can be seen that the cutting temperature decreases slightly with the cutting speed in the range of 145 - 185 m/min. When the cutting speed is further increased, the rate of temperature change increases drastically.

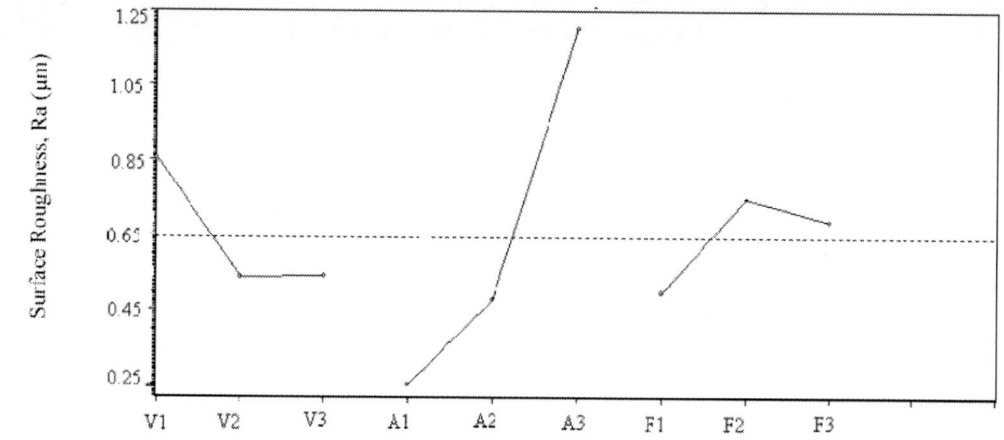

a) Average responses for surface roughness (Ra)

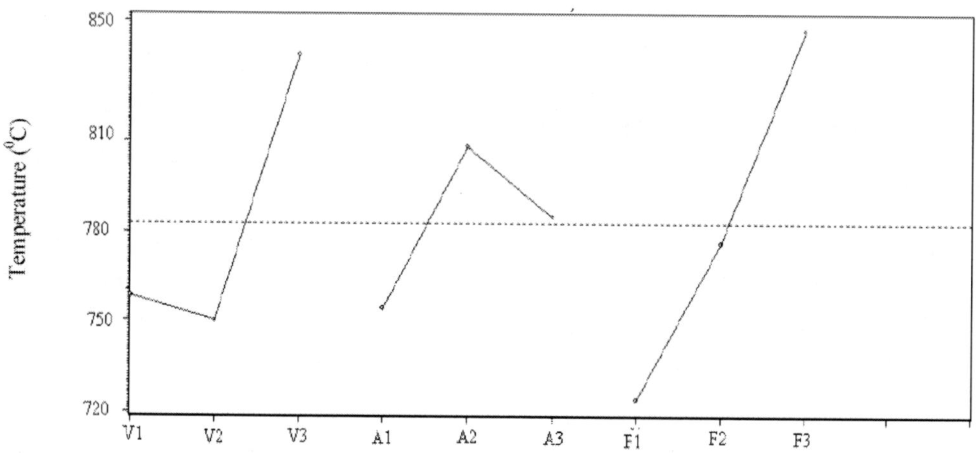

b) Average responses for tool temperature

Figure 1. Continued on next page.

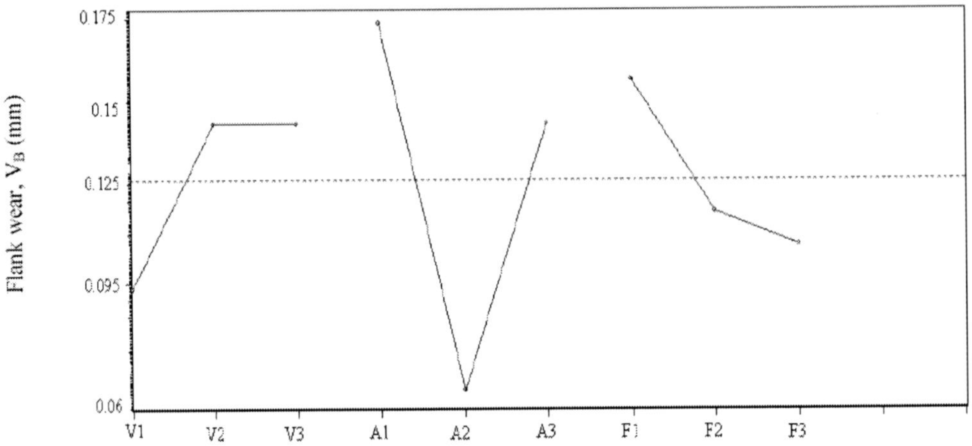

c) Average responses for flank wear (V_B)

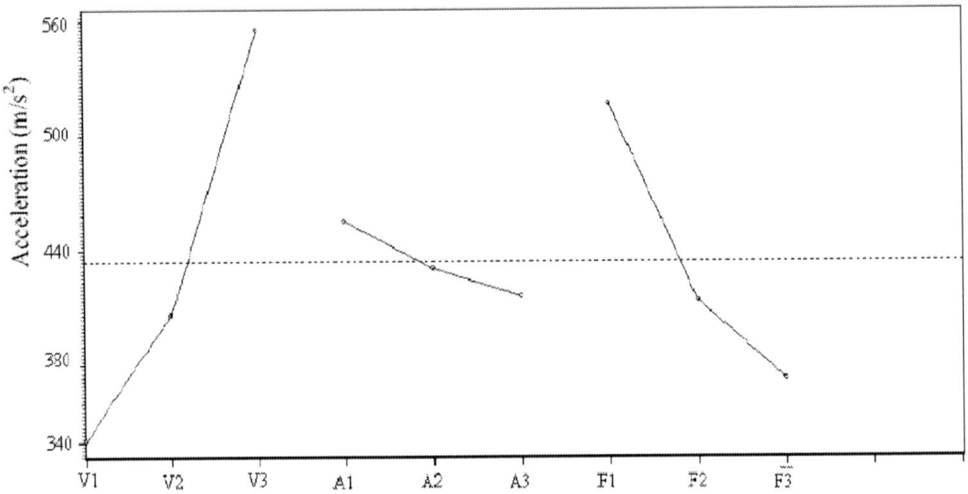

d) Average responses for vibration acceleration

Figure 1. Main factor plots. (Cutting Speed: V1 = 145 m/min, V2 = 185 m/min, V3 = 225 m/min, Depth of cut: a1 = 0.05 mm, a2 = 0.10 mm, a3 = 0.15 mm, Feed rate: f1 = 0.05 mm/rev, f2 =0.10 mm/rev, f3 = 0.16 mm/rev).

When the cutting speed is between 145 m/min and 185 m/min, reduction of shear energy due to the appearance of saw tooth chip is responsible for the changes in the cutting temperature. At higher cutting speeds, the increase of heat flux is far greater than the reduction of shear energy and hence a drastic rise of cutting temperature is observed. The cutting temperature increases from 754^0C to 808^0C, when depth of cut increases from 0.05 mm to 0.1 mm. The reason for this increase in temperature is that the cutting force increases with the depth of cut, thereby increasing the cutting temperature. At increased depth of cut of 0.15 mm, the cutting temperature decreases as shown in the Figure 1(b). This is because, more heat is carried away by the chips produced in this condition. It can be seen from the figure that the tool temperature is higher when feed rate is large than when the feed rate is small. The temperature ahead of cutting edge increases from about 724 to 847 °C when feed rate increases from 0.05 to 0.16 mm/rev. This increase in temperature by increased feed rate is due to the increase of material removal rate since the cutting energy increases with the material removal rate. For getting minimum tool temperature, the optimum cutting conditions are cutting speed 185 m/min, depth of cut 0.05 mm and feed rate 0.05 mm/rev.

Figure 1(c) shows the effect of cutting speed on tool wear V_B. Tool wear increases with increase in cutting speed. High cutting temperature produced during high cutting speed in turning promotes increased tool wear. Increase in depth of cut from 0.05 mm to 0.10 mm results in a decrease of tool wear from 0.17 mm to 0.06 mm. A further increase in depth of cut leads to increase in tool wear. Increase in feed rate results in decreased tool wear as shown in Figure 1(c). Therefore the optimum cutting conditions for minimizing the tool wear are cutting speed 145 m/min, depth of cut 0.1 mm and feed rate 0.16 mm/rev.

From Figure 1(d), it can be seen that there is an increasing trend of RMS (Root Mean Squares) acceleration with respect to the cutting speed. However, RMS acceleration decreases as depth of cut and feed rate increase. For getting minimum RMS acceleration cutting speed should be 145 m/min, depth of cut 0.15 mm and feed rate should be 0.16 mm/rev.

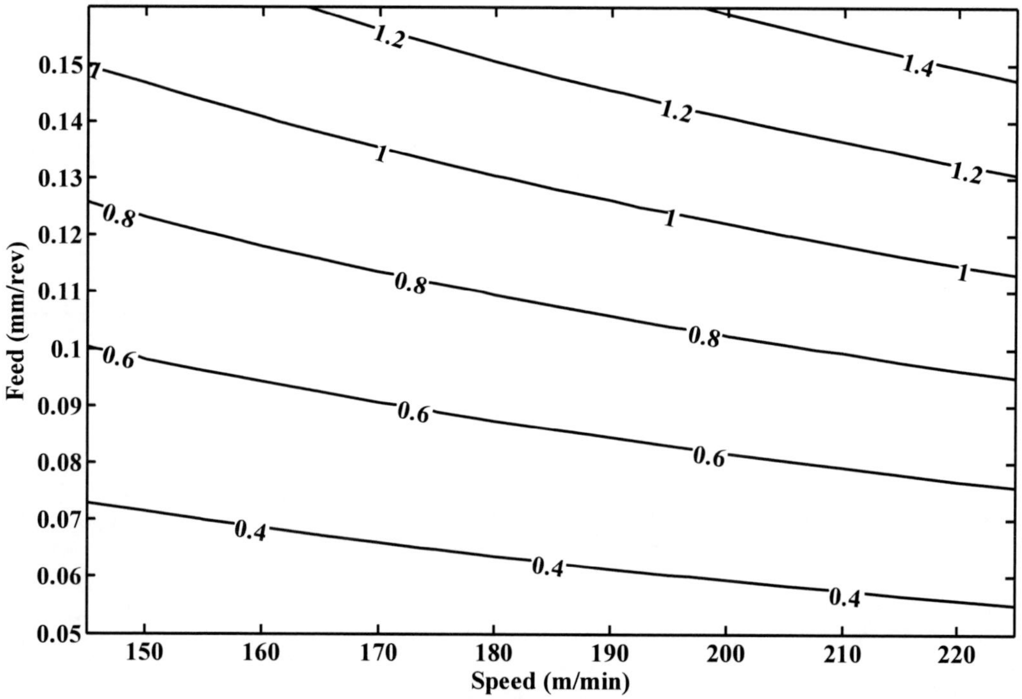

(i) Depth of cut = 0.05 mm

Figure 2. Continued on next page.

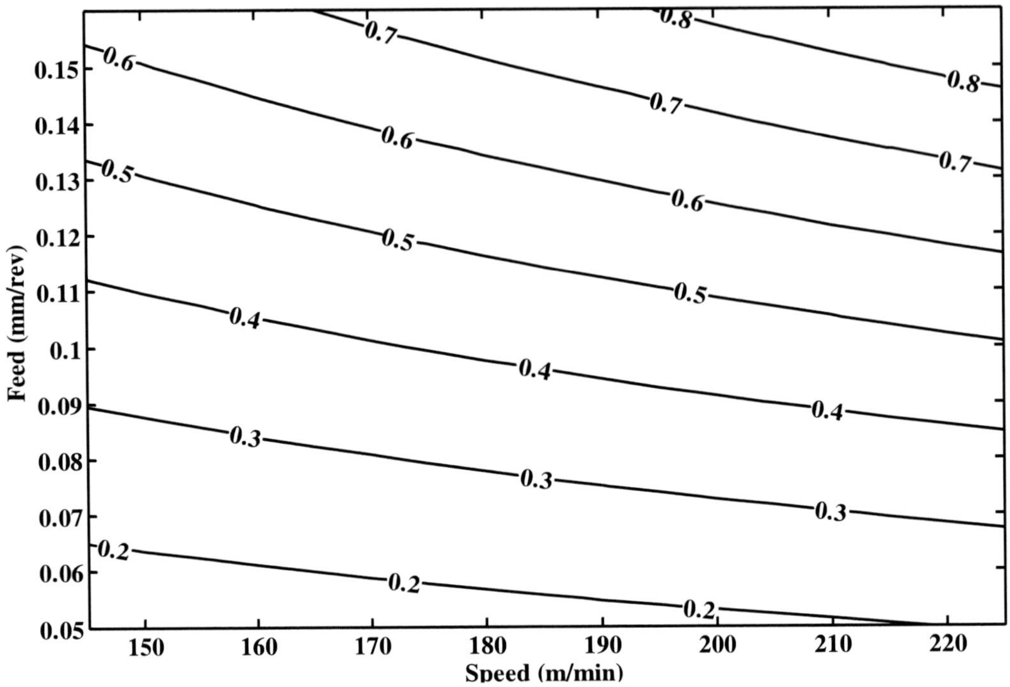

(ii) Depth of cut = 0.15 mm

(a) Contour plots for R_a

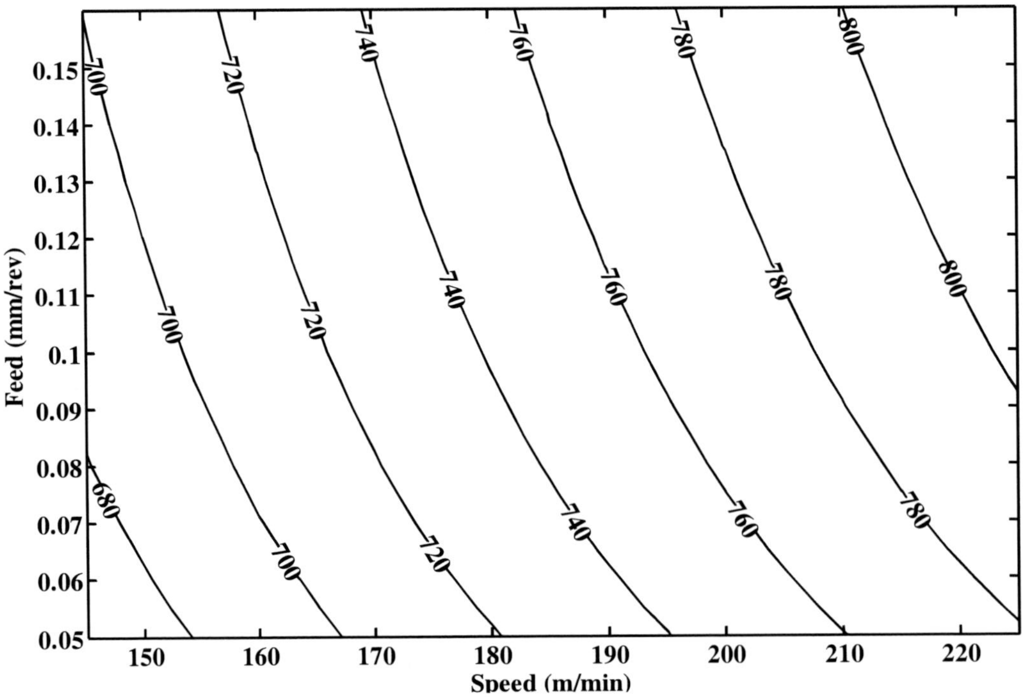

(i) Depth of cut = 0.05 mm.

Figure 2. Continued on next page.

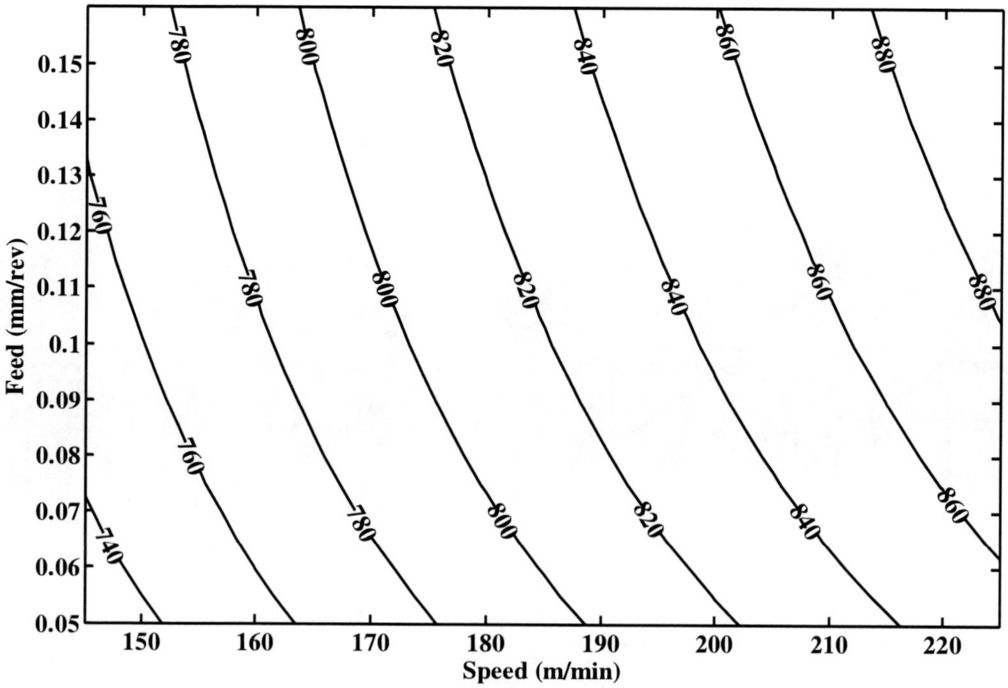

(ii) Depth of cut at 0.15 mm.

(b) Contour plots for tool temperature.

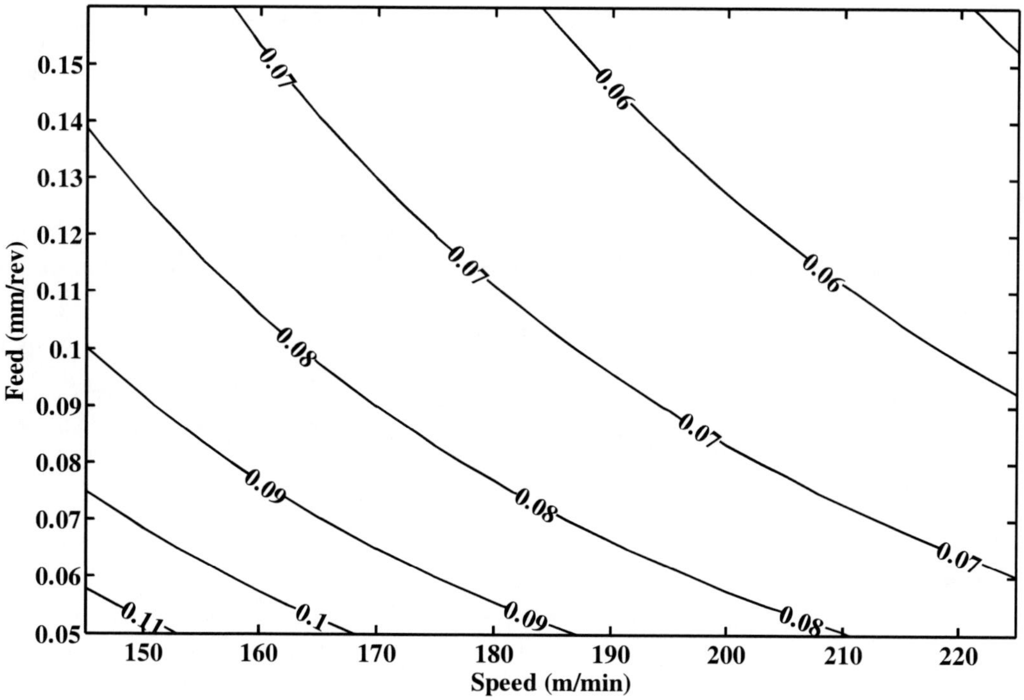

(i) Depth of cut = 0.05 mm

Figure 2. Continued on next page.

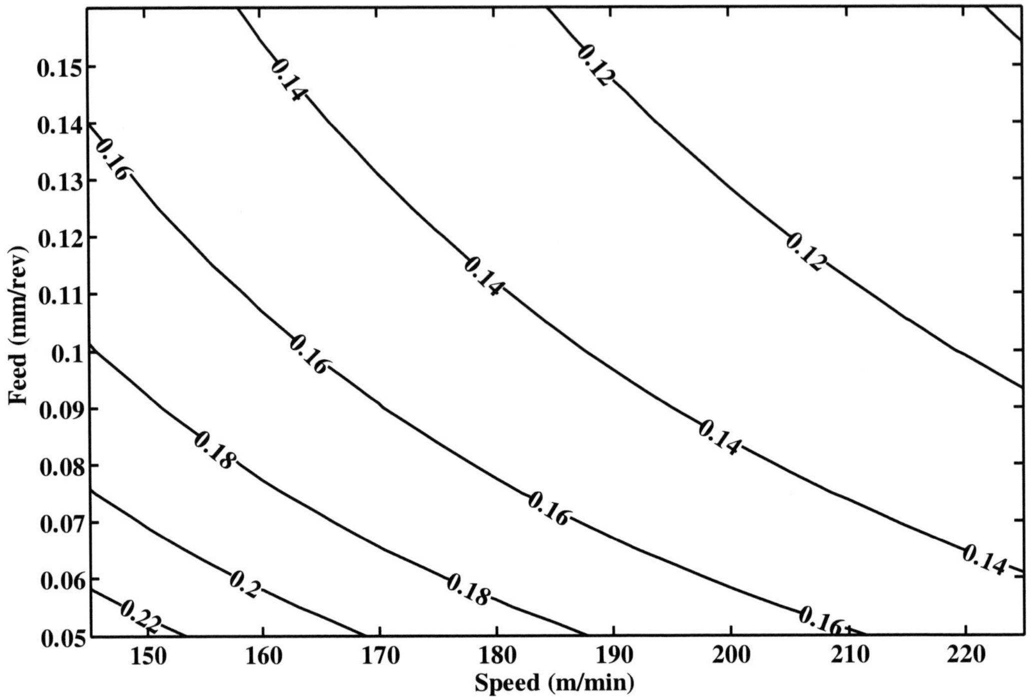

(ii) Depth of cut = 0.15 mm

(c) Contour plots for flank wear.

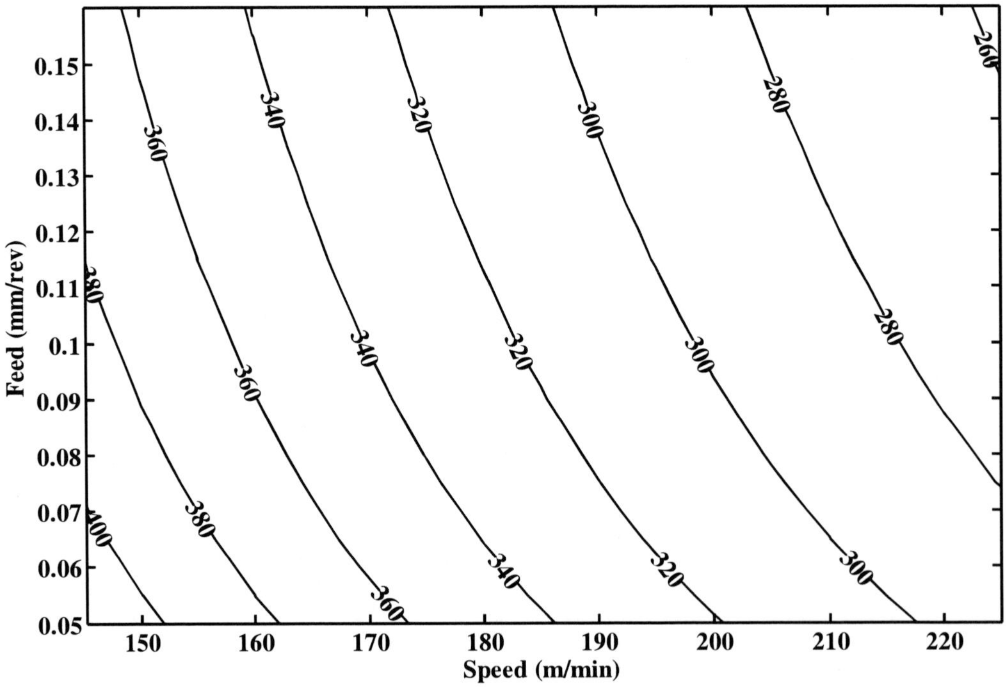

(i) Depth of cut = 0.05 mm

Figure 2. Continued on next page.

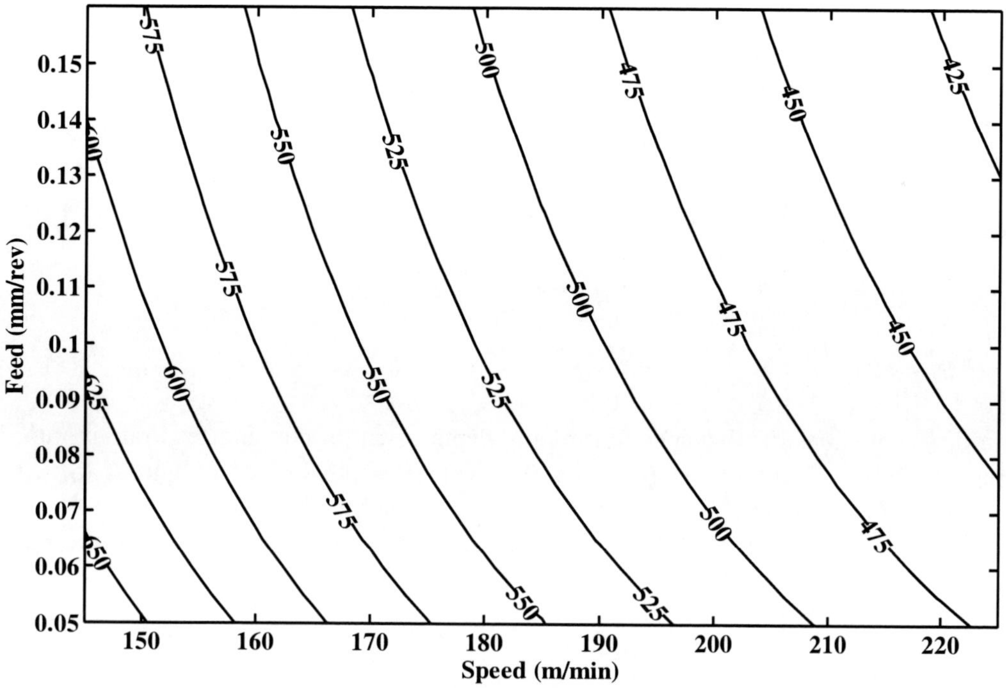

(ii) Depth of cut = 0.15 mm

(d) Contour plots for vibration acceleration.

Figure 2. Predicted performances.

3.3. Response Equations

The relationship between the factors and the performance measures are established from the experimental data in the power form given below:

$$y = cV^p a^q f^r \tag{1}$$

where *SR*, *TT*, *TW* and *ACCN* are represented by *y*. The terms *V*, *a* and *f* represent speed, depth of cut and feed respectively.

Taking log on both sides,

$$ln(y) = ln(c) + (p)ln(V) + (q)ln(a) + (r)ln(f) \tag{2}$$

In a more general form, the above equation can be rewritten as

$$Y = \beta_0 + \beta_1 X_1 + \beta_2 X_2 + \dots \dots + \beta_k X_k + \varepsilon \tag{3}$$

This regression model is represented in matrix form as

$$Y = X\beta + \varepsilon \tag{4}$$

where the least square estimate of β is given by $(X^TX)^{-1}(X^TY)$.

Taking anti-log of β_0, value of c in Eqn. 1 is obtained. Other β values represent the exponents and the equations in the power form are obtained as given below:

$$SR = 0.044628\ (V)^{0.81398}\ (a)^{-0.49724}\ (f)^{1.2734}$$

$$TT = 163.37\ (V)^{0.35844}\ (a)^{0.08233}\ (f)^{0.044473}$$

$$TW = 36.365\ (V)^{-0.99189}\ (a)^{0.63377}\ (f)^{-0.36333}$$

$$ACCN = 60054\ (V)^{-0.80226}\ (a)^{0.43478}\ (f)^{-0.1073} \quad (5)$$

In these equations, cutting speed in m/min, depth of cut in mm and feed rate in mm/rev are represented by V, a and f respectively. The values of the surface roughness (*SR*), tool temperature (*TT*), tool wear (*TW*) and vibration acceleration (*ACCN*) are given by these equations for any combination of factor levels given that the levels are within the ranges indicated in Table 2. Figure 2 shows predicted values in the form of contour plots for low and high values of depth of cut.

4. Experiments with Simulated Flank Wear

Second part of the work presents details of the performance of cBN tool during finish-turning of hardened steel under the optimal cutting conditions identified in Sec.3.2. In the optimal modes, cBN inserts might take very long time to wear out and reach a failure criterion. In order to simulate different flank wear condition, cBN inserts were subjected to accelerated cutting conditions. The flank wear was measured and each insert was identified with a pre-measured flank wear value.

Experiments were conducted with tools with different flank wear and the resultant surface roughness, tool temperature and vibration of the cutting tool were measured. The results obtained for the optimum cutting conditions are reported below.

4.1. Effect on Surface Roughness

Variation of surface roughness parameters with the tool wear is shown in Figure 3 under the optimal conditions, namely cutting speed of 185 m/min, 0.05 mm depth of cut and feed rate of 0.05 mm/rev. The roughness parameters, R_a, R_z and R_k, tend to increase significantly with tool wear as shown in figure. Usually $V_B = 0.3$ mm is considered as criterion for the end of the tool life. It can bee seen the there is a significant increase in the surface roughness parameters after this tool wear. From Figure 3, it is evident that the good replication of the tool profile implies that changes in its topography with wear will contribute to the work piece roughness.

Examination of the roughness profile with the tool wear in Figure 4 reveals that tool wear process has a strong influence on surface roughness.

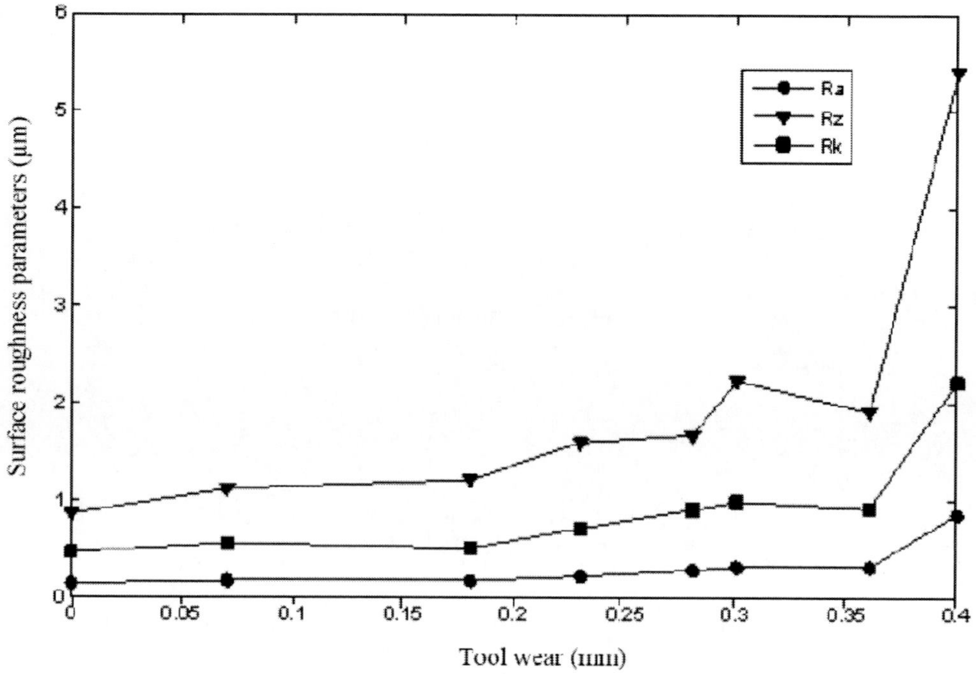

Figure 3. Variation of surface roughness parameters with flank wear.(Cutting speed: 185 m/min, Depth of cut: 0.05 mm, Feed rate: 0.05 mm/rev.)

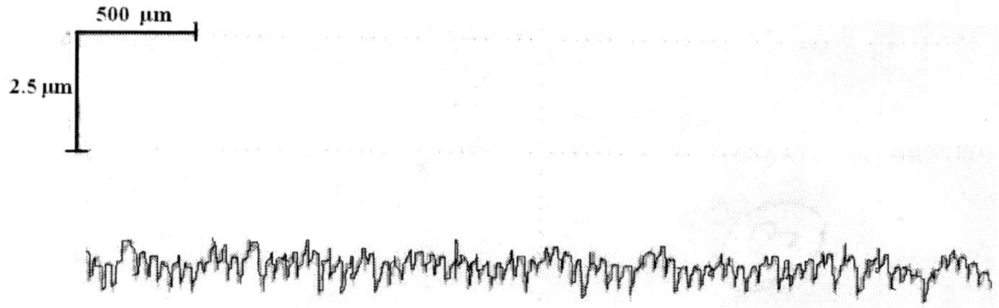

(a) Flank wear, V_B = 0.07 mm (Ra = 0.192 μm)

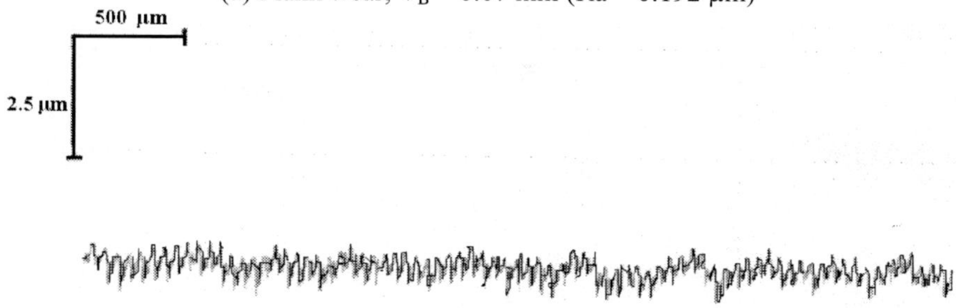

(b) Flank wear, V_B = 0.23 mm (Ra = 0.207 μm)

Figure 4. Continued on next page.

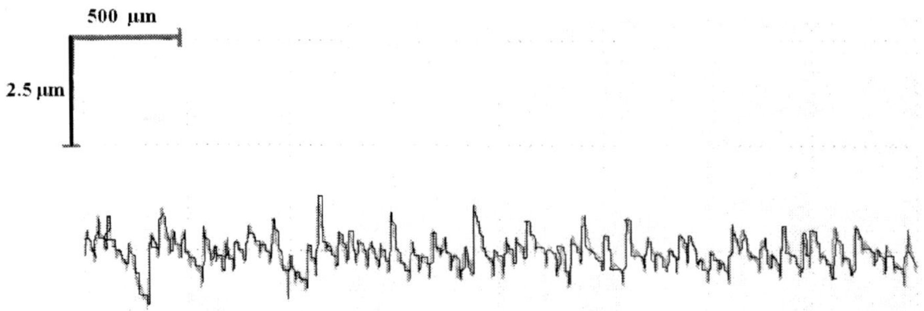

(c) Flank wear, V_B = 0.30 mm (Ra = 0.283 μm)

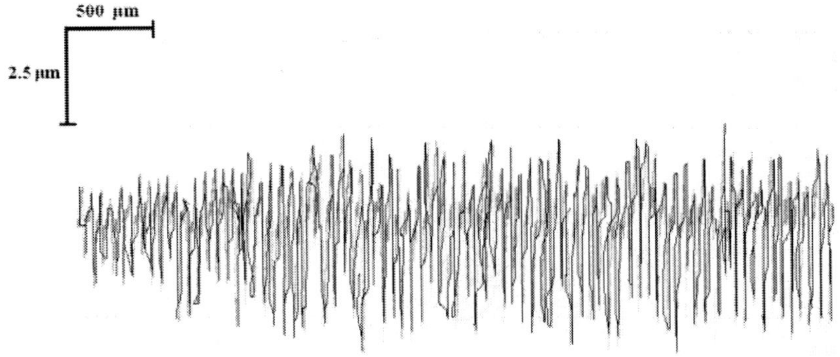

(d) Flank wear, V_B = 0.40 mm (Ra = 0.875 μm)

Figure 4. Surface profiles obtained for different tool wear. (Cutting speed: 185 m/min, Depth of cut: 0.05 mm, Feed rate: 0.05 mm/rev).

For a fresh tool, the classical turning trace similar to a very fine thread can be expected. In case of a tool with a flank wear of 0.07 mm, the surface profile as shown in Figure 4(a) is obtained. It can be seen that peaks of the profile generated by the tool feed tend to become higher as the tool wear increases as shown in Figure 4(a) to 4(c) and it increases significantly after the tool wear reaches V_B = 0.3 mm as shown Figure 4(d).

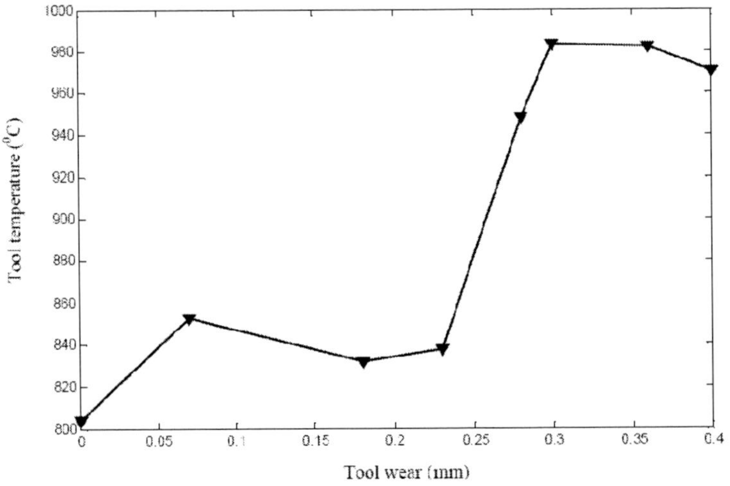

Figure 5. Variation of tool temperature with Flank wear. (Cutting speed: 185 m/min, Depth of cut: 0.05 mm, Feed rate: 0.05 mm/rev.).

4.2. Effect on Tool Temperature

Figure 5 shows that the variation of tool temperature with tool wear for optimal cutting speed of 185 m/min, depth of cut of 0.05 mm and feed rate of 0.05 mm/rev generally increases with the increase in tool flank wear. With increase in tool wear, the amount of energy flowing into the tool at the cutting edge increases. This is due to the increase in contact area at the tool – work piece interface in the presence of tool flank wear. The overall efficiency of the cutting process is thus reduced as tool wear increases; because the amount of energy required to perform the cutting process also increases. The increase in energy level is translated as higher cutting tool temperature.

4.3. Effect on Tool Vibration

The vibration accelerations in the direction of main cutting force for different tool wear are shown in Figure 6 for the optimal cutting speed of 145 m/min, depth of cut of 0.15 mm and feed rate of 0.16 mm/rev. It shows an increase in vibration acceleration with the increase in tool wear. Since the time domain plots could not lead to clear and meaningful interpretation, frequency analysis was carried out.

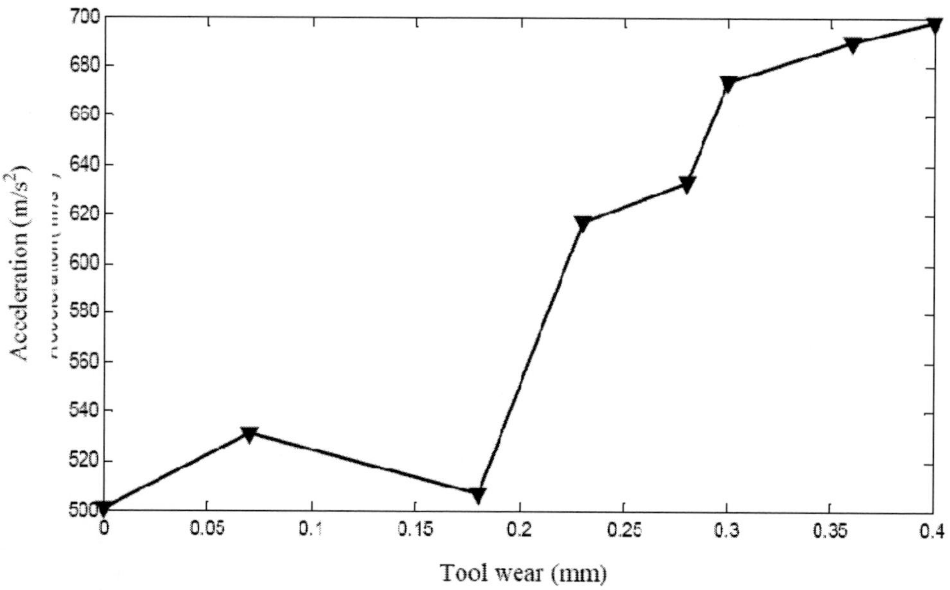

Figure 6. Variation of vibration acceleration with flank wear. (Cutting speed: 145 m/min, Depth of cut: 0.15 mm, Feed rate: 0.16 mm.).

Frequency analysis involved performing a forward FFT on the data, and the results interpreted through power spectra. With spectra plots, it is possible to visualize significant amplitude variation and their corresponding frequencies. Figure 7 shows the power spectra computed for the vibration signals generated during turning operations with different values of tool flank wear for optimal cutting conditions of 145 m/min cutting speed, 0.15 mm depth of cut and 0.16 mm/rev feed rate.

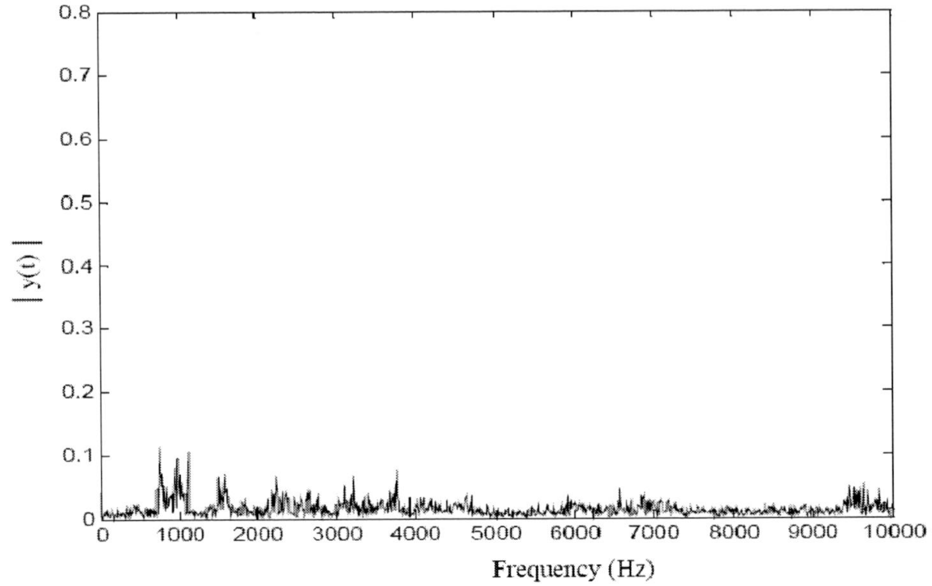

(a) Flank wear, $V_B = 0.07$ mm

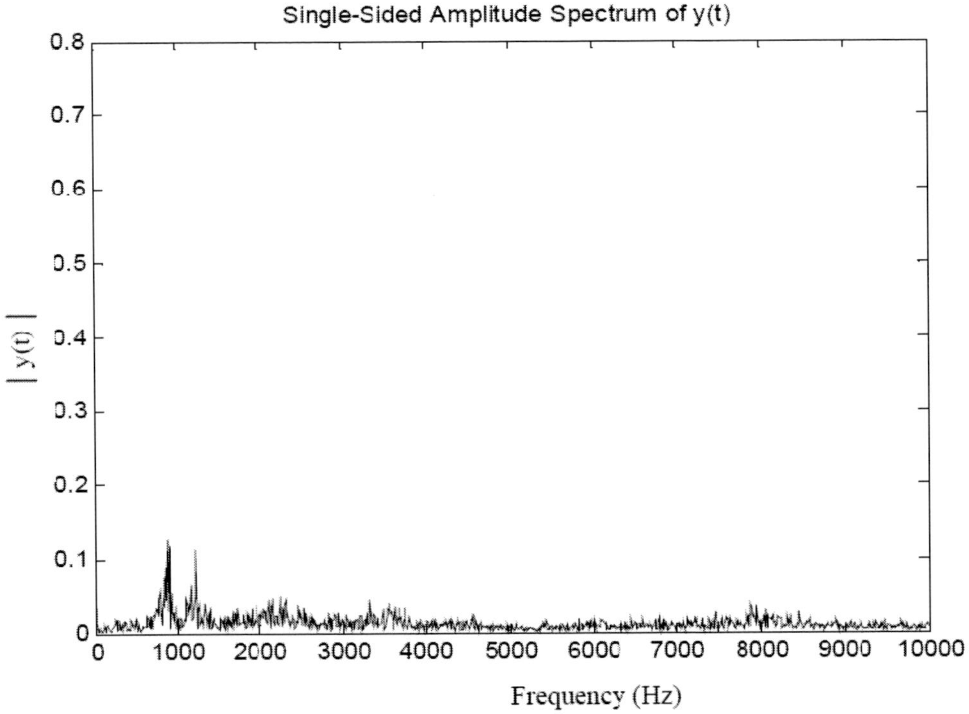

(b) Flank wear, $V_B = 0.23$mm

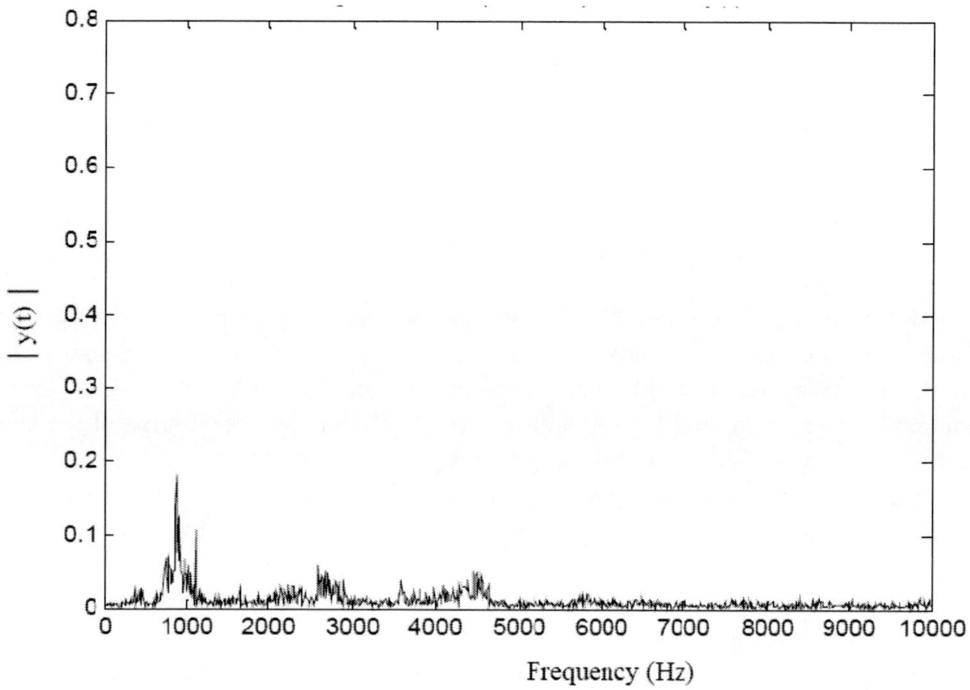

(c) Flank wear, $V_B = 0.30$ mm

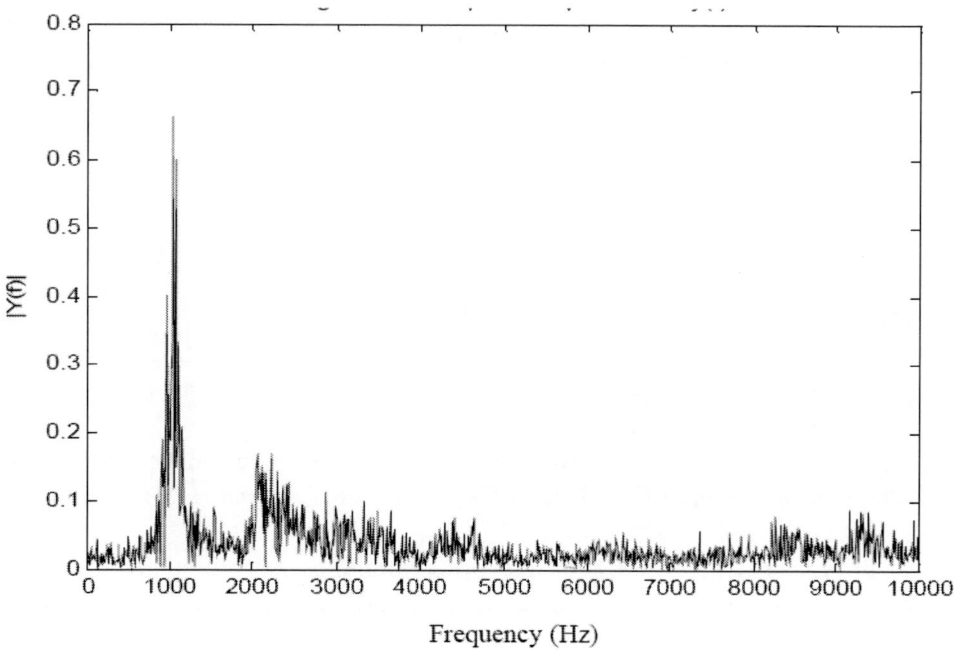

(d) Flank wear, $V_B = 0.40$ mm)

Figure 7. Power spectra with different tool wear. (Cutting speed: 145 m/min, Depth of cut: 0.15 mm, Feed rate: 0.16 mm/rev).

Though spectra plots show several peaks, for comparison sake, one can look at three peaks at frequencies of approximately 1 kHz, 3 kHz and 9.5 kHz. The amplitude of 1 kHz (

the fundamental resonant frequency of the tool holder) peak increases gradually as the flank wear progresses as shown in Figure 7(a) to 7(d). There is a significant increase in peak after tool wear reaches V_B = 0.3 mm as shown in Figure 7(d). This plots shows that the vibration signal is sensitive to flank wear.

CONCLUSIONS

The finish-turning experiments have been carried out on bearing steel hardened to 55 HRC using cBN inserts. From statistical analysis, the parameters found to be significant at 95% confidence level are, the depth of cut for surface roughness and tool flank wear, feed rate for tool temperature, and cutting speed and feed rate for vibration acceleration of the cutting tool, within the selected range of cutting conditions.

The relations between the machining parameters, namely cutting speed, feed rate and depth of cut, and the performance measures such as surface roughness, flank wear, tool temperature, and vibration acceleration established by response analysis can be used to estimate the expected performance level for any combination of parameters within the ranges used in the present investigation.

The surface roughness, tool temperature and vibration on the cutting tool show an increase with increase of flank wear for machining under optimal conditions. The increase in these parameters is very high after the tool wear reaches V_B = 0.3 mm, indicating the end of the tool life criterion. The vibration monitoring can be very effectively employed to control the nature of the surface in finish hard-turning.

REFERENCES

Abukhshim N.A., Mativenga P.T., Sheikh M.A., (2005) "Investigation of heat partition in high speed turning of high strength alloy steel", *International Journal of Machine Tools and Manufacture*, 45 : 1687–1695.

Aslan E., Camuscu N., Birgoren B., (2007) "Design optimization of cutting parameters when turning hardened AISI40 steel (63 HRC) with Al2O3 + TiCN Ceramic tool", *Materials and Design,* 28: 1618 - 1622.

Costes J.P., Guillet Y., Poulachon G., Dessoly M., (2007) "Tool life and wear mechanisms of CBN tools in machining of Inconel 718", *International Journal of Machine tools and Manufacture,* 47:1081 – 1087.

Dimla D.E. Sr., Lister P.M., (2000) "On-line metal cutting tool condition monitoring.1: force and vibration analyses.," *International Journal of Machine Tools and Manufacture*, 40:739-768.

Ghani A.K., Choudhury I.A., Husni., (2002) "Study of tool life , surface roughness, and vibration in machining nodular cast iron with ceramic tool", *Journal of Materials Processing Technology,* 127: 17-22.

Lahiff C., Gordon S., Phelan P., (2007) "PCBN tool wear modes and mechanisms in finish hard turning", *Robotics and Computer-Integrated Manufacturing*, 23 : 638–644.

Liu X.L., Wen D.H., Li Z.J., Xiao L., Yan F.G., (2002) "Cutting temperature and tool wear of hard turning of hardened bearing steel", *Journal Of Material Processing Technology*, 129: 200 – 206.

Luo S.Y., Liao Y.S., Tsai Y.Y., (1999) " Wear characteristics in turning high hardness alloy steel by ceramic and CBN tools", *Journal of Materials Processing Technology*, 88 :114–121.

Montgomery D.C, (1991) *Design and Analysis of Experiments*. John Wiley and Sons Inc, Singapore.

Muller B., Renz U., (2003) " Time resolved temperature measurements in manufacturing" *Journal of Measurement*, 34 : 363 – 370.

Pavel R., Marinescu I., Deis M, Pillar J., (2005) "Effect of tool wear on surface finish for a case of continuous and interrupted hard turning", *Journal of Materials Processing Technology*, 170: 341- 349.

Penalva M.L., Arizmendi M., Diaz F, Fernandez J., (2007) "Effect of tool wear on roughness in hard turning", *Annals of CIRP*, 51: 57 – 60.

Scheffer C., Heyns P.S., (2001) " Wear monitoring in turning operations using vibration and strain measurements", *Mechanical Systems and Signal Processing*, 15(6): 1185-1202.

Sullivan D.O, Cotterall M., (2001) " Temperature measurement in single point turning" *Journal of Material Processing Technology*, 118: 301 – 308.

In: Machining and Forming Technologies, Volume 2
Editor: J. Paulo Davim

ISBN: 978-1-61122-820-5

Chapter 5

ANALYSIS OF VIBRATION SIGNAL CHARACTERISTICS AND ITS CORRELATION WITH SURFACE ROUGHNESS IN TURNING

T. Alwarsamy* and N. Sangeetha
Department of Mechanical Engineering, Government College of Technology, Coimbatore - 641013, Tamil Nadu, India

ABSTRACT

Hard machining is a competitive finishing operation in precision manufacturing. Hard machining applications have been remaining slow due to the existence of surface damage such as chatter marks. The chatter marks on a machined component surface could be found only after machining. This post-process scenario imposes a potential danger to subsequent product performance such as fatigue life. Therefore, real-time monitoring of hard machining operation has significant economical and durability importance. Hard turning creates high pulsating pressures on the insert seat and shim. The shim acts as a shock absorber for the insert and workpiece. The shim protects both insert and tool holder from damage due to high cutting forces. In this research, the shim material is replaced with different damping materials, and cutting tests were conducted to find out the improvement in surface finish. The improvement in quality of surface finish is investigated by real time observations using the vibration data collector and MCMe software. The experimental results show that RMS and frequency shift may be used to monitor the surface integrity factors. The surface finish of the workpiece is significantly improved after replacing Carbide shim with Brass shim.

Keywords: Chatter, Damping, Machining, Vibration signal, Surface roughness.

* Corresponding author, E-mail : *alwar_samy@yahoo.co.in*

1. Introduction

In the metal-cutting process, oscillations of the tool may be caused by stick slip friction which occurs as the chip slides past the tool. The frequency of these oscillations has been measured for a range of cutting variables, including the degree of bluntness of the tool, the approach angle, and the amount of tool overhang. (Scrutton R. F., 1972). This implies several economical as well as environmental disadvantages.

The quality of machined component is evaluated by how closely they adhere to set product specifications of length, width, surface finish etc. Among them surface finish is the central to determining the quality of workpiece. In practice, predicting and controlling the roughness is difficult, which due to the fact that many variables are affecting the process. Some of these factors can be controlled and some cannot. Controllable process parameters include feed, cutting speed, tool geometry and tool setup. Other factors, such as tool wear and degradation, material variability of tool and workpiece cannot be controlled as easily (Caldeirani Filho J et.al., 2002; Dr.Mandar 1999). Many of these variables are non-linear, interdependent or difficult to quantify with crisp numeric precision.

2. Literature Review

2.1. Vibration in Machining Process

Chatter during cutting process is a complex phenomena, in which uncontrolled can lead to premature tool failure, inferior surface finish, etc. This is particularly important with brittle tool materials such as Carbide and Polycrystalline diamond (PCD). In addition, vibration affects the machining surface and its integrity. Furthermore, the noise generated may be objectionable, particularly if it is at a high frequency. Basically, there are two types of vibration in machining.

1. *Forced vibration*: This type of vibration generally caused by some periodic force present in the machine such as, that comes from gear drive, imbalance of the machine tool components etc,. The essential efforts here is to minimize the vibration amplitude, since it causes inferior surface finish of the workpiece, and changing its frequency far away from the natural frequency of the system to prevent the probability of resonance occurrence. The tool rake angles, depth of cut, ronouts (both tool and workpiece) are directly affect forced vibration occurred in machining.
2. *Self – Excited vibration*: These vibrations called "chatter", happen due to the interaction of the dynamics of the chip removal process and the structural dynamics of the machine tool. Chatter which accompanies some times the cutting process, being manifested by auto excited vibrations appearing between tool and workpiece, is to be explained by the regeneration of these vibrations from one cutting cycle to another. Regeneration of vibrations is caused by the modification of the cutting force, as a result of the variation of the cut of material thickness. George Balan (1996) proposed a new general nonlinear system for the study of the phenomenon.

Vibration measurements are frequently carried out on rotating and reciprocating machinery for analysis, design and trouble shooting purposes. Much knowledge has been gained in the recent years and computer solutions of various vibration problems have been developed (Hassui A 2003; Taejun Choi et al 2003). Often it becomes essential to make actual measurements of vibration characteristics by test during development, either on the machine itself or on its prototype, because of the fact that it is difficult to build a perfect mathematical model with all its interrelationship and complexity.

2.2. Signal Processing

Quality evaluation and control is an almost universal concern in across all manufacturing sectors. Statistical methods play an important role, as do both automated and human inspection. Many sensing modalities including electrical, physical, optical, x-ray, acoustical, ultrasound, chemical measurements (M.P.S. Krishnan Kiran et al.,2007) applied in inspection procedures; many of these measurements already contain, or could be enhanced by, signal conditioning and processing. Many signal analysis methods are applied in selected manufacturing and machine monitoring applications. These include auto-regressive (AR) models, time-domain signal envelope analysis and other time –domain methods such as kurtosis or root-mean-square (RMS) signal power (Alexandria,VA 1996). The constant false alarm rate (CFAR), ratio of power (ROP), Kurtosis and correlation of AE have been presented more sensitive than root mean square (RMS) (Paulo R .Aguiar et al., 2006). Many paper described the practical aspect of vibration phenomena and measurement requirements of a general monitoring system consisting of data collection with data reports in digital manner, followed by the acquisition phase calculating the statistical values and functions in time and frequency domain with integrated data reduction by fault and operational pattern (Wilfried Reimche et al., 2003; Warren Liao T et al., 2007; Steven Y.Liang et al., 2004). AE RMS, frequency and count rate have good correlation with white layer formation and thus, may be used to monitor surface integrity factors (Guo. Y.B. et al., 2005). The tool wear forms were correlated to features in the vibration signals in the time and frequency domains (Dimla D.E 2002). Signal parameters characterizing acoustic emission (AE) detected during metal cutting have been theoretically correlated in a simple manner, to the work material properties, cutting conditions and tool geometry (Keraita J.N. et al., 2000 and Richard Y. Chiou et al., 2000). Devillez et al (2006) were discussed the problem of vibration occurring during a machining operation and signals of tool micro movement were correlated with surface finish. Surface roughness has received serious attentions for many years. A considerable number of studies have been investigated the general effects of the speed, feed, and depth of cut, nose radius and others on the surface roughness. In the present paper, the effects of Carbide shim, Brass shim and Aluminum shim on surface roughness in turning process are also studied. Vibration signals are collected using accelerometer during machining process for different dampers. The surface roughness of the machined workpieces is measured using surface tester. Features of the amplitude spectrum and probability density curve are correlated with the measured roughness values.

2.3. Surface Roughness in Machining

Surface measuring techniques are grouped into contact and non contact methods. An amplified stylus profilometer is most popular and prevalent contact instrument used to measure surface roughness in industry and research laboratories because it is fast, repeatable, easy to interpret and relatively inexpensive. However, this stylus profilometer is limited because it requires an excessive amount of time to scan large areas, it has limited range of use on non flat surface and it is restricted to off-line use (Mandar D. Savage et al., 1999). Monitoring can be performed in-process or off-line using direct or indirect methods. The in-process designation for a sensing methods means that it is performed while metal is being removed without interrupting process. However monitoring or measurement conducted in-process or off-line would not be possible without the use of sensory devices. Sensor technologies can measure surface characteristics either directly or indirectly. Direct measurement methods using sensors include optical, electromagnetic and ultrasonic methods. (Rodriguez E, 2003). Direct sensors scan the workpiece surface directly and obtain surface roughness information as well as workpiece dimensions. However, these processes are limited because the chips and /or cutting fluid block the line of sight. Indirect methods have been successfully used to extrapolate the surface condition form vibration signals measured by accelerometer (Murad S, et al., 2005; Daniel Kirby E, et al., and 2004; Joseph C.Chen et al., 1999) or dynamometer. Indirect measurement is not impeded by the presence of chips during cutting. For that reason, the present research employs an accelerometer sensor to measure vibration data to correlate the surface roughness. As suggested by Paulo R. Aguiar et al. (2007), the vibration sensors are inexpensive, easy to mount and analog to signal processing is comparatively simple, but good techniques for extracting reliable process information from the signals are still lacking. J. Paulo Davim, and Luís Figueira (2006), made an investigation using ceramic cutting tools on cold work tool steel D2 (AISI). This study shows that with an appropriate cutting parameters choice is possible to obtain a surface roughness ($Ra < 0.8$ μm) that allows to eliminate cylindrical grinding operations. Literature review in this field shows a good scope for further research in hard machining.

3. Experimental Procedure, Materials and Equipments

Experimental procedure adopted and the details of workpiece, tool and dampers and the equipments used for the experiments are discussed in this section.

3.1. Machine Tool, Workpiece, Cutting Tool and Dampers

The CNC machine tool of ACE designers was identified for conducting experiments. Tool and die steel of grade OHNS (Oil Hardened Non Shrinking) steel which is used for making punches of forging machines were selected as workpiece. Carbide insert SNMM 120408 and PSBNR 2525M12 tool holder with shim were used for machining OHNS steel. Shock absorbing material Brass and Aluminum were identified and new shims were prepared to the standard Carbide shim size.

3.2. Cutting Tests

The experimental setup consists of an accelerometer and vibration data collector are shown in Figure 1. The details of cutting tests are given in the Table 1.

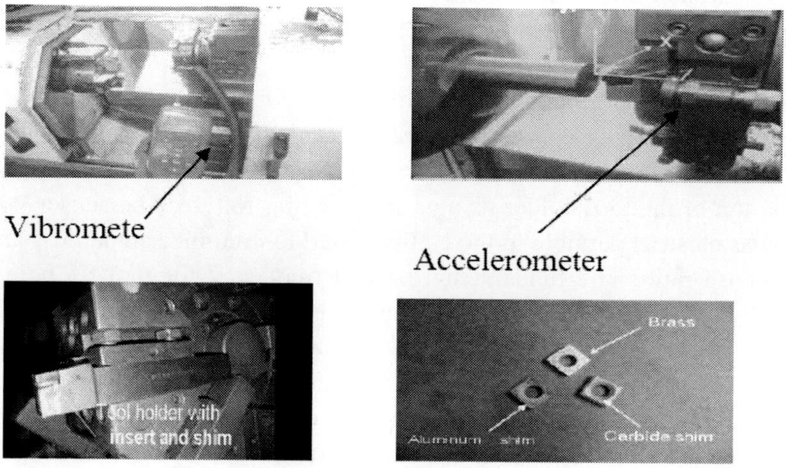

Figure 1. Experimental setup.

Table 1. Experimental conditions

Machine tool:	CNC Lathe Machine (M/s Lakshmi Machine Tool, Coimbatore, India)
	Make: ACE Designers (Bangalore)
	Model: Classic 20T
	Capacity: Ø350X630L
Workpiece:	
	Materials: OHNS AISI 01 (C=0.90%, Cr=.50%, Mn=1.00%, W=0.5%),
	Hardness =228 Br
	Size: 100mm length and 30 mm diameter.
Cutting tool:	
	Cutting insert: Carbide, SNMM 120408 (P-30 ISO specification), Drillco
	Tool holder: PSBNR 2525M12(ISO specification), Drillco
	Working tool geometry: -6, -6, 6, 6, 15, 75, 0.8 (mm)
Cutting condition:	
	Cutting speed: 63, 80, m/min,
	Feed rate: 0.10 mm/rev,
	Depth of cut, 0.5 mm and 1.0 mm
Accelerometer:	
	Type: SENDIG L-15 Accelerometer
	Connector type: TNC
	Connector location: Head
	Frequency range: 0.2 Hz-5 kHz (+10%)
	Resonant frequency: 18 kHz
	Sensitivity: 40-50 PC/g

Table 1. (Continued)

Surface	roughness tester: Mitutoyo, surf test SJ-400
Sensor:	Piezoelectric sensor, Frequency range: 0.2 Hz-5 kHz (+10%)
Dampers:	Carbide, Brass, Aluminum

The signals were post-processed using software MCMe 2.0 of SENDIG.

In the first experimental group, 12 experiments were carried out using 3 different shims and for each shim four cutting conditions were selected. Each workpiece is 100 mm length and 30 mm diameter. A piezoelectric sensor of Beijing SENDIG Technology Co was mounted on the top of the tool holder using a magnet ring fixture. The sensor was placed in Z-axial position as close as possible to the cutting insert to minimize signal loss and achieve a good signal-to-noise ratio. The fundamental natural frequency for tool tip point is 190 Hz (Claudiu Bisu, et al.,)Data base were created to collect signal up to 500 Hz frequency span (Kumarasamy S, 2002) using MCMe 2.0 software package of SENDIG and the data base were transferred to vibration data collector before machining. The signals were collected by vibration data collector 911 through the piezoelectric sensor during the cutting test. After machining the collected vibration data and signals were uploaded to software package MCMe 2.0.

4. Discussion of Results

The raw data obtained from a vibration sensor mounted on a tool holder are in time domain. The vibration signal in time domain is useful to the extent of finding out the overall vibration level. While the time domain analysis is direct and does not provide useful information, it has also served limitation that it does not provide desired information about all frequencies present in the response from the structure. Frequency spectrum analysis or simply spectrum analysis is a powerful technique of signal analysis. The spectrum signals have been discussed in the section 4.1 and probability distribution in section 4.2. The surface roughness of machined component discussed in section 4.3.

4.1. Analysis of Amplitude Spectrum Signals

The amplitude spectrum signals for four different cutting conditions were shown in the Figure 2.1 to Figure 2.4. In all cutting conditions the spectrum signals of Carbide shim are shown in Figure 2.1 (a), 2.2 (a), 2.3 (a) and 2.4 (a). The comparison signal of Aluminum with Carbide is shown in Figure 2.1 (b), 2.2 (b), 2.3 (b) and 2.4 (b) and the comparison signal of Brass with Carbide is shown in Figure 2.1 (c), 2.2 (c), 2.3 (c) and 2.4 (c). In comparison signals continuous lines indicates spectrum of Carbide shim and dotted line indicates the spectrum of Aluminum shim and Brass shim. The spectrum signals give amplitudes values in 500 Hz frequency span. Right side of each waveform contains first 10 peak values of amplitude and corresponding frequencies.

From the Figure 2.1 (a) the amplitude peak value is 0.29 mm and the corresponding frequency 83.8 Hz recorded for Carbide shim in the first cutting condition. For the same cutting condition the amplitude peak value is reduced to 0.18 mm and frequency shifted to 118 Hz when Carbide shim is replaced with the Aluminum shim shown in Figure 2.1 (b). Similarly for Brass shim amplitude is reduced to 0.27 mm and frequency is shifted to 131.2 Hz which is shown in Figure 2.1 (c).

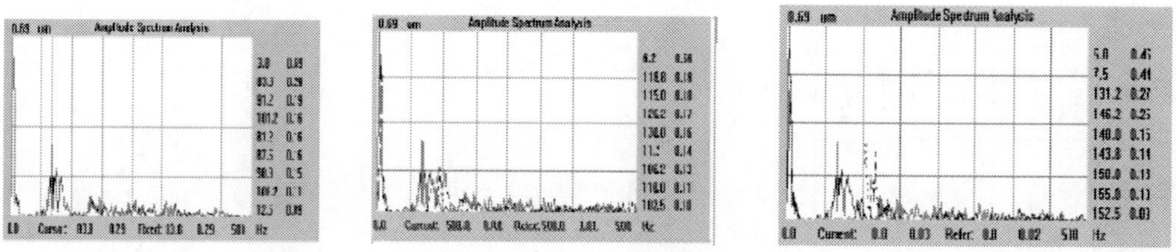

a b c

a) Carbide shim b) Comparison of Carbide and Aluminum shims c) Comparison of Carbide and Brass shims.

Figure 2.1. Spectrum signals at speed 750 rpm and depth of cut 0.5 mm.

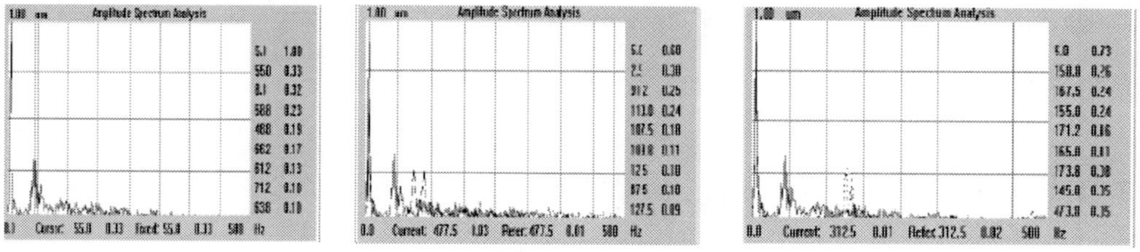

a b c

a) Carbide shim b) Comparison of Carbide and Aluminum shims c) Comparison of Carbide and Brass shims.

Figure 2.2. Spectrum signals at speed 750 rpm and depth of cut 1.0 mm.

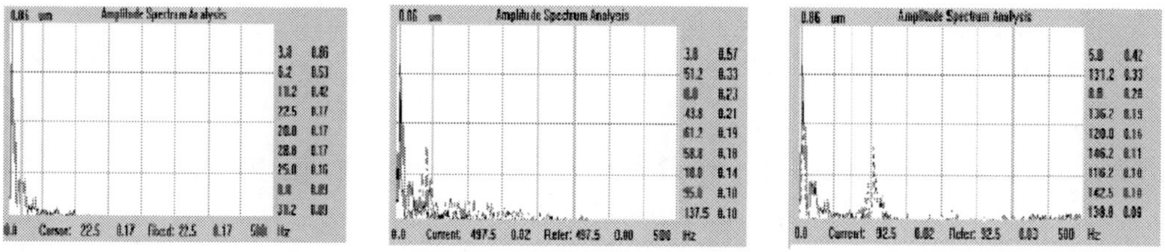

a b c

a) Carbide shim b) Comparison of Carbide and Aluminum shims c) Comparison of Carbide and Brass shims.

Figure 2.3. Spectrum signals at speed 900 rpm ad depth of cut 0.5 mm.

Table 2. Spectrum data (amplitude peak and frequency)

Cutting condition / Material of shim	Speed 750 rpm and depth of cut 0.5 mm (I)		Speed 750 rpm and depth of cut 1.0 mm (II)		Speed 900 rpm and depth of cut 0.5 mm(III)		Speed 900 rpm and depth of cut 1. mm (IV)	
	Amplitude peak	Frequency (Hz)	Amplitude peak	Frequency (Hz)	Amplitude peak	Frequency (Hz)	Amplitude peak	Frequency (Hz)
Carbide shim	0.29	83.8	0.33	55	0.42	11.2	0.32	17.5
Aluminum shim	0.18	118.8	0.25	91.2	0.33	51.2	0.29	52.2
Brass shim	0.27	131.2	0.26	158.8	0.33	131.2	0.27	163.8

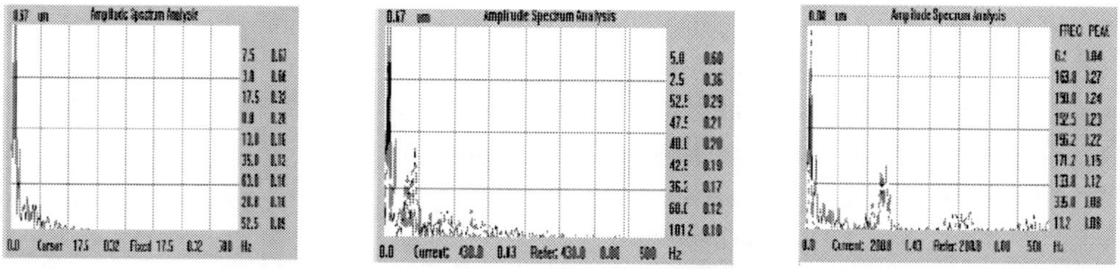

a b c

a) Carbide shim b) Comparison of Carbide and Aluminum shims c) Comparison of Carbide and Brass shims.

Figure 2.4. Spectrum signals at speed 900 rpm and depth of cut 0.5 mm

The amplitude value of second peak and the corresponding frequency values are taken from spectrum signal and tabulated in Table.2 for the all four cutting conditions.

The amplitude reduction and frequency shift of aluminum and brass are clearly shown in Figure 3.1 and Figure 3.2 for the four different machining conditions.

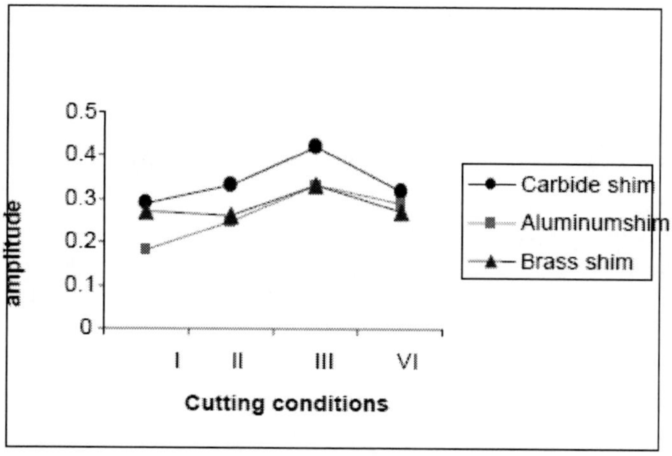

Figure 3.1. Reduction of amplitude.

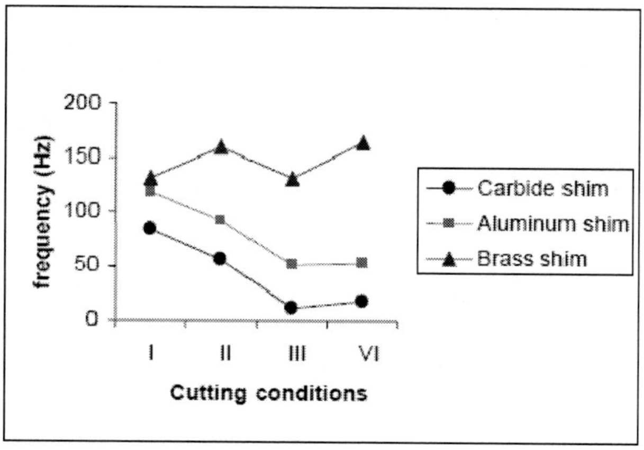

Figure 3.2. Frequency shift.

From the Figure 3.1 and Figure 3.2, it is observed that the amplitude second peak is reduced and the frequency is shifted to the right hand side of the spectrum graph for aluminum and brass when compared with carbide shim. Changes in shim material lead to a change not only in energy distribution of signals but also in the location of peaks in the frequency domain.

4.2. Surface Characterization by Probability Distribution

One of several quantities that could be used to describe the vibration effects is peak value either it is displacement, velocity, or acceleration form of vibration, on the other hand, more complex vibrations are being studied other descriptive quantities may be preferred. A further descriptive quantity that does take the time history into account is the Root Mean Square (RMS) value and could be formulated as

$X_{RMS} = \sqrt{1/T \int_0^T X^2(t)\, dt}$

X_{RMS}: The RMS value of the vibration signal.

T: Period of vibration signal.

t: Time (sec).

A measured surface profile representing a stationary random function with ergodic properties may be comprehensively described by probability distribution function p(x) described by probability density curve (Daniel Kirby E et al., 2004). This probability curves were obtained from the time domain waveform collected during machining. This wave form gives probability function for the various amplitudes and which gives the R.M.S of amplitudes, skew-index and kurtosis-index values.

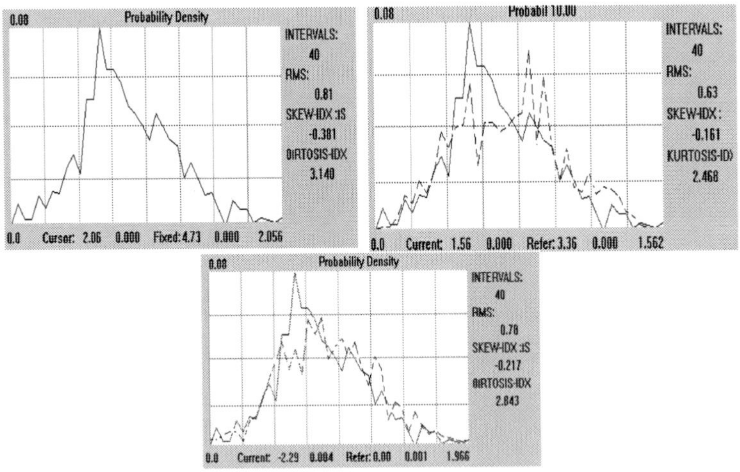

a b c

a) Carbide shim b) Comparison of Carbide and Aluminum shims c) Comparison of Carbide and Brass shims.

Figure 4.1. Probability density plots at speed 700 rpm and depth of cut 0.5 mm.

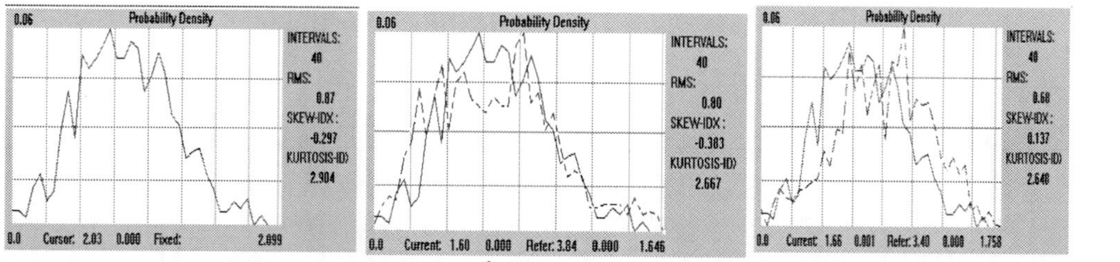

a b c

a) Carbide shim b) Comparison of Carbide and Aluminum shims c) Comparison of Carbide and Brass shims.

Figure 4.2. Probability density plots at speed 700 rpm and depth of cut 1.0 mm.

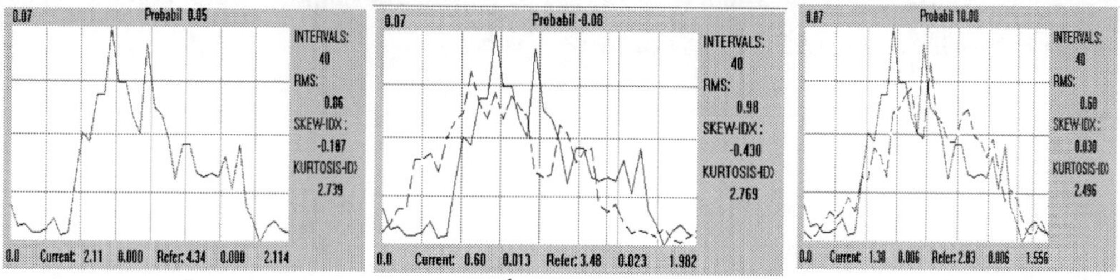

a b c

a) Carbide shim b) Comparison of Carbide and Aluminum shims c) Comparison of Carbide and Brass shims.

Figure 4.3. Probability density plots at speed 900 rpm and depth of cut 0.5 mm.

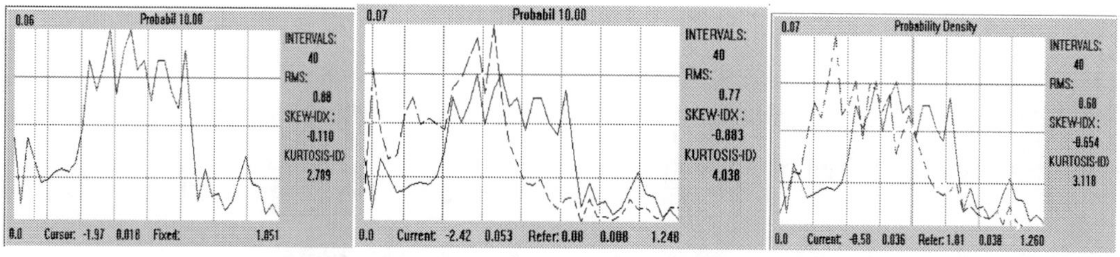

a b c

a) Carbide shim b) Comparison of Carbide and Aluminum shims c) Comparison of Carbide and Brass shims.

Figure 4.4. Probability density plots at speed 900 rpm and depth of cut 1.0 mm.

The probability density plots are shown in Figure 4.1 to Figure 4.4. In the first cutting condition the root mean square value of the amplitudes is 0.81 for Carbide shim as shown in Figure 4.1 (a). The RMS value is reduced to 0.63 and 0.78 as the Aluminum and Brass shims are used as shown in Figure 4.1 (b) and Figure 4.1 (c). The RMS value of vibration signals are taken from Probability density plots and tabulated in Table.3 for the all four cutting conditions.

Table 3 RMS value from probability density plot

Cutting condition / Material of shim	RMS			
	Speed 750 rpm and depth of cut 0.5 mm (I)	Speed 750 rpm and depth of cut 1.0 mm (II)	Speed 900 rpm and depth of cut 0.5 mm (III)	Speed 900 rpm and depth of cut1.0 mm (IV)
Carbide Shim	0.81	0.87	0.86	0.88
Aluminum shim	0.63	0.80	0.98	0.77
Brass shim	0.78	0.68	0.68	0.68

The RMS values for carbide, aluminum and brass shims for different four machining conditions are compared in the Figure 5.

From the Figure 5 it is observed that, the RMS values are reduced when components are machined with Aluminum and Brass shim respectively.

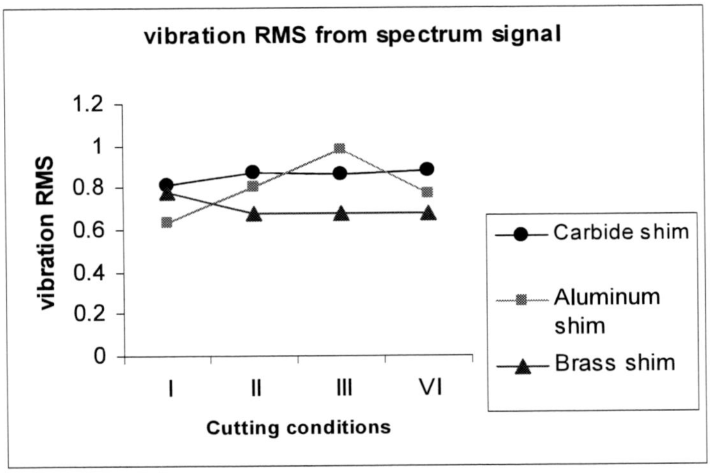

Figure 5. Amplitude RMS of different shim material.

Figure 6. Surface roughness tester.

4.3. Surface Roughness

After machining, surface roughness was measured using surface roughness tester SJ-400. The standard type stylus was used for measurements. After zero adjustment the roughness values were recorded for the length of 4 mm and which gives the value of Ra, Rz and Rq. The average roughness Ra is the most commonly used parameter in surface finish measurement. The roughness tester SJ-400 are shown Figure 6. Surface roughness Ra in μm measured by the surface roughness tester and RMS value from vibration signals for different cutting conditions are given in the Table 4 .1 and Table 4.2.

Table 4.1. Speed 750 rpm, feed rate 0.1 mm/s

Sl. No	Machining Condition	Depth of cut 0.5 mm		Depth of cut 1.0 mm	
		RMS	Ra (μm)	RMS	Ra (μm)
1	Carbide Shim	0.81	2.54	0.87	2.03
2	Aluminum Shim	0.78	1.88	0.80	1.92
3	Brass Shim	0.63	1.53	0.68	1.20

Table 4.2. Speed 900 rpm, feed rate 0.1 mm/s

Sl. No	Machining Condition	Depth of cut 0.5 mm		Depth of cut 1.0 mm	
		RMS	Ra (μm)	RMS	Ra (μm)
1	Carbide Shim	0.86	2.70	0.88	2.73
2	Aluminum Shim	0.98	2.26	0.72	3.04
3	Brass Shim	0.60	1.08	0.68	2.73

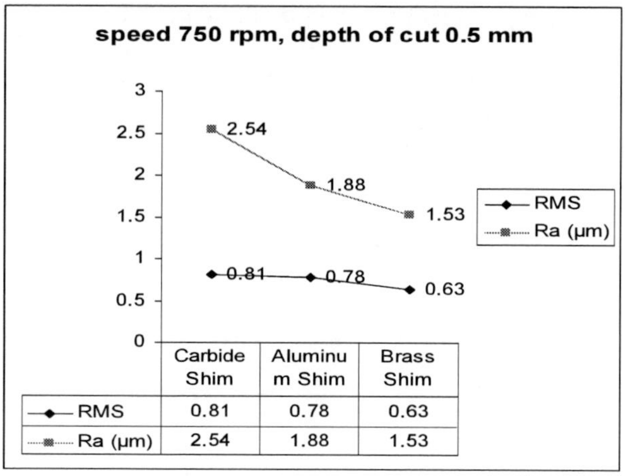

Figure 7.1. Speed 750 rpm and depth of cut 0.5 mm.

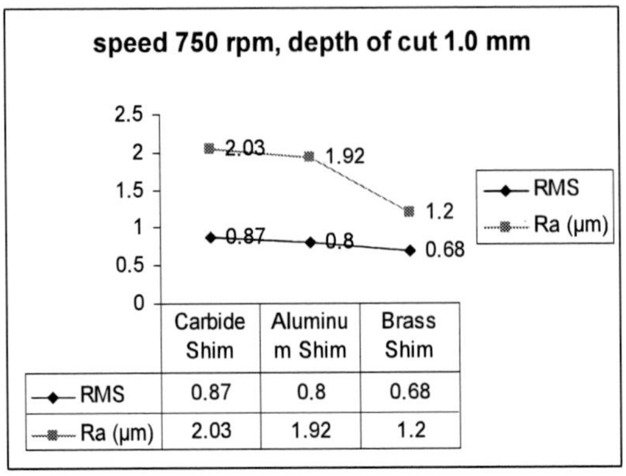

Figure 7.2. Speed 750 rpm and depth of cut 1.0 mm.

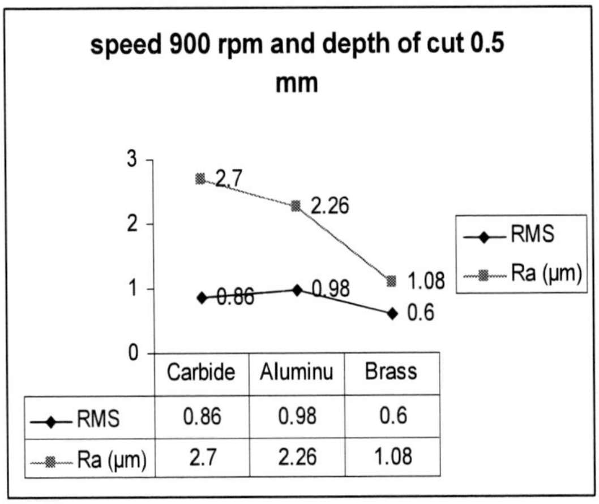

Figure 7.3. Speed 900 rpm and depth of cut 0.5 mm.

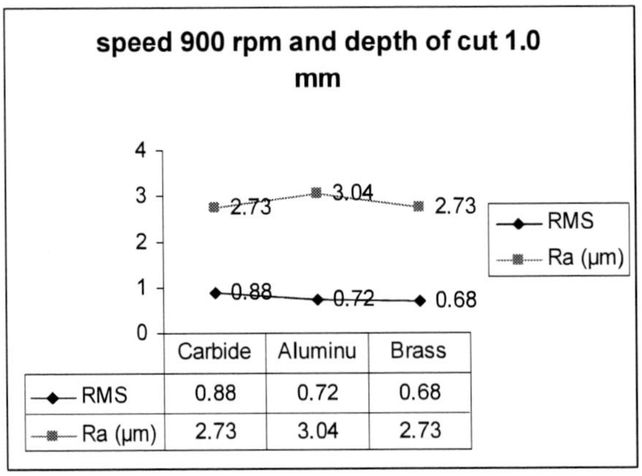

Figure 7.4. Speed 900 rpm and depth of cut 1.0 mm.

The comparison of Ra and RMS values for different cutting condition are clearly shown in graphical representation in Figure 7.1 to Figure 7.4. The Figure 7.1 shows that the average roughness of the machined component with Carbide shim is 2.54 μm and this value is reduced as 1.88 μm and 1.53 μm when component machined with Aluminum and Brass shim respectively for the first machining condition. Similarly in all other three cutting conditions, it is observed that average roughness value is reduced when component machined with Aluminum and Brass shim as shown in Figure.7.2 to Figure 7.4.

From the Figure 7.1 to Figure 7.4, it is observed that the vibration signal parameter amplitude RMS is proportional to roughness (Ra) value.

CONCLUSION

The following results are obtained by introducing passive dampers for the tool inserts. The dynamic characteristics for turning of OHNS material with various shims are studied by using vibration data collector and MCMe software.

1. RMS value of vibration signal and the average surface roughness (Ra) value for aluminum shim and brass shim were reduced when compared with the carbide shim.

(i) Reduction of RMS value for Aluminum shim is 3.7 % and Brass shim is 22.7 % for the cutting condition of 750 rpm and 0.5 mm depth of cut, for second cutting conditions the RMS is reduced to 8 % for Aluminum shim and 21 .3 % for Brass shim, similarly for the cutting conditions at the speed of 900 rpm an depth of cut of 1.0 mm the RMS is reduced to 18 % for Aluminum shim and 22.72 % for Brass shim.

(ii) Reduction of Ra for Aluminum shim is 25.98 % and Brass shim is 39.76 % for the cutting condition of 750 rpm and 0.5 mm depth of cut, for second cutting condition, Ra is reduced to 5.4 % for Aluminum and 21.83 % for Brass shim, similarly for the cutting conditions at the speed of 900 rpm and depth of cut of 0.5 mm, the Ra is reduced to 20.74 % for Aluminum shim and 18.6 % for Brass shim.

2. It is observed that the RMS value of Brass shim is reduced by 21 % and the Aluminum shim by 10 % when compared with Carbide shim and it is also observed that the Ra value of Brass shim reduced by 45 % and Aluminum shim by 17 % when compared with Carbide shim.

3. Based on the observations made, Brass shim is more suitable for machining OHNS steel which gives better surface finish. Vibration signal features can be easily realized during machining and it can also be used to detect state of surface roughness.

REFERENCES

Alexandria,V.A., (1996),"*Signal Processing for manufacturing and machine monitoring*", National Science Foundation Workshop, Les atlas, University of Washington and Douglas Jones, University of Illinois, pp. 1-8.

Alexej V. Barkov, Natalia A. Barkova, (1999), "The Artificial Intelligent systems for Machine condition monitoring and diagnostics by vibration", *Proceedings of the Saint Petersburg Post-graduate Institute of the Russian federation Power industry and vibration Institute,* USA, volume 9.

Caldeirani Filho, J. Diniz A.E, (2002), " Influence of cutting conditions on tool life, tool wear and surface finish in the face milling process", *Journal of the Brazilian Society of Mechanical Sciences* Vol.24, No.1, pp. 10-14.

Change-Xue (Jack) Feng, Xianfeng Wang, (2002), "Development of empirical Models for surface roughness prediction in finish turning", *International Journal of Advanced Manufacturing Technology*, Vol.20, No.5, pp. 348-356.

Claudiu Bisu, Alain Gérard, Jean-Yves K'nevez, Raynald Laheurte, Olivier Cahuc"Self-excited vibrations in turning: forces torsor analysis" *The International Journal of Advanced Manufacturing Technology.*

Daniel Kirby E., Ms. Zhe Zhang and Dr. Joseph.C.chen, (2004), "Development of an accelerometer-Based surface roughness system in turning operations using multiple regression techniques" *Journal of Industrial Technology*, Vol. 20, No.4, pp. 2 – 8.

Devillez Arnaud and Dudzinski Daniel, (2006), "Tool Vibration with eddy current sensors in machining process and computation of stability lobes using fuzzy classifiers", *Journal of Mechanical Systems and Signal Processing*, Vol.21, No 1, pp. 441-456.

Dimla D.E., (2002), "The Correlation of vibration signal features to cutting tool wear in metal turning operation", *The International Journal of Advances Manufacturing Technology,* Vol. 19, No.10, pp. 705-713.

Hassui A. and Diniz A.E, (2003) "Correlating surface roughness and vibration on plunge cylindrical grinding of steel", *International Journal of Machine Tools and Manufacturing,* Vol. 43, No. 8, pp. 855-862.

George Balan (1996), "*New Results in the Study of Machine – Tools Vibrations* Part 1: Critique look on a Current state in the Study of Chatter phenomenon" FASCICULA V, Dunarea de Jos, Galati University ISSN 1221-4566, pp. 21-24.

Ghani A.K., Choudhury I.A. and Husni, (2003), "Study of the tool life roughness and vibration in machining nodular cast iron with Carbide tool", *Journal of material processing and technology*, Vol. 127, No. 1, pp. 17-22.

Guo Y.B and Ammula S.C., (2005), "Real-time acoustic emission monitoring for surface damage in hard machining", *International Journal of Machine tools and manufacture,* Vol. 45, pp.1622-1627.

Joseph C.Chen, luke H.Huang, Ashley X.Lan and Samson Lee (1999), "Analysis of an effective Sensing Location for an In-process surface recognition system in turning Operations", *Journal of Industrial Technology*, Vol.15, No.3, pp.1-6.

Julie Z.Zhang, Joseph Chen C., Danial Kirby E., (2007), "Surface roughness optimization in an end-milling operation using the Tauguchi design method", *Journal of material processing Technology*, Vol.184, pp. 233-239.

Keraita J.N., Oyango H.J. and Misoi G.K, "Lathe Stability charts via acoustic emission monitoring", *African Journal of Science and Technology, Science and engineering series,* Vol.2, No. 2, pp. 81-93.

Kozochkin M.P., Kochinev N.A. and Sabirov F.S, (2006), "Diagonstics and monitoring of complex production processes using measurement of vibration-acoustic signals", *Journal of Measurement Techniques*, Springer New York, Vol. 49, No.7, pp. 672-678.

Krishnan Kiran M.P.S., Suhas S.Joshi (2007), "Modelling of Surface Roughness and Role of Debries in Micro-EDM", *Journal of Manufacturing Science and Engineering*, Vol.129, pp. 265-273.

Kumarasamy. S, Rakesh. J. and Amol kumar nalavade (2002), "Standardization of Absolute Vibration Level and damping factors for Machinery Health Monitoring", *Proceedings of International Conference,* VETOMAC-2, pp.16-18.

Mandar, D.Savage and Dr.Joseph C.Chen, (1999), "Effects of Tool Diameter Variations in on-line Surface Roughness Recognition system", *Journal of Industrial Technology,* Vol.15, No.4, pp. 1-6.

Murad S. Samhouri and Brain W.Surgenor, (2005), "Surface Roughness in grinding: on-line prediction with adaptive Nero –fuzzy inference system", *Transaction of NAMRI /SME* Vol. 33, pp 57 – 64.

Paulo Davim J., and Luís Figueira, (2007), "Machinability evaluation in hard turning of cold work tool steel (D2) with ceramic tools using statistical techniques" *Materials and Design*, Elsevier sc.,, Vol. 28, Issue 4, pp.1186-1191.

Paulo R .Aguiar, Carlos E.D.Cruz, Wallace.F.paula, Eduardo C. Binachi,etal, (2007), "Neural network Approach for Surface roughness prediction in surface grinding", *Proceedings of the 25th IASTED International Conference on Artificial Intelligent and Applications*, pp. 96 – 101.

Paulo R.Aguiar, Paulo J.A.Serni, Fabio R.L.Dotto, Eduardo C. Bianchi, (2006), "In-Process grinding monitoring through acoustic emission", *Journal of the Brazilian Society of Mechanical Sciences and Engineering,* Vol.28, No.1, pp.118-124.

Richard Y. Chiou, Steven Y. Liang, (2000), "Analysis of acoustic emission in chatter vibration with tool wear effect in turning", *International Journal of Machine Tools and Manufacture,* Vol. 40, pp. 927-941.

Rodriguez E. , Fraudita F.,.Amorer Y.L, (2003), "Effect of surface roughness in Ultrasonic testing (Pulse-echo by direct contact) in AISI/SAE 4340 steel samples", Paper Presented At The 8th Ecndt, Barcelona Vol.8, No.11.

Sadettin Orhan, Ali Osman Er, Necip Camuscu and Ersan Aslan, (2007), "Tool wear evaluation by vibration analysis during end milling of AISI D3 cold work tool steel with 35 HRC hardness", NDT and E International, ISSN 0963 - 8695, Vol. 40, No.2, pp. 121-126.

Scrutton R. F., (1972), "Frictional Vibrations in Metal Cutting" *International Journal of Production Research*, Vol. 10, Issue 3 July, pp. 195 – 204.

Steven Y. Liang, Regelio L.Hecker and Robert g Landers, (2004), "Machining Process monitoring and Control: The State-of-the-Art", *Journal of manufacturing Science and Engineering*, Vol.126, pp. 297-307.

Taejun Choi, Young C. Shin, (2003), "On-line Chatter detection wavelet-based Parameter Estimation", *Journal of Manufacturing Science and Engineering*, Vol 125, Issue 1, pp. 21-28.

Warren Liao T., Chi-Fen Ting, and Blau.P.J, (2007), "A wavelet-based methodology for grinding wheel condition monitoring", *International Journal of Machine tool and Manufacturing,* Vol. 47, No.3-4, pp. 580-592.

Wilfried Reimche, Ulrich Sijdmersen, Oliver Pietch, Christian Scheer Fiedrich-Wilhelm Bach (2003), "Basics of Vibration monitoring for fault detection and process control" 3rd Pan-*American Conference for Nondestructive Testing*-PANNDT, pp. 1-10.

In: Machining and Forming Technologies, Volume 2
Editor: J. Paulo Davim

ISBN: 978-1-61122-820-5

Chapter 6

FINITE ELEMENT ANALYSIS FOR THE EFFECTS OF FLANK WEAR AND SOME CUTTING PARAMETERS ON CUTTING STRESSES IN TURNING OF AISI 1060 STEEL

Bekir Yalçin* and Nihat Yilmaz
Technical Education Faculty, S. Demirel University,
Cunur Campus, Isparta, Turkey

ABSTRACT

Cutting forces affecting cutting tool in metal cutting are one of the important parameters necessarily to know tool failure and to optimize the cutting parameters. It is known that flank wear describing the tool life is formed with mechanical and thermal interaction between cutting tool and workpiece. In present study, the cutting stress analyses were carried out from two aspects. Firstly, changing in cutting stress distribution with wear band width is determined with modeling of between 0.3 mm and 0.9 mm wear band values by finite element method using Ansys 10.0 programme. Secondly, the effect of the some important cutting parameters such as cutting depth, feed rate, and approach angle on the cutting stress was performed in present analyses. The results showed that stresses increase to parallel an increase with flank wear value. The cutting depth has more a significant effect on the tool stresses than feed rate parameter in respect of von-Mises stresses. Also, cutting stresses decreased with an increase in the approach angle.

Keywords: Flank wear, tool life, finite element analyses, stress, approach angle.

---

* Corresponding author, bekir@tef.sdu.edu.tr

1. Introduction

The turning operation is a reference machining process used in modeling the cutting temperature and cutting forces. Tool wear in turning has an important effect on machining costs and quality components. Therefore, the experimental and mathematical researches relating with machining focused on the tool wear, cutting forces and cutting temperature. Some of these work (Kishwey et. al., 2005; Shimada et. al., 2004; Takeyama et. al., 1963; Astakhov, 2004) focused on the empirical modeling of tool wear, the others carried out on the estimation and simulation of tool wear as mathematically (Attanasio et al.,2008; Valiorgue et al., 2008). Also, others reported (Buryta et al., 1994; Trent, 1991) that the thermal and mechanical loads on the tool flank surface and chip surface are deteriorated during cutting. One approach emphases in report, cutting temperatures reach to 500–1000 ^{o}C and cutting stresses reach to average 2GPa on flank and chip surfaces during cutting. Kose et al. (2008) determined that cutting tool stresses in the machining of Inconel 718 increase with an increase in the feed rate by finite element analyses using ANSYS programme.

The main tool wear in turning inserts is formed with two phenomena. One is the crater wear occurred with chip sliding from tool chip surface, another is the flank wear occurred with friction between machined surface and chip. The contact tool edge with machined surfaces affecting the surface quality and cutting force is an important phenomena and then flank wear take into consideration in tool life equation of Taylor's (Köse, et al., 2008). The flank wear type is the measurable directly according to ANSI/ASME B94.55M-1985 standard (Astakhov, 2006).

During cutting operations, contact stress between tool chip surface and chip can cause the friction between tool flank surface and machined surface. Therefore, this phenomenon causes the extremely friction. Hence, all wear type depending on wear mechanisms are occurred on the tool chip surface, flank surface, and tool tip. The main wear types can be appeared in Figure 1. The flank wear is formed on the tool flank surface with abrasive wear mechanism. Especially, worn tool may causes phenomena such as increasing the cutting forces, deterioration of workpiece surface and dimension quality, decreasing the productivity, increasing the cutting temperature and vibration.

In this paper, effects of the flank wear and some cutting parameters on the main cutting stresses during the machining of AISI 1060 steel were investigated by finite element method. The present study consists from two analyses. In first analyses, tool cutting edge geometry together with wear band width (flank wear) was modeled by Solidwork pack sent to Ansys in order to investigate of the effects of flank wear value and other cutting parameters. Flank wear values (Vb:0mm, 0.3 mm, 0.9 mm) were given to tool cutting edge and flank wear as suited to literature (ASM Handbook, 1998) in finite element models.

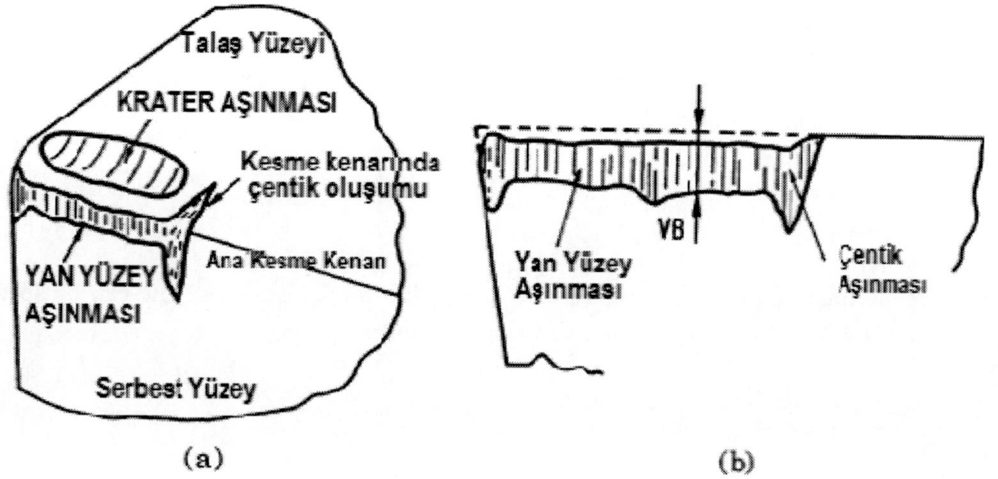

Figure 1. Wear type at turning tool.

In second analyses, the effects of the machining parameters such feed rate, approach angle, cutting depth on cutting stresses in machining of SAE 1060 steel were determined with finite element method. Some results were obtained with total thirty three finite element solutions performed in present study.

1.1. The Main Cutting Force in Turning

The mechanical properties such as elastic module and poisson ratio of high speed steel tool were used in finite element analyses. Tool wear and stresses are a function of the main cutting force (Fc) and cutting temperature. Also, it is known that cutting forces increase with increases in the flank wear. Cutting mechanics can be seen in Figure 2.

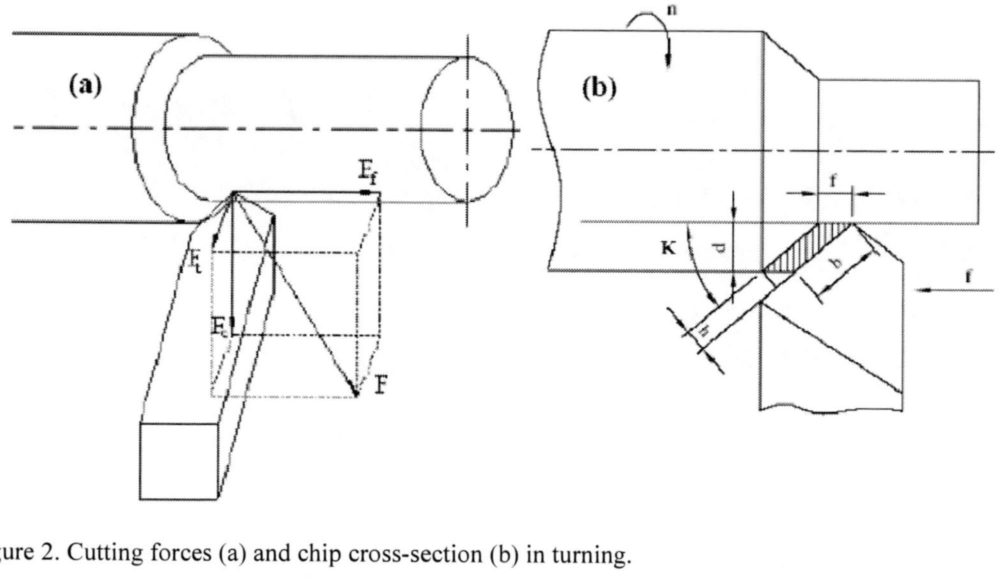

Figure 2. Cutting forces (a) and chip cross-section (b) in turning.

The cutting force is calculated with some parameters such as cutting depth (d), chip cross-section (Ac) and feed rate (f), approach angle (K). The cutting force shown in Figure 3 are depended on the relation uncut chip cross-section (Ac) and the specific cutting power. The following equations are given;

$$Fc = A_c k_s N \tag{1}$$

$$Ac = b.h = d.f \tag{2}$$

where, b is the uncut chip width and h is the uncut chip thickness. This state can be seen in Figure 3-b. b and h are calculated with following equations:

$$b = \frac{d}{\sin \chi}, \text{ and } h = f \times \sin K \tag{3}$$

But, specific cutting force, ks is effected by factors such as tool and workpiece materials, cutting depth, chip angle, specific cutting force (ks) obtained by taking into consideration these correction factors;

$$ks = k_{11}\, h^{-z}.kv.k\gamma.kt \tag{4}$$

In equation 4, k_{11} is an average specific cutting force for workpiece material. h^{-z} is a increasing the specific cutting force as function of cutting depth. Also, kv is a cutting speed correction factor, kγ is a chip angle factor ($k\gamma = \frac{(C - 1{,}5\gamma)}{100}$. C for steel is 100) (Sağlam et. al., 2007), and kt is removal ration and tool material factor. As a result, the main cutting force is calculated with Eq. 5 if all factors are taken into consideration:

$$Fc = d \text{ x } f \text{ x } (k_{11}\, h^{-z}.kv.k\gamma.kt) \tag{5}$$

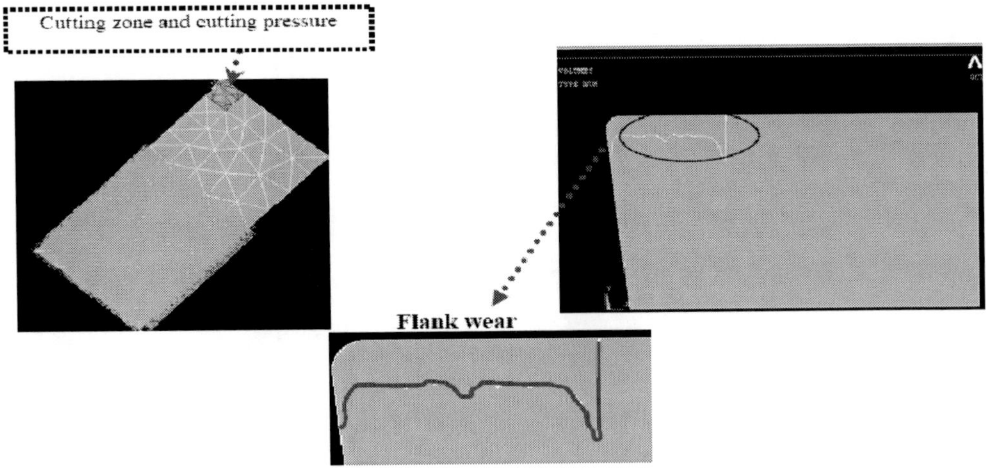

Figure 3. a) Tool displacement (boundary condition) at all dof and meshing. b) Worn tool (flank wear).

2. Materials and Method

2.1. The Cutting Tool Modelling

A geometric property of the cutting tool model was performed by solid model formed on the Solidworks 2007 pack sent to ANSYS in “iges” format in each two analyses. In first analyses, the main cutting forces applied in finite element analyses were calculated for one imaginative turning operation (Table 1) with Equation 5. The main cutting force is calculated by Eq 5 as 4390 N for AISI 1060 steel according to machining condition given in Table 1. Cutting pressure occurred with the main cutting force at cutting zone was calculated as 2693 MPa with Pc=Fc/Ac equation and then this pressure value is applied to cutting zone.

The cutting zone area was accepted as chip cross-section area. The calculated pressure values were applied to there different model with Vb:0 mm (new tool). Vb: 0.3 mm and Vb: 0.9 mm. respectively. In first analyses, cutting tool was modeled as not only new tool form but worn tools with 0.3 mm and 0.9 mm flank wear value. Finite element model for cutting tool is shown in Figure 3. In first analyses, element type was selected as Solid 97, elasticity modulus, poisson ration are 210 GPa and 0.28.

In second analyses, flank wear value was fixed as 0 mm (new tool). The main cutting forces are calculated by Eq. 5 depended on changing of the cutting depth, feed rate, and the approach angle. The main cutting forces were applied on the chip section at -z direction.. With this way, the effects of important cutting parameters on cutting pressure were investigated in these analyses. Table 2 shows the changing the cutting parameters.

Table 1. Fixed cutting parameters for changing the flank wear value

Imaginative Turning Parameters	Values
Cutting depth, d (mm)	4
Chip angle, γ	6
Feed rate, f (mm/dev)	0.4
For steel, k_{11} (N/mm^2	2260
Approach angle. (K)	60
Specific cutting force exponent, z	0.3
Cutting speed, Vc (m/min)	170
Yield stress for HSS tool, (MPa)	2200
Elastic Module for HSS tool, (MPa)	210000
Poisson ratio for HSS tool, (v)	0.3

Table 2. Changing the cutting parameters for second analyses

Changing the Turning Parameters	Values
First analyses: Changing the cutting stress with flank wear values	
Fixed	Vb: 0mm; Vb:0.3 mm; Vb:0.9 mm
Second Analyses: Changing the cutting stress with important cutting parameters	
Cutting Depth , d.(mm)	1, 2, 3, 4
Feed rate, f (mm/rev)	0.1, 0.2, 0.3, 0.4
Approach angle. K	30, 60, 90
Flank wear value, Vb (mm)	New tool

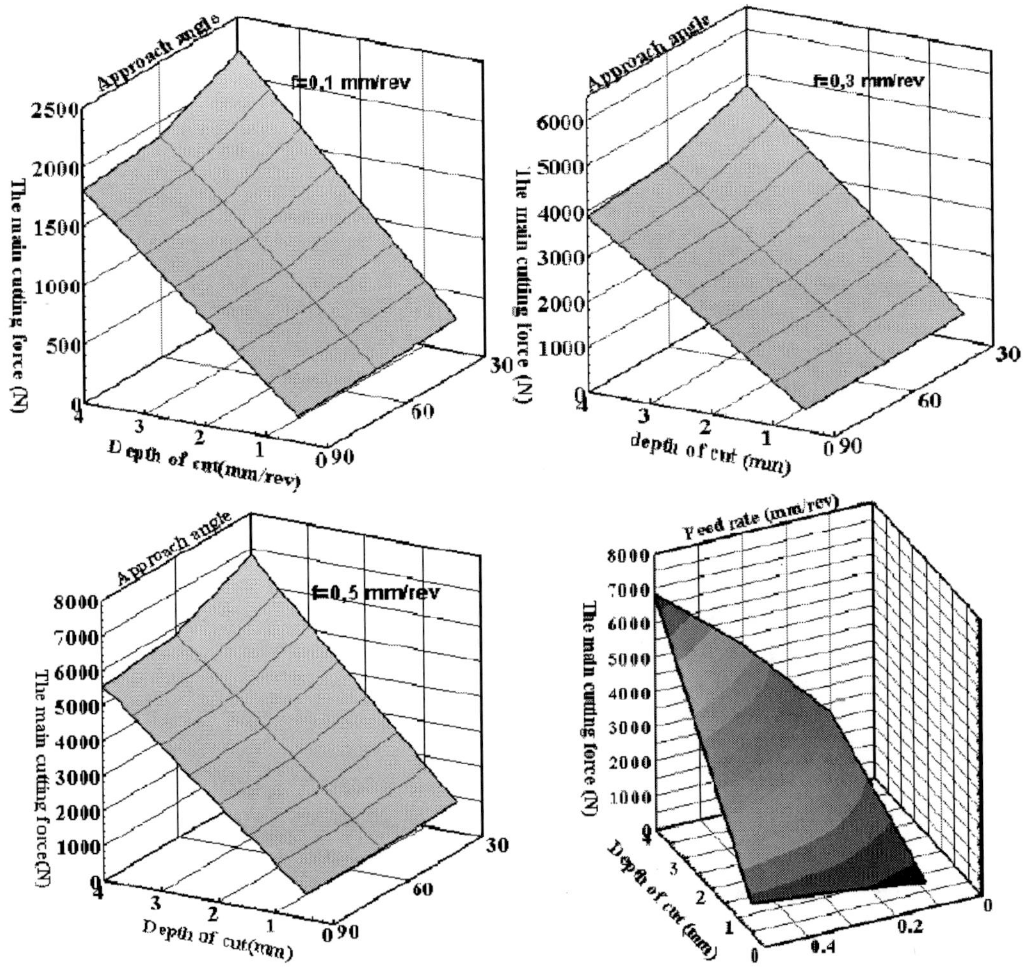

Figure 4. Variables calculated as mathematically for changing of the main cutting force according to some cutting parameters.

Besides, element type was selected as Solid 97, elasticity modulus and poisson ration are 210 GPa and 0.28 in FEM model. 86630 elements and 129945 nodes were produced in this FEM model. Tool was fixed at all dof from tool fixed area.

In each two analyses, the main cutting forces were calculated by taken chip cross-section area and according to changing of the cutting parameters. Fc force, chip cross-section area and the main cutting pressure calculated as mathematical for changing of each cutting parameter are given in Figure 4.

The main cutting forces obtained as mathematically by the Eq. 5 according to change in the feed rate, approach angle, and cutting depth values were shown in Figure 4. Generally, the main cutting force increase with an increase in the feed rate, cutting depth; on the other hand, the main cutting force decreases with an increase in approach angle. This state can be seen in Figure 4. The calculated main cutting force is extremely higher value in cutting condition such as minimum approach angle (30°) and 0.5 mm/rev of feed rate and 4 mm of cutting depth than other cutting condition. The cutting depth has more a significant effect on the main cutting forces than other parameters. This tendency is shown in Figure 4.

The calculated main cutting force value for the 4 mm cutting depth are higher than that of 1 mm cutting depth. Also, the main cutting force for 0.5 mm the feed rate is lower than 0.1 mm the feed rate.

Second analyses were carried out for all thirty cutting condition with regardless of cutting parameters the stated in Table 2. Cutting speed is fixed for 150 m/min. The calculated main cutting forces applied to cutting zone are given in Figure 3, and then S_{EQV}-von Mises stresses are obtained from finite element analyses according to change in the feed rate, cutting depth and approach angle. The finite element models for second analyses are given in Figure 3.

3. Results

In first stress analyses, the cutting stresses obtained from the finite element analyses according to the change in the flank wear values were shown in Figures 5 and 6. The von-Misses stresses, S_{QEV}, increased with an increase in the flank wear value. Stresses value for new tool (Vb: 0 mm) were obtained as lower than for the other flank wear value. Also, stresses at new tool during metal cutting are lower than the yield stress of tool. Therefore, any fatigue cracks don't expect in new tool. But, the obtained S_{QEV} stresses of worn tool with 0.3 mm flank wear were obtained close to the yield stress of tool. The 2046 MPa stress value is lower than the 2200 MPa the tool yield stress. Therefore, any fatigue cracks for the 0.3 mm flank wear value may not occur under cutting condition given in Table 1. But, the 0.3 mm flank wear is a critical value due to 2046 MPa stress value closing the yield stress (2200 MPa). Then, the 0.3 mm flank wear may cause the initial fatigue cracks on the flank wear notch. Because, stresses of tool with 0.3 mm flank wear value were focused on the notch of flank wear. On the other hand, stresses in the new tool focuses on the chip surface. This state can be seen in Figures 6-b-c.

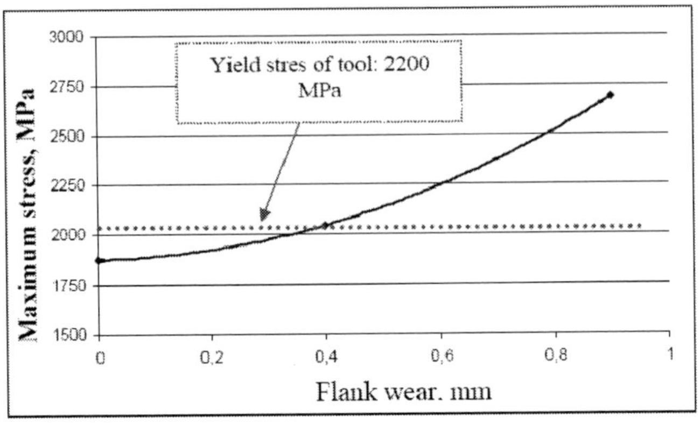

Figure 5. σmax obtained by finite element analyses according to wear value (Vb).

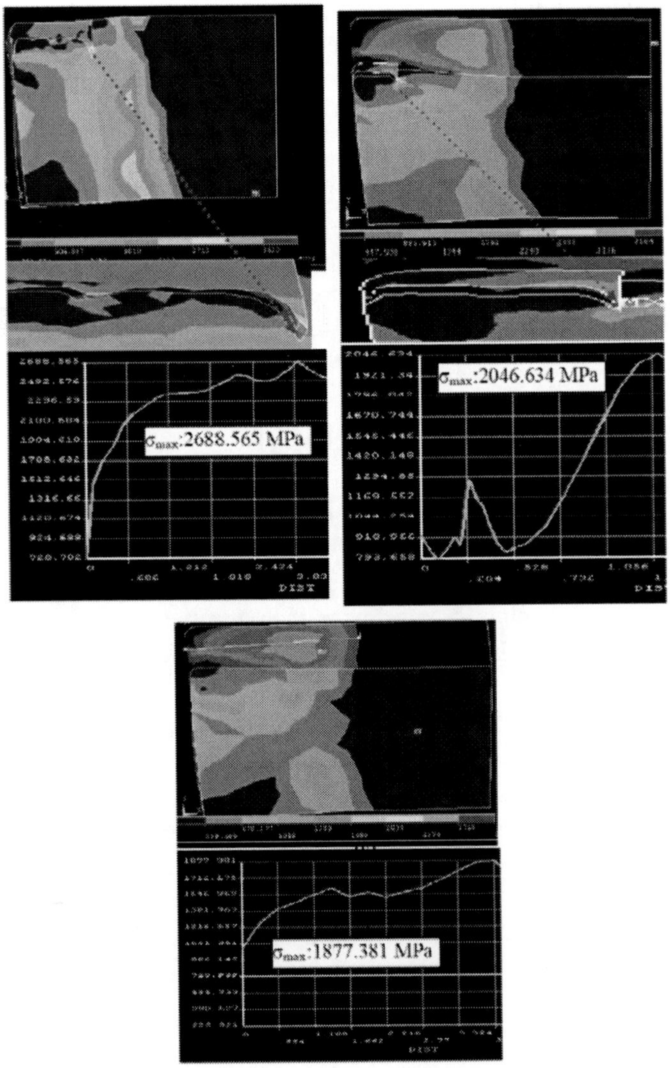

Figure 6. a) von-Mises stresses at Vb: 0.9 mm tool, b) von-Mises stresses at Vb: 0.3 mm tool, c) von-Mises stresses at Vb: 0 mm tool (new tool).

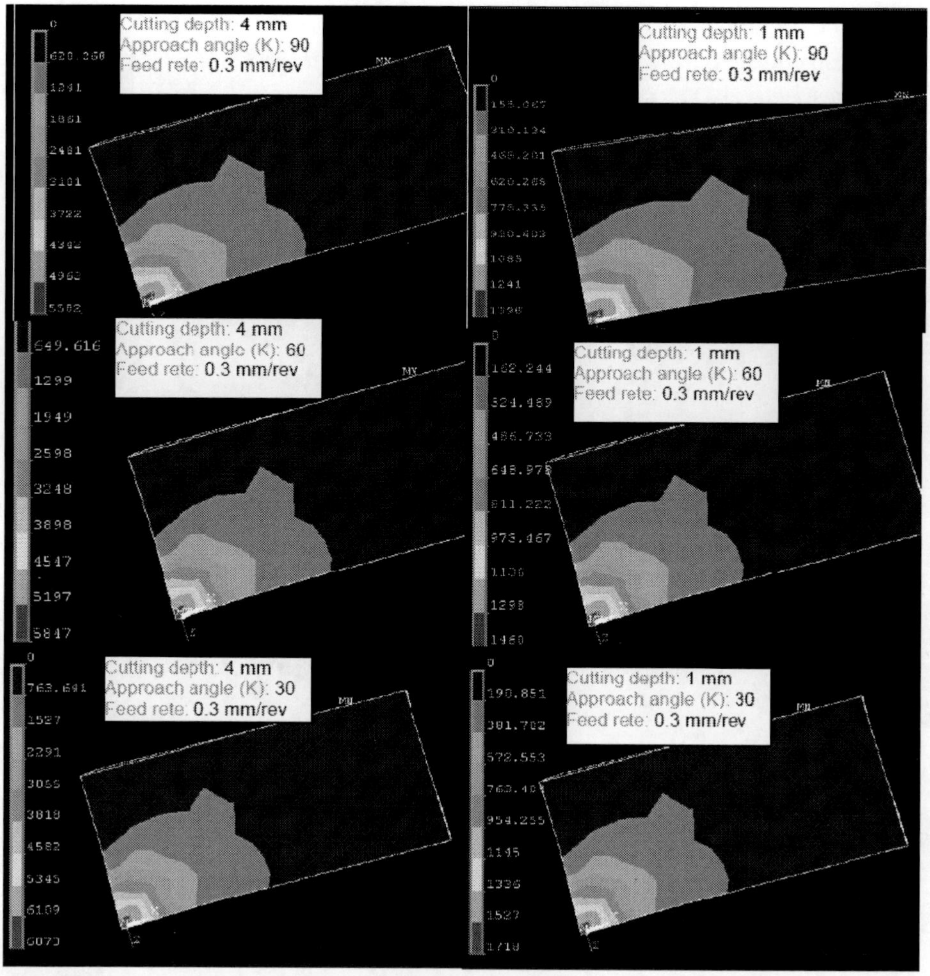

Figure 7. The effect of the approach angle and cutting depth on the von Mises stresses.

Stresses of tool with 0.9 mm flank wear are extremely higher than other new and 0.3 mm worn tools. Also, the 2688 MPa stress value for worn tool (Vb:0.9 mm) is extremely higher than the yield stress of cutting tool. Therefore, the fatigue crack leading final failure develops. The fatigue cracks for the 0.9 mm flank wear occurs due to extremely stresses focused on the notch of flank wear. This can be shown in Figure 6-a. It is concluded that von-Mises stresses increased with an increase the flank wear value in first analyses. Fatigue cracks may develop on the wear notch with an increase in the flank wear. This phenomenon may gradually leads to final tool fracture. Especially, initial fatigue cracks may start at below of the 0.3 mm flank wear value. Hence, 0.3mm flank wear value is a critical value for starting of the tool initial crack. Because of this, the worn HSS tool with below of the 0.3 mm flank wear value may breakage in any metal cutting due to higher cutting stresses than tool yield stress.

In the second analyses, stresses analyses were performed with changing of the cutting parameters such as approach angle (30°, 60 ^{o}, 90 ^{o}), feed rate (0.1, 0.2, 0.3, 0.4 mm/rev.) and cutting depth (1, 2, 3, 4 mm). The cutting speed was fixed as 150 m/min. The von-Mises stresses, S_{QEV} and shear stresses were obtained from the second solutions. Analyses were carried for all 30 cutting conditions according to the stated loading condition.

Figure 7 and Figure 8 show the stress distributions for 150 m/min cutting speed, and changing in some cutting parameters. Shear stresses and von Mises stresses were expressed as S_{XY} and S_{EQV}, respectively.

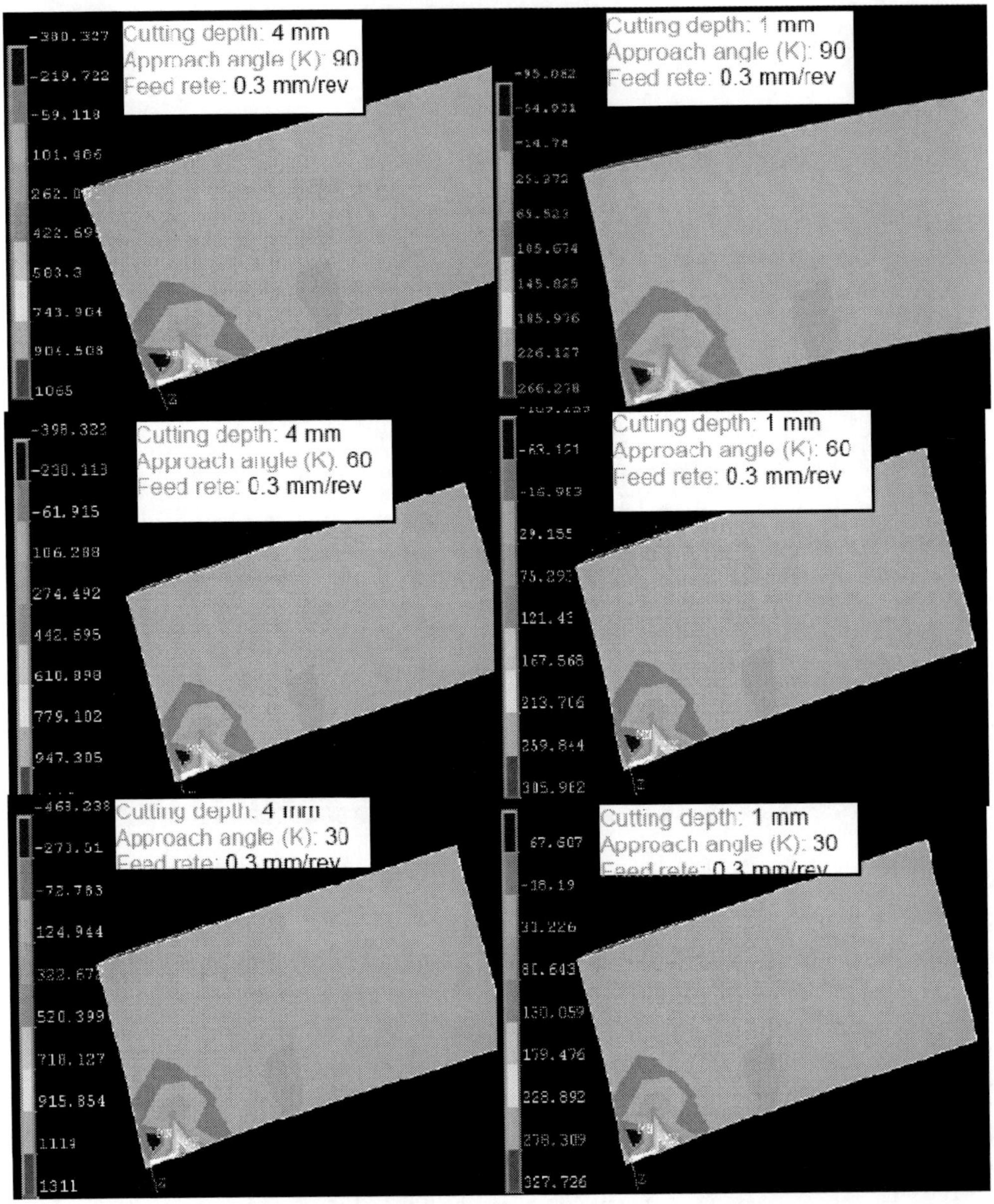

Figure 8. The effect of the approach angle and cutting depth on the S_{XY} shear stresses.

The critical zone in term of the cutting tool for second analyses is determined as notch area, cutting edge, and nose area. It is shown that all von Mises stresses values increase in parallel to a decrease the approach angle and to an increase the cutting depth. The highest value of the von Mises stress and S_{XY} shear stress are 6873 MPa and 1311 MPa for feed rate 0.3 mm/ rev and 4 mm cutting depth, respectively. The largest stress (S_{EQV}) 6873 MPa value may cause the initial fatigue cracks in cutting tool nose area due to higher than tool yield stress. Figure 8 shows that the S_{XY} shear stresses focused on the cutting edge contacted with the machined surface. Thence, possible wear mode will be almost on the flank surface. On the other hand, possible second wear mode will be almost as nose wear due to fact that S_{EQV} stresses intensified on the cutting zone.

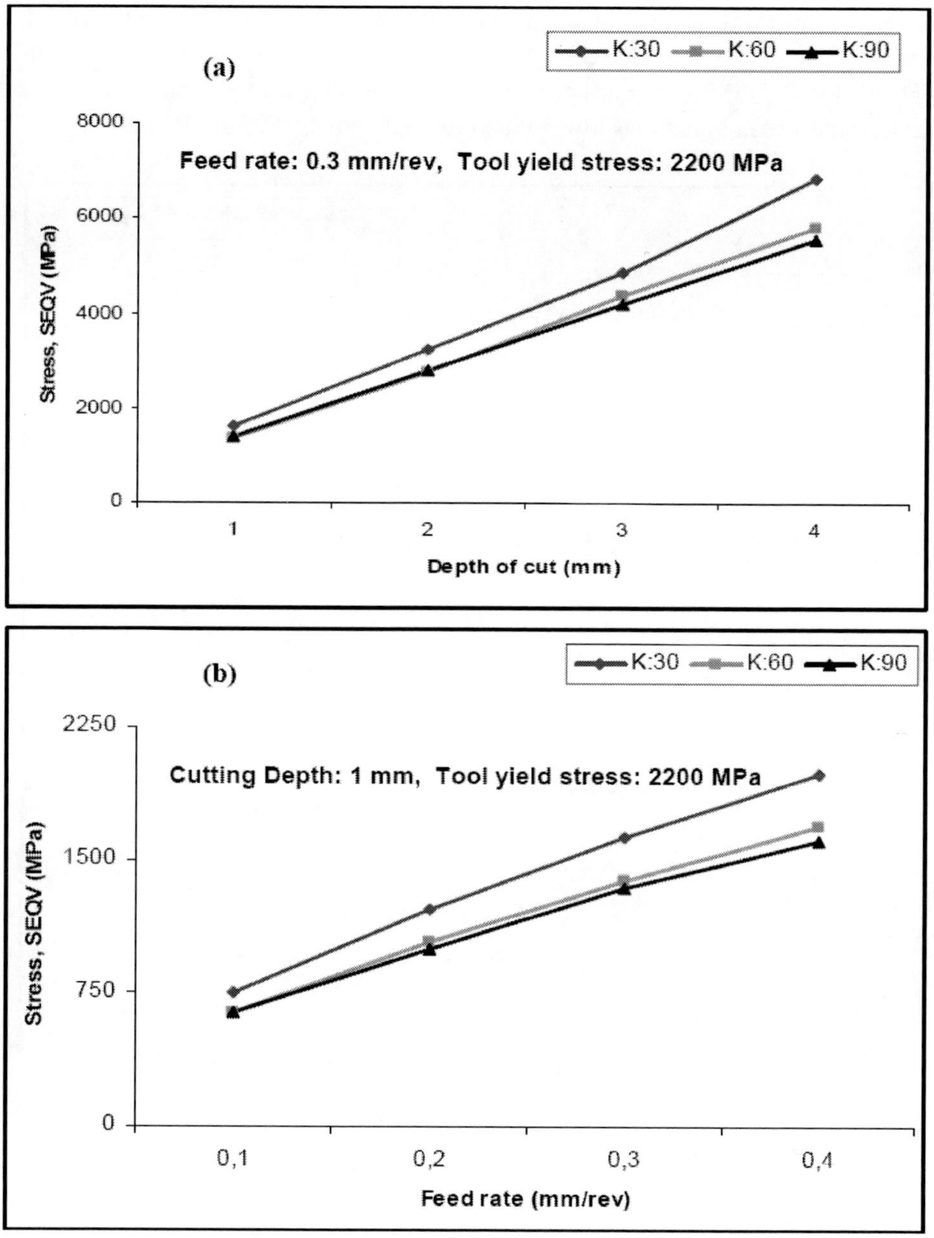

Figure 9. The effect of the cutting depth (a) and feed rate (b) on the von Mises stresses (SEQV).

Also, Figure 7 shows that the von Mises stresses increase in parallel to an increase the cutting depth. The effects of cutting depth on cutting stresses are higher than effects of the feed rate parameter. 3 mm and 4 mm cutting depth values and 0.3 mm feed rate value and 150 m/min cutting speed value for turning of the SAE 1060 steel may cause the any initial fatigue cracks due to the fact that cutting stresses are obtained as higher than tool yield stresses from analyses. According to finite element analyses, 0.3 mm/rev feed rate value and 1-2 mm cutting depth, approach angle between 60 and 90 values are optimal turning condition for AISI 1060 steel with HSS tools. Furthermore, the cutting stresses for 1 mm cutting depth and all feed rate values was obtained as lower than tool yield stress (Figures 9 and 10).

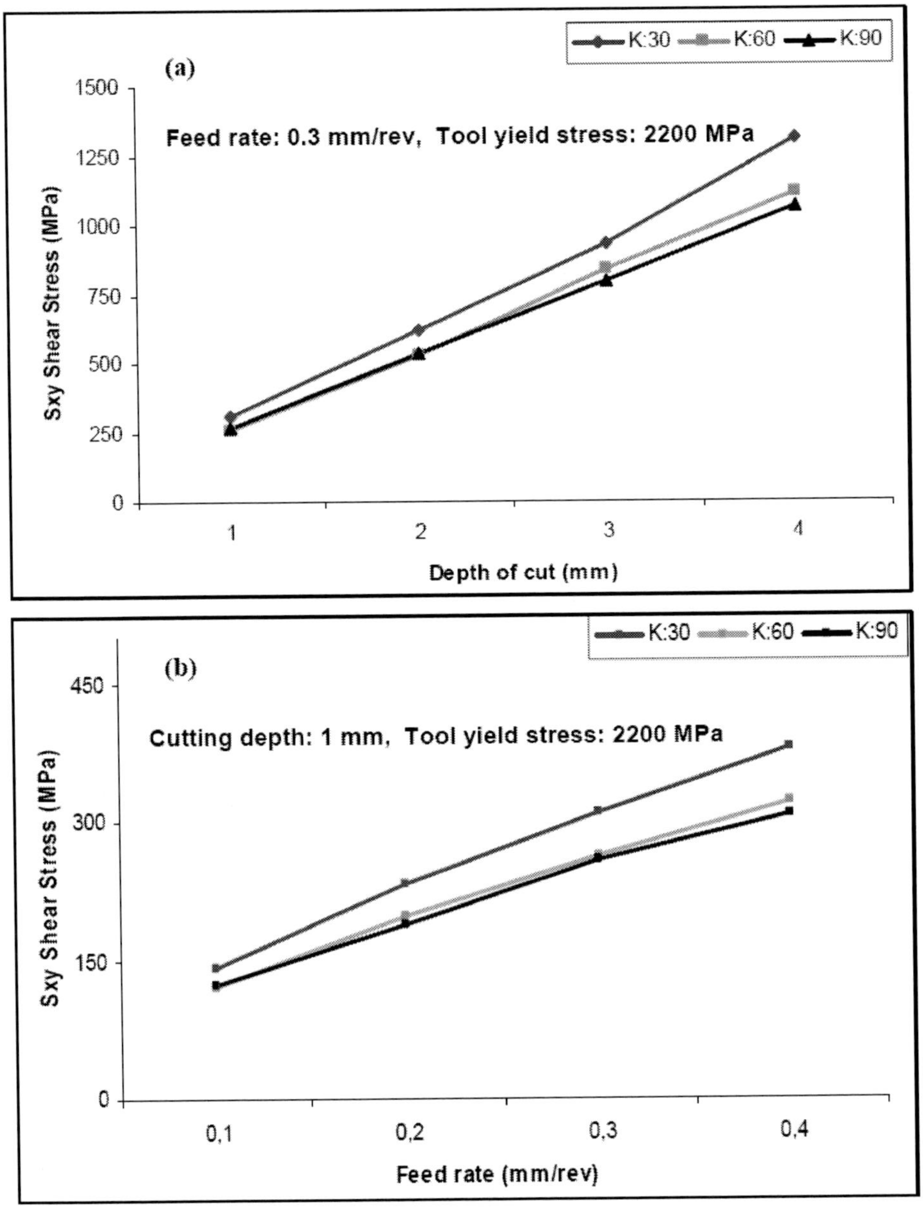

Figure 10. a) The effect of the cutting depth, b) The effect of feed rate on the shear stresses (S_{XY}).

The highest von Mises Stress for 1 mm cutting depth and 0.4 mm/rev feed rate value was obtained as 2000 MPa. Any initial fatigue cracks will not expect at tool edge because of the fact that this value is lower than tool yield stress.

Furthermore, shear stresses increase along with an increase in the feed rate and cutting depth (Figure 10). The largest shear stress has occurred as 1311 MPa on cutting edge for 0.3 mm/rev feed rate, 4 mm cutting depth and 30 degree approach angle. The lower shear stresses have occurred as between 120 MPa and 300 MPa for 1 mm cutting depth, and 60-90 approach angles. Besides, the largest von Mises Stress has occurred as 6866 MPa on the tool nose or cutting zone for 4 mm cutting depth, 0.3 mm/rev feed rate and 30 degree approach angle (Figure 9). The minimum von Mises stresses were obtained as 120 MPa for 1 mm cutting depth, 0.1 mm/rev feed rate, between 60 and 90 approach angles.

CONCLUSION

With the help of the mathematical studies performed on turning of the AISI 1060 steel workpiece using the high speed steel cutting tool, the results of the analyses obtained ANSYS propgramme based on finite element method are follows:

1. Firstly, the present studies give the preinformation before turning of the AISI 1060 steel. Thence, the obtained variables help to cutting parameters selection.
2. The stresses on the HSS tool increase with an increase the flank wear value. Then, 0.3 mm flank wear value is a critical for starting of the fatigue cracks.
3. The tool with 0.9 mm flank wear will cause the tool breakage due to high stresses focused on the notch of flank wear. Because, the stresses are obtained as higher than tool yield stress from this analyses.
4. 0.3 mm flank wear value may be limit in the turning of the AISI 1060 steel with HSS tool.
5. The stresses on the new HSS tool increase with an increase the feed rate and cutting depth. Also, between 60 and 90 degree approach angles, between 1 mm and 2 mm cutting depths must be selected to turning of the AISI 1060 steel.
6. The cutting depth has more a significant effect on the tool stresses than feed rate parameter in respect of von-Mises stresses.
7. As a result of the cutting tool stresses analyses, especially from von Mises stress and shear stress distributions, wear mode will be on the surfaces focused areas such as cutting edge, tool cutting zone and nose area. Especially, shear stresses focused on the cutting edge will cause the flank wear on the cutting edge (Figure 10).

REFERENCES

ASM Handbook. (1998) *Machining,* 16: ASM International, USA.

Astakhov VP.(2004) The Assessment of Cutting Tool Wear. *International Journal of Machine Tools and Manufacture* 44: 637–647.

Attanasio, A.; E. Ceretti; S. Rizzuti, D; Umbrello, F; Micari. (2008) 3D *Finite Element Analysis of Tool Wear in Machining, CIRP Annals - Manufacturing Technology* 57: 61–64.

Buryta, D.; R. Sowerby and I. Yellowley. (1994) Stress Distributions on the Rake Face during Orthogonal Machining, *Int. J. Mach. Tools Manuf.* 35 (5): 721–739.

Erdoğan Köse; Abdullah Kurt; Ulvi Şeker (2008) The effects of the feed rate on the cutting tool stresses in machining of Inconel 718. *Journal of Materials Processing Technology,* 196: 165-173.

Astakhov, VP. (2006) *Tribology of Metal Cutting, Tribology and Interface Engineering Series,* No 52.

Haci Saglam; Suleyman Yaldiz; Faruk Unsacar.(2007) The effect of tool geometry and cutting speed on main cutting force and tool tip temperature, *Materials and Design* 28: 101–111.

Kishawy HA; Kannan S; Balazinski M.(2005) Analytical Modeling of Tool Wear Progression During Turning Particulate Reinforced Metal Matrix Composites. *Annals of CIRP* 54(1):55–58.

Shimada S; Tanaka H; Higuchi M; Yamaguchi T; Honda S; Obata K.(2004) Thermo-chemical Wear Mechanism of Diamond Tool in Machining of Ferrous Metals. *Annals of CIRP* 53(1):57–60.

Takeyama H; Murata T.(1963) Basic Investigations on Tool Wear. Transactions of the *ASME-Journal of Engineering for Industry* 85:33–38.

Trent, E.M. (1991) *Metal Cutting. Butterworth Heinemann*, ISBN-10: 0750610689.

Valiorgue, F.; J. Rech; H. Hamdi; C. Bonnet; P. Gilles; J.M. Bergheau. (2008) Modelling of Friction Phenomena in Material Removal Processes. *Journal of Materials Processing Technology,* Volume 201, 1-3, 450-45.

In: Machining and Forming Technologies, Volume 2
Editor: J. Paulo Davim
ISBN: 978-1-61122-820-5

Chapter 7

MULTI-ATTRIBUTE OPTIMIZATION OF CNC END MILLING PROCESS PARAMETERS USING UTILITY BASED TAGUCHI PHILOSOPHY

Sanjit Moshat[1], Saurav Datta*[2], Asish Bandyopadhyay[3] and Pradip Kumar Pal[4]

[1, 3, 4]Department of Mechanical Engineering,
Jadavpur University, Raja S. C. Mallik Road, Kolkata- 700032, West Bengal, India
[2]Department of Mechanical Engineering,
National Institute of Technology, Rourkela, Orissa-769008, India

ABSTRACT

In the present work an attempt has been made to predict surface roughness and material removal rate (MRR) in CNC end milling process considering spindle speed, feed rate and depth of cut as machining parameters. The experimental work has focused on determination of optimum process environment for achieving expected quality as well as high productivity by ensuring minimization of surface roughness and maximization of material removal rate. Optimization of the process parameters for the desired surface roughness and material removal rate has been carried out by utility based Taguchi approach. The aforesaid method has been proved very helpful to optimize multiple performance responses for continuous quality and productivity improvement of the process/product.

Keywords: CNC end milling, Utility based Taguchi approach.

---

* Corresponding author, E-mail: sdatta@nitrkl.ac.in

1. Introduction

Milling is one of the most widely used metal removal processes. Among different types of milling processes, end milling is one of the most vital and common metal cutting operations used for machining parts because of its capability to remove materials at faster rate with a reasonably good surface quality. Also, it is capable of producing a variety of configurations using end milling cutter. In recent times, computer numerically controlled (CNC) machine tools have been adopted to make the milling process fully automated. It provides greater improvements in productivity, increases the quality of the machined parts and requires less operator input. For these reasons, CNC end milling process has been recently proved to be very versatile and useful machining operation in most of the modern manufacturing industries. Only the implementation of automation in end milling process is not the last achievement. It is also necessary to improve the machining process and machining performances continuously for effective machining and also for the fulfillment of varied requirements of the industries.

Surface roughness is a key factor in the machining process while considers machining performance and that is why in many cases, industries are looking for maintaining the good surface quality of the machined parts. Surface roughness is a measure of the technological quality of a product and a factor that greatly influences manufacturing cost and quality. It describes the geometry of the machined surface and combined with the surface texture, it can play an important role on the operational characteristics of the part. It also influences several functional attributes of a part, such as light reflection, heat transmission, coating characteristics, surface friction, fatigue resistance etc. However, the mechanism behind the formation of surface roughness is very dynamic, complicated and process dependent; therefore it is very difficult to calculate its value through analytical formulae. Various theoretical models that have been proposed in literature are not accurate enough and apply only to a limited range of processes and cutting conditions. Therefore, machine operators usually use "trial and error" approaches to set-up milling machine cutting conditions in order to achieve the desired surface roughness. Obviously, the "trial and error" method is not effective and efficient. Also, this method is very much time consuming. Therefore, the dynamic nature and widespread usage of milling operations in practice have raised a need for seeking a systematic approach that can help to set-up milling operations and also to help in achieving desired surface roughness quality.

On the other hand, material removal rate (MRR), which indicates processing time of the work piece, is another important factor that greatly influences production rate and cost. It is also necessary to study the material removal rate along with surface roughness in CNC end milling process. So, as a whole, there is a need for a tool that will allow the evaluation of the surface roughness and material removal rate (MRR) before the machining of the part and which, at the same time, can be easily used in the production-floor environment for contributing to the minimization of required time and cost and the production of desired surface quality. Both the surface roughness and material removal rate greatly vary with the change of cutting process parameters. That is why proper selection of cutting process parameters is also essential along with its prediction to obtain good surface finish (lower R_a value) and higher material removal rate in CNC end milling process.

Around 1980, the first CNC milling machine was developed. Since then, many researches had been done continuously to improve the system and process, machining performances of CNC milling machine. Alauddin *et al.* (1995) developed two mathematical polynomial models in end milling of 190 BHN steel considering the factors - cutting speed, feed and axial depth of cut. Response surface contours were constructed based on these models. Baek *et al.* (1997) suggested that both static and dynamic variables of the milling process should be included in the surface roughness model. Kim and Chu (1999) used the texture superposition method in the case of three- or multi-axis milling. The development of a generalized cutter model, which used the fillet radius as a variable, in conjunction with the scallop height and the geometric inaccuracy of the cutter allowed for a three-dimensional formation of the surface topography. In the research made by Kuang-Hua and Chiu-Fu (1995), Analysis of variance had been adopted in order to determine the most influential parameters concerning surface quality. It was shown that the tool nose radius and feed influence surface roughness. Lee *et al.* (2001), developed simulation algorithms to predict the machined surface in end milling. The spindle's acceleration signal, in particular its frequency and its amplitude, was used to monitor the vibration state of the milling process and to construct an algorithm that would calculate the surface roughness and plot the machined surface. Tsai *et al.* (1999) used neural networks for in-process prediction of surface roughness in milling operations. They had considered spindle speed, feed, depth of cut, and vibration "intensity" per revolution as parameters which would influence surface roughness. The interesting point of this work was the use of the neural network in real time. For the same purpose, hybrid techniques had been employed by the same authors, i.e. neural networks combined with fuzzy logic (1997). Similarly, fuzzy Petri nets had been used in the same context [Hanna et al. (1996)]. Benardos and Vosniakos (2002) explored a neural network modeling approach for the prediction of surface roughness (R_a) in CNC face milling. They had considered the factors - depth of cut, the feed rate per tooth, the cutting speed, the engagement and wear of the cutting tool, the use of cutting fluid and the three components of the cutting force. Using feed-forward artificial neural networks (ANNs) trained with the Levenberg-Marquardt algorithm, the most influential of the factors were determined also surface roughness had been predicted. Ozcelik and Bayramoglu (2006) developed a statistical model for surface roughness estimation in a high-speed flat end milling process under wet cutting conditions, using machining variables such as spindle speed, feed rate, depth of cut, and step over. First- and second-order models were developed using experimental results of a central composite design, and assessed by means of various statistical tests. The highest coefficient of correlation was obtained with a 10-parameter second-order model. Meanwhile, a time trend was observed in residual values between model predictions and experimental data, reflecting the probable effect of the tool wear on surface roughness. Thus, in order to enhance the estimation capability of the model, another independent variable was included into the model to account for the effect of the tool wear, and the total operating time of the tool was selected as the most suitable variable for this purpose. Back *et al.* (2001) studied the effects of the insert run-out errors and the variation of the feed rate on the surface roughness and the dimensional accuracy in a face-milling operation using a surface roughness model. The model was developed to predict the machined surface roughness from the information of the insert run-outs and the cutting parameters. From the estimated surface roughness value, the optimal feed rate that gave a maximum material removal rate under the given surface roughness constraint could be selected by a bisection method. Alauddin *et al.* (1997) presented a study of the development of mathematical models for tool life in end milling steel (190 BHN) using high-speed steel slot drills under dry conditions. The predictive models for tool life had been developed in terms of primary machining variables such as cutting speed, feed and axial depth of cut by using response surface methodology. Response surface contours were constructed in speed-feed planes and used for determining the optimum cutting conditions for a required tool life. The adequacy of the predictive models was tested by analysis of variance and found to be adequate. Yang and Chen (2001) demonstrated a systematic procedure of using Taguchi parameter design in process control of individual milling machines. The Taguchi parameter design had been done in order to identify the optimum surface roughness performance with a particular combination of cutting parameters in an end-milling operation. Ghani *et al.* (2004) applied Taguchi optimization methodology to optimize cutting parameters in end milling while machining hardened steel with TiN coated carbide insert tool under semi-finishing and finishing conditions of high speed cutting considering the milling parameters - cutting speed, feed rate and depth of cut. From the analysis of the result, the optimal combination of process parameters for low resultant cutting force and good surface finish was determined. Mansour and Abdalla (2002) developed a surface roughness model for the end milling EN32M (a semi-free cutting carbon casehardening steel with improved merchantability). The mathematical model for the surface roughness had been developed, in terms of cutting speed, feed rate, and axial depth of cut. The effect of these cutting parameters on the surface roughness had been carried out using design of experiments and the response surface methodology (RSM).

Reddy *et al.* (2006) investigated the role of solid lubricant assisted machining with graphite and molybdenum disulphide lubricants on surface quality, cutting forces and specific energy while machining AISI 1045 steel using cutting tools of different tool geometry (radial rake angle and nose radius). The performance of solid lubricant assisted machining had been studied in comparison with that of wet machining. The results indicated that there was a considerable improvement in the process performance with solid lubricant assisted machining as compared to that of machining with cutting fluids.

Colak *et al.* (2007) used genetic expression programming method for predicting surface roughness of milling surface under varying cutting parameters. Cutting speed, feed and depth of cut of end milling operations were collected for predicting surface roughness. A linear equation was predicted for surface roughness estimation in this study.

Oktem *et al.* (2005) focused on the development of an effective methodology to determine the optimum cutting conditions leading to minimum surface roughness in milling of mold surfaces by coupling response surface methodology (RSM) with a developed genetic algorithm (GA). RSM was utilized to create an efficient analytical model for surface roughness in terms of cutting parameters: feed, cutting speed, axial depth of cut, radial depth of cut and machining tolerance. Lou *et al.* (1999) used multiple regression method to predict surface roughness for CNC end milling considering feed, speed and depth of cut as input process parameters. Fuh and Chang (1997) presented a dimensional-accuracy model for the peripheral milling of aluminum alloys under dry and down-milling conditions. The model was postulated as a second-order equation and developed in terms of the Brinell hardness of the work piece, cutting speed, feed, radial and axial depths of cut. To build the quadratic model and minimize the number of experiments for the design parameters, response surface methodology (RSM) with an orthogonal central composite design was used. By means of variance analyses and additional experiments, the adequacy of this model was confirmed. Bayoumi *et al.* (1993) developed a method combining a mechanistic model and empirical relationships for helical end milling operations. Computation of cutting tool engagements and flank wear width occurred in real time as the cutting progressed using averaged force data. Response surface methodology had been used to investigate the effects of the process and operating variables on the process-dependent parameters; a second order relationship was obtained. The approach can also determine variation of wear land width along the length of the tool as axial engagement changes. Kansal *et al.* (2005) made a study to optimize the process parameters of powder mixed electrical discharge machining (PMEDM). Response surface methodology had been used to plan and analyze the experiments. Pulse on time, duty cycle, peak current and concentration of the silicon powder added into the dielectric fluid of EDM were chosen as variables to study the process performance in terms of material removal rate and surface roughness. The results identified the most important parameters to maximize material removal rate and minimize surface roughness.

The recommended optimal process conditions had been verified by conducting confirmation experiments. Ebeid *et al.* (2004) developed the mathematical models for correlating the inter relationships of various machining parameters such as applied voltage, feed rate, back pressure and vibration amplitude on overcut and conicity for achieving high controlled accuracy based on response surface methodology approach.

Literature depicts that previous investigators focused on various aspects of modeling and simulation of surface roughness in various milling operations. Predictive models were developed to represent correlations among surface roughness and various milling parameters. Compared to surface roughness, study on MRR has been found done to a limited extent. Parametric optimization of milling process responses have been addressed too. But it has been observed that optimization has highlighted a single objective function.

In a multi-response process, it may happens that optimization of a single response may cause severe quality loss for rest of the responses. Therefore, for solving a multi-criteria optimization problem; it is advised to convert individual objectives into an equivalent single objective function.

In view of the above consideration the present study has been dealt with application of utility based Taguchi method for optimization of surface roughness and MRR in end milling operation.

2. INTRODUCTION TO TAGUCHI METHOD

Taguchi Method is developed by Dr. Genichi Taguchi, a Japanese quality management consultant. The method explores the concept of quadratic quality loss function and uses a statistical measure of performance called Signal-to-Noise (S/N) ratio, [Antony and Antony (2001)]. It is the ratio of the mean (Signal) to the standard deviation (Noise). The ratio depends on the quality characteristics of the product/process to be optimized. The standard S/N ratios generally used are as follows (Equation 1-3): - Higher the Better (HB), Lower the Better (LB) and Nominal is Best (NB). The optimal setting is the parameter combination, which has the highest S/N ratio.

Higher-the-better (HB)

$$\text{S/N ratio} = -10\log_{10}\left(\frac{1}{n}\sum_{i=1}^{n}\frac{1}{y_i^2}\right) \quad (1)$$

where n = number of replications and y is the observed data

This is applied for problems where maximization of the performance characteristic of interest is desired. This is referred to as the larger-the-better type problem.

Lower-the-better (LB)

$$\text{S/N ratio} = -10\log_{10}\left(\frac{1}{n}\sum_{i=1}^{n}y_i^2\right) \quad (2)$$

This is applied for problems where minimization of the performance characteristics is intended. This is termed as smaller-the-better type problem.

Nominal-the-best (NB)

$$\text{S/N ratio} = -10\log_{10}\left(\frac{\mu^2}{\sigma^2}\right) \quad (3)$$

Here, μ = mean and σ = Standard deviation

Based on the signal-to-noise (S/N) analysis, the signal-to-noise (S/N) ratio for each level of process parameters are computed. Larger S/N ratio corresponds to better performance characteristics, regardless of their category of performance. It means that the level of process parameters with the highest S/N ratio corresponds to the optimum level of process parameters. Finally, a confirmatory experiment is conducted to verify the optimal processing parameters obtained from the parameter design.

However, traditional Taguchi approach cannot solve a multi-response optimization problem. It is therefore, necessary to convert these multiple objectives (criteria or attributes) into an equivalent single objective function which is generally treated as the representative index for all quality responses/attributes. In this context application of utility concept deserves mention. This technique can transform multi-quality responses into a single quality index (utility index); which is optimized by Taguchi method. The present work aims at applying utility concept coupled with Taguchi method for multi-response optimization through a case study in SAW. Utility concept combines the utility measure of individual quality attributes into a single utility index, which serves as the objective function for optimization in Taguchi philosophy.

3. Introduction to Utility Concept

A product or a process is generally evaluated on the basis of certain quality attributes, may be conflicting in nature. Therefore, a combined measure is required to estimate its overall performance, which must take into account the relative contribution of individual quality characteristics. In this work, a methodology based on utility concept coupled with Taguchi method has been applied for determining the optimal settings of process or parameters for multi-response/multiple quality characteristic of the weld in this section.

Utility can be defined as the usefulness of a product or a process in reference to the levels of expectations to the consumers, [Kumar [42]]. The overall usefulness of a process/product can be represented by a unified index termed as utility which is the summation of the individual utilities of various quality characteristics. The methodological basis for utility approach is to transform the estimated response of each quality characteristic into a common index, [Walia et al. [43]].

If X_i is the measure of effectiveness of an attribute (or quality characteristics) i and there are n attributes evaluating the outcome space, then the joint utility function can be expressed as:

$$U(X_1, X_2, ..., X_n) = f(U_1(X_1), U_2(X_2), ..., U_n(X_n)) \quad (4)$$

Here $U_i(X_i)$ is the utility of the i_{th} attribute.

The overall utility function is the sum of individual utilities if the attributes are independent, and is given as follows:

$$U(X_1, X_2, ..., X_n) = \sum_{i=1}^{n} U_i(X_i) \quad (5)$$

The attributes may be assigned weights depending upon the relative importance or priorities of the characteristics. The overall utility function after assigning weights to the attributes can be expressed as:

$$U(X_1, X_2, ..., X_n) = \sum_{i=1}^{n} W_i . U_i(X_i) \quad (6)$$

Here W_i is the weight assigned to the attribute i. The sum of the weights for all the attributes must be equal to 1. A preference scale for each quality characteristic is constructed for determining its utility value. Two arbitrary numerical values (preference number) 0 and 9 are assigned to the just acceptable and the best value of the quality characteristic respectively. The preference number P_i can be expressed on a logarithmic scale as follows:

$$P_i = A \times \log\left(\frac{X_i}{X_i^{'}}\right) \quad (7)$$

Here X_i is the value of any quality characteristic or attribute i , $X_i^{'}$ is just acceptable value of quality characteristic or attribute i and A is a constant. The value A can be found by the condition that if $X_i = X^*$ (where X^* is the optimal or best value), then $P_i = 9$.

Therefore,

$$A = \frac{9}{\log \frac{X^*}{X_i^{'}}} \quad (8)$$

The overall utility can be expressed as follows:

$$U = \sum_{i=1}^{n} W_i . P_i \quad (9)$$

Subject to the condition: $\sum_{i=1}^{n} W_i = 1$ (10)

Among various quality characteristics types, viz. Lower-the-Better, Higher-the-Better, and Nominal-the-Best suggested by Taguchi, the utility function would be Higher-the-Better type. Therefore, if the quality function is maximized, the quality characteristics considered for its evaluation will automatically be optimized (maximized or minimized as the case may be).

4. Experimental Setup and Procedure

Details of experimental plan are given below.

(a) Checking and preparing the Vertical CNC milling machine system ready for performing the machining operation.
(b) Preparing rectangular aluminum plates of size 95 mm × 75 mm × 10 mm in shaping machine for performing CNC end milling.
(c) Calculating weight of each plate by the high precision digital balance meter before machining.
(d) Creating CNC part programs for tool paths with specific commands using different levels of spindle speed, feed and depth of cut and then performing end milling operation.
(e) Calculating weight of each machined plate again by the digital balance meter.
(f) Measuring surface roughness and surface profile with the help of a portable stylus-type profilometer, *Talysurf* (Taylor Hobson, Surtronic 3+, UK)

The present experimental work seeks to evaluate the optimal result for selection of spindle speed (S), feed rate (*f*) and depth of cut (*d*) in order to achieve good surface roughness

(R_a value) and high material removal rate (MRR) during the CNC end milling process. Table 1 represents selected process control parameters used during the experiments. These parameters have been allowed to vary in three different levels. The values of coded and actual value of each parameter used in this work are listed in Table 1. The experimental matrix adopted as per Taguchi's L_9 Orthogonal Array (OA) design.

It is obvious that prediction accuracy depends on sample size. More is the data more accurate will be the prediction. However, this statement is true if we want to develop mathematical models of process phenomena. That's why full factorial design is recommended. It is also fact that if number of factors (as well as their levels) is increased; number of experimental runs will increase exponentially.

Table 1. Process control parameters and their limits

Coded levels	Spindle speed, S (rpm)	Feed rate, f (mm/min)	Depth of cut, d (mm)
-1	300	30	0.2
0	450	50	0.5
+1	600	70	0.8

Intensive experiments lead increase in experimental cost and consumes considerable amount of time. That's why people go for fractional factorial, central composite design. This is for modeling aspects.

The present work aims at investigating optimal process environment which can simultaneously satisfy contradicting requirements of surface roughness and MRR. In this context Taguchi's L_9 Orthogonal Array has been selected because Taguchi's robust design has been recommended long back and widely applied to various fields only due to its efficacy to predict acceptable optima by exploring minimum (limited) experimental runs. Moreover, in the present case, it has been assumed that there is no interaction effect of process parameters.

The machine used for the milling is a 'DYNA V4.5' CNC vertical milling machine having the control system SINUMERIK 802 D with a vertical milling head.

For generating the milled surfaces, CNC part programs for tool paths have been created with specific commands. The photograph of DYNA V4.5 CNC milling machine has been shown in Figure 1.

Figure 1. CNC vertical milling machine used for experiment.

The compressed coolant servo-cut has been used as cutting environment.

Commercially available CVD coated carbide tools have been used. The tools used are flat end mill cutters produced by WIDIA (EM-TiAlN). The tools are coated with TiAlN coating. For each material a new cutter of same specification has been used. The details of the end milling cutter are given below:

Cutter diameter = 8 mm
Overall length = 108 mm
Fluted length = 38 mm
Helix angle = 30^0
Hardness = 1570 HV
Density = 14.5 g/cc
Transverse rupture strength =3800 N/mm^2

The test work pieces are made of Aluminum of size 95 mm x 75 mm x 10mm rectangular plate. Different plates of same dimension and material are used for each experimental run. Material properties are given below:

Density: 2600-2800 kg/m^3
Melting Point: 660 °C
Elastic Modulus: 70-79 GPa
Tensile Strength: 230-570 MPa
Yield Strength: 215-505 MPa

Roughness measurement has been done using a portable stylus-type profilometer, *Talysurf* (Taylor Hobson, Surtronic 3+, UK) [Surface texture (of some samples) has been shown in Appendix; Figures A1 to A6].

Material removal rate (MRR) has been calculated from the difference of weight of work piece before and after experiment.

$$MRR = \frac{W_i - W_f}{\rho_s t} mm^3.\min^{-1} \tag{11}$$

Here, W_i is the initial weight of work piece in g; W_f is the final weight of work piece in g; t is the machining time in minutes; ρ_s is the density of aluminum (2.7 x 10^{-3} g/mm^3).

The weight of the work piece has been measured in a high precision digital balance meter (Model: DHD – 200 Macro single pan DIGITAL reading electrically operated analytical balance made by Dhona Instruments), which can measure up to the accuracy of 10^{-4} g and thus eliminates the possibility of large error while calculating material removal rate (MRR) in CNC end milling process.

5. Optimization of Surface Roughness and MRR in End Milling Process

Experimental data regarding R_a value and MRR (material removal rate) corresponding to L_9 Orthogonal Array (OA) design of experiment has been tabulated below (Table 2).

Experimental data have been explored to calculate utility values of individual quality attributes by using equations (7-8). For MRR (HB) criterion and for surface roughness (LB) has been used (equation (1) and (2) respectively). In case of MRR (corresponding to HB criteria) its minimum observed value (from Table 2) has been considered as just acceptable value; whereas maximum observed value has been treated as the most expected value. For surface roughness, the maximum of entries of Table 2 has been considered as just acceptable value; whereas minimum observed value has been treated as the best (desired) value. These individual utility measures have been furnished in Table 3. The overall utility index has been computed using equation (9); tabulated in Table 4 with their corresponding S/N ratio. In this computation it has been assumed that all quality features are equally important (same priority weightage).

Table 2. Experimental design and collected response data

Sl. No.	Parametric combination (Design of experiment)			Response features	
	Spindle Speed (rpm)	Feed rate (mm/min)	Depth of Cut (mm)	R_a (μm)	MRR (mm^3/min)
1	300	30	0.2	1.167	62.500
2	300	50	0.5	1.544	229.743
3	300	70	0.8	1.720	428.803
4	450	30	0.5	0.9696	133.076
5	450	50	0.8	1.541	304.770
6	450	70	0.2	1.210	148.200
7	600	30	0.8	1.250	172.811
8	600	50	0.2	1.000	116.400
9	600	70	0.5	1.44	338.680

Table 3. Utility values related to individual responses

Sl. No.	Utility measures of individual attributes	
	Surface roughness	MRR
1	6.0904	0.0000
2	1.6949	6.0837
3	0.0000	9.0000
4	9.0000	3.5319
5	1.7255	7.4043
6	5.5223	4.0349
7	5.0116	4.7529
8	8.5153	2.9062
9	2.7898	7.8974

Figure 2 reflects S/N ratio plot for overall utility index; S/N ratio being computed using equation (1). Optimal parameter setting has been evaluated from this plot. The optimal setting should confirm highest utility index (HB criterion). The predicted optimal setting becomes *S3 f3 d2*.

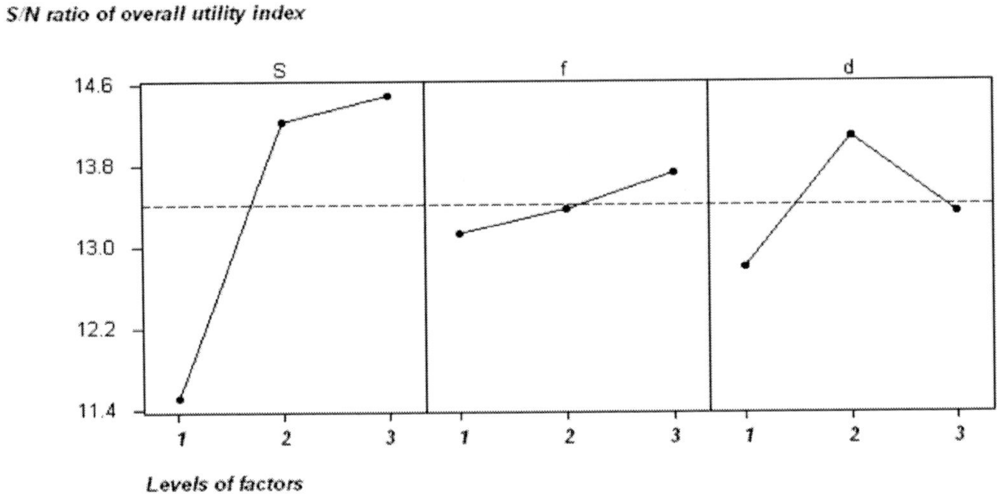

Figure 2. S/N ratio plot for evaluation of optimal setting.

Table 4. Overall utility index

Sl. No.	Overall utility index	Corresponding S/N ratio
1	3.0452	9.6723
2	3.8893	11.7974
3	4.5000	13.0643
4	6.2659	15.9397
5	4.5649	13.1886
6	4.7786	13.5860
7	4.8822	13.7723
8	5.7107	15.1338
9	5.3436	14.5567

Table 5. Results of confirmatory experiment

	Optimal setting	
	Prediction	Experiment
Level of factors	S3 f3 d2	S3 f3 d2
S/N ratio of Overall utility index	15.4965	15.1930
Overall utility index	5.9542	5.7498

After evaluating the optimal parameter settings, the next step is to predict and verify the optimal result using the confirmatory test. Table 5 reflects the satisfactory result of confirmatory experiment.

CONCLUSIONS

In literature related to CNC end milling operation; optimization problem has been solved using traditional Taguchi method. This method fails to solve multi-objective optimization problems. To overcome this present work proposes utility based hybrid Taguchi approach for simultaneous optimization of conflicting multi-quality characteristics. Compared to grey Taguchi or Taguchi based desirability function approach, utility based Taguchi method is very straight forward without complexity of understanding as well as computation.

Based on the scope and limitations of the present work, the following conclusions may be drawn from the results of the experiments and analysis of the experimental data in connection with multi-response optimization in CNC end milling process.

(1) Utility based Taguchi method has been found fruitful for evaluating the optimum parameter setting. It has been found that spindle speed of 600 rpm, feed rate of 70 mm/min and depth of cut of 0.5 mm is the optimal parametric combination capable of producing desired surface finish and high material removal rate.
(2) This approach is efficient enough to solve a multi-response optimization problem.
(3) Confirmatory test has validated the parametric setting determined by utility based Taguchi method.
(4) The said approach can be recommended for continuous quality improvement and off-line quality control of a process/product.

APPENDIX

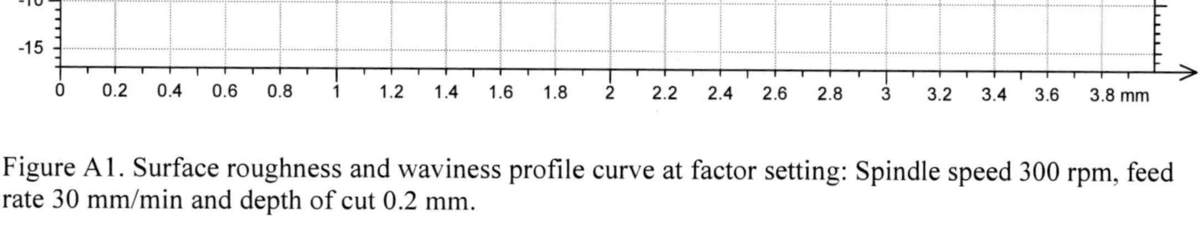

Figure A1. Surface roughness and waviness profile curve at factor setting: Spindle speed 300 rpm, feed rate 30 mm/min and depth of cut 0.2 mm.

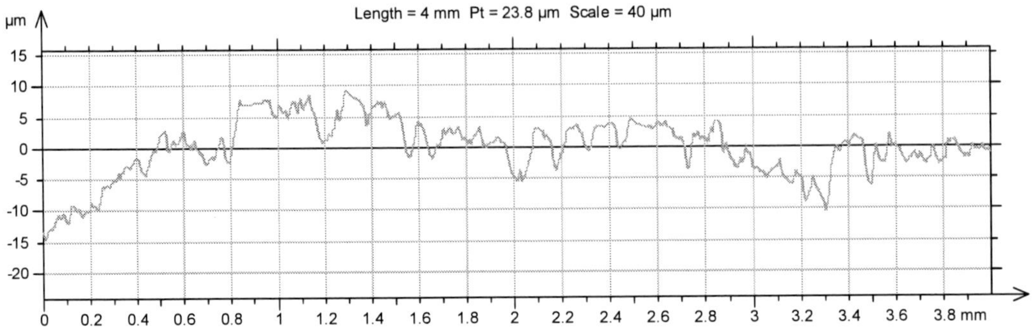

Figure A2. Surface roughness and waviness profile curve at factor setting: Spindle speed 300 rpm, feed rate 50 mm/min and depth of cut 0.5 mm.

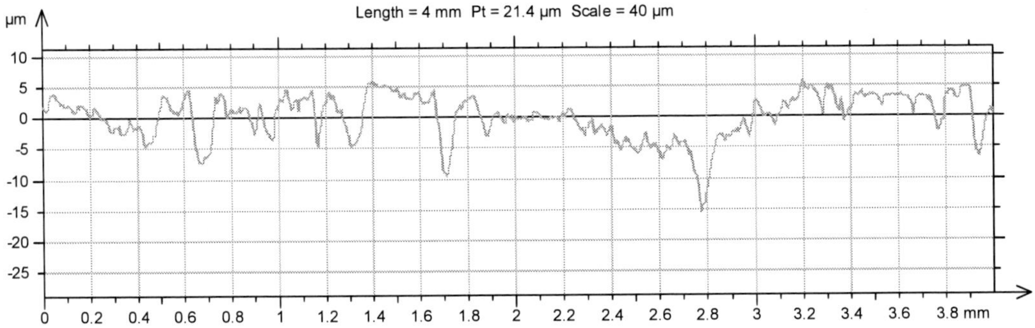

Figure A3. Surface roughness and waviness profile curve at factor setting: Spindle speed 300 rpm, feed rate 70 mm/min and depth of cut 0.8 mm.

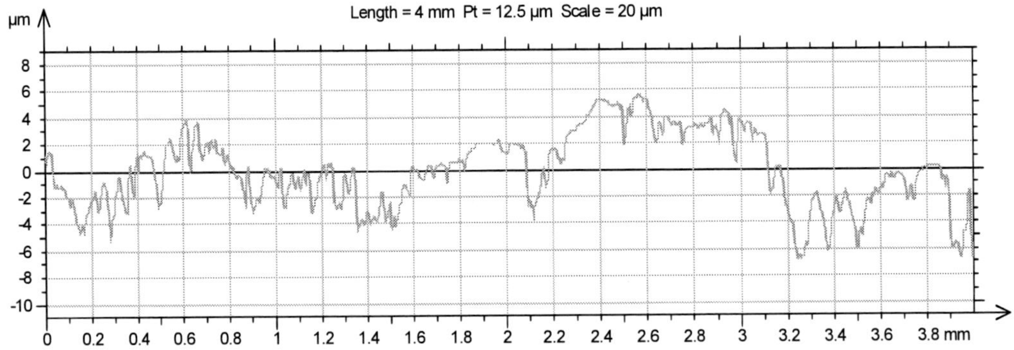

Figure A4. Surface roughness and waviness profile curve at factor setting: Spindle speed 450 rpm, feed rate 30 mm/min and depth of cut 0.5 mm.

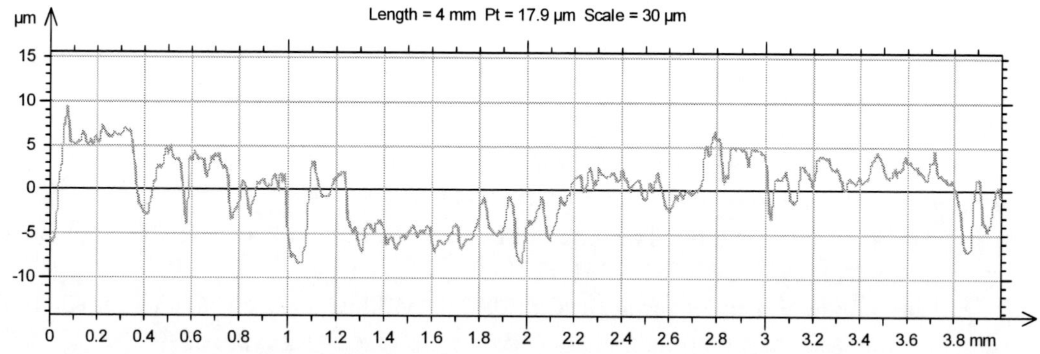

Figure A5. Surface roughness and waviness profile curve at factor setting: Spindle speed 450 rpm, feed rate 50 mm/min and depth of cut 0.8 mm.

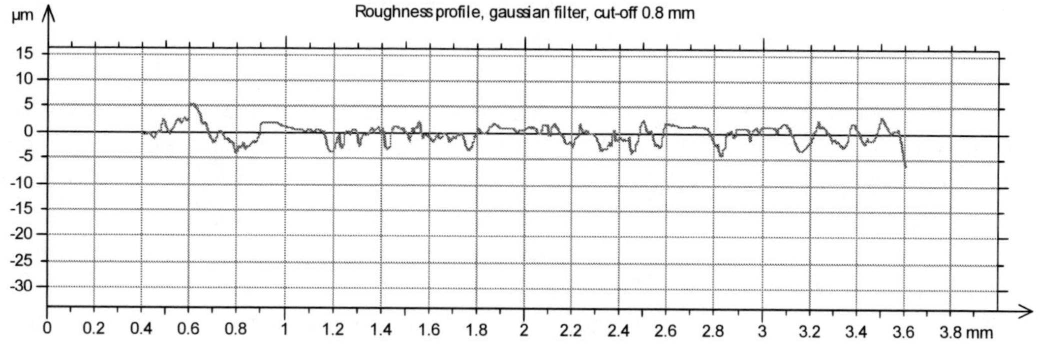

Figure A6. (confirmatory test): Surface roughness and waviness profile curve at factor setting: Spindle speed 600 rpm, feed rate 70 mm/min and depth of cut 0.5 mm.

REFERENCES

Allen, D. M., (1971), "Mean Square Error of Prediction as a Criterion for Selecting Variables," *Proceedings of Technometrics*, 13, 469-475.

Allen, D. M., (1974), "The Relationship between Variable Selection and Data Augmentation and a Method for Prediction," *Proceedings of Technometrics*, 16, 125-127.

Alauddin, M., El Baradie M. A. and Hashmi M. S. J., (1995), "Computer aided analysis of a surface roughness model for end milling", *Journal of Material Processing Technology*, 55, 123-127.

Alauddin, M., El Baradie, M. A. and Hashmi, M. S. J., (1997), "Prediction of tool life in end milling by response surface methodology", *Journal of Material Processing Technology*, 71, 456-465.

Box, G. E. P. and Draper, N. R., (1987), "Empirical Model-Building and Response Surfaces," *Proceedings of Jon Wiley and Sons*, New York.

Bayoumi, A. E., Barnwal, S. and Hutton, D. V., (1993), "Prediction of flank wear and engagements from force measurements in end milling operations", *Wear*, 170, 255-266.

Back, D. K., Ko, T. J. and Kim, H. S., (2001), "Optimization of feed rate in a face milling operation using a surface roughness model", *International Journal of Machine Tools and Manufacturing*, 41, 451-462.

Benardos, P. G. and Vosniakos, G. C., (2002), "Prediction of surface roughness in CNC face milling using neural networks and Taguchi's design of experiments", *Robotics and Computer Integrated Manufacturing*, 18, 343 - 354.

Bhattacharyya, A., "Metal Cutting: Theory and Practice", Reprinted 2006, New Central Book Agency, Page No. 495 – 501. ISBN: 81-7381-228-4.

Colak, O., Kurbanoglu C. and Kayacan M. C., (2007), "Milling surface roughness prediction using evolutionary programming methods", *Materials and Design*, 28, 657-666.

Dae Kyun Baek, Tae Jo Ko and Hee Sool Kim, (1997) "A dynamic surface roughness model for face milling", Precision Engineering, 20, 171-178.

DIN 4760: Form deviations; concepts; classification system. Deutsches Institute Fuer Normung, e.V., 1982.

Ebeid, S. J., Hewidy, M. S., El-Taweel, T. A. and Youssef, A. H., (2004), "Towards higher accuracy for ECM hybridized with low-frequency vibrations using response surface methodology", *Journal of Material Processing Technology*, 149, 432-438.

Fuh Kuang-Hua and Wu Chiu-Fu, (1995), "A proposed statistical model for surface quality prediction in end milling of Al alloy", *International Journal of Machine Tools and Manufacturing*, 35(8), 1187-1200.

Fuh, K. H. and Chang, H. Y., (1997), "An accuracy model for the peripheral milling of aluminum alloys using response surface design", *Journal of Material Processing Technology*, 72, 42-47.

Ghani, J. A., Choudhury, I. A. and Hassan, H. H., (2004), "Application of Taguchi method in the optimization of end milling parameters", *Journal of Material Processing Technology*, 145, 84-92.

Groover, Mikell P., (2007). "Theory of Metal Machining", Fundamentals of Modern Manufacturing, 3rd Ed, John Wiley and Sons, Inc. ISBN 0471744859.

Hanna, M.M., Buck, A and Smith, R., (1996), "Fuzzy Petri nets with neural networks to model product quality from a CNC milling machining center", IEEE Trans Systems Man Cybernet Part A, 26(5), 638-645.

Kaczmarek J. (1983). "Principles of machining by cutting", abrasion and erosion. London: Peter Peregrinus.

Khuri, A. I. and Cornell, J.A., (1996), "Response Surfaces," 2nd edition, Marcel Dekker. New York.

Kim, B. H. and Chu, C. N., (1999), "Texture prediction of milled surfaces using texture superposition method", *Computer Aided Design*, 31, 485-494.

Kalpakjian S. and Schmid Steven R. (2000), "Manufacturing Engineering and Technology", 4th Ed, Pearson Education Asia. ISBN 81-7808-157-1.

Kumar, P., Barua, P. B. And Gaindhar, J. L., (2000), Quality Optimization (Multi-Characteristics) Through Taguchi's Technique and Utility Concept, *Quality and Reliability Engineering International*, 16, 475-485.

Ki Yong Lee, Myeong Chang Kang, Yung Ho Jeong, Deuk Woo Lee and Jeong Suk Kim, (2001), "Simulation of surface roughness and profile in high-speed end milling", *Journal of Material Processing Technology*, 113, 410-415.

Kansal, H. K., Singh, S. and Kumar, P., (2005), "Parametric optimization of powder mixed electrical discharge machining by response surface methodology", *Journal of Material Processing Technology*, 169, 427 – 436.

Kopac, J. and Krajnik, P., (2007), "Robust design of flank milling parameters based on grey-Taguchi method", *Journal of Material Processing Technology*, 1-4.

Lou Shi-Jer and Chen Joseph C., (1997), "In-process surface recognition of a CNC milling machine using the fuzzy nets method", *Computer Integrated Engineering*, 33(l-2), 401-404.

Lou, M. S., Chen, J. C. and Li, C. M., (1999), "Surface Roughness Prediction Technique for CNC End-Milling", *Journal of Industrial Technology*, 15 (1), 1-6.

Myers, R. H., (1990), "Classical and Modern Regression with Applications," 2nd edition, Duxbury Press, Boston.

Montgomery, D. C., Peck, E. A. and Vining, G. G., (2001), "Introduction to Linear Regression Analysis," 3rd edition, John Willey and Sons, New York.

Mansour, A. and Abdalla, H., (2002), "Surface roughness model for end milling: a semi-free cutting carbon casehardening steel (EN32) in dry condition", *Journal of Material Processing Technology*, 124, 183-191.

Myers, Raymond H. and Montgomery, B.C., (2002), "Response Surface Methodology: process and product optimization using designed experiment", Proceedings of A Wiley-Interscience Publication.

Oktem, H., Erzurumlu, T. and Kurtaran, H., (2005), "Application of response surface methodology in the optimization of cutting conditions for surface roughness", *Journal of Material Processing Technology*, 170, 11-16.

Ozcelik, B. and Bayramoglu, M., (2006), "The statistical modeling of surface roughness in high-speed flat end milling", *International Journal of Machine Tools and Manufacturing*, 46, 1395-1402.

Rao, P. N., (2001). "Manufacturing Technology – Metal Cutting and Machine Tools", First reprint 2001, Tata McGraw-Hill. ISBN 0-07-463843-2.

Reddy, N., Kumar, S. and Rao, P. Venkateswara, (2006), "Experimental investigation to study the effect of solid lubricants on cutting forces and surface quality in end milling", *International Journal of Machine Tools and Manufacturing*, 46, 189-198.

Taguchi, G., (1986), "Introduction to Quality Engineering," Proceedings of Asian Productivity Organization, UNIPUB, White Plains, NY.

Taguchi, G., (1987), "System of Experimental Design: Engineering Methods to Optimize Quality and Minimize Cost," Proceedings of UNIPUB/Kraus International, White Plains, NY.

Tsai Yu-Hsuan, Chen Joseph C. and Lou Shi-Jer, (1999), "An in-process surface recognition system based on neural networks in end milling cutting operations" *International Journal of Machine Tools and Manufacturing*, 39, 583-605.

Tosun, N., (2006), "Determination of optimum parameters for multi-performance characteristics in drilling by using grey relational analysis", *International Journal Advanced Manufacturing Technology*, 28, 450-455.

Walia, R. S., Shan, H. S. and Kumar, P., (2006), Multi-Response Optimization of CFAAFM Process Through Taguchi Method and Utility Concept, *Materials and Manufacturing Processes*, 21, 907-914.

Yang, L. John and Chen, C. Joseph, (2001), "A Systematic Approach for Identifying Optimum Surface Roughness Performance in End-Milling Operations", *Journal of Industrial Technology*, 17, 1-8.

In: Machining and Forming Technologies, Volume 2
Editor: J. Paulo Davim
ISBN: 978-1-61122-820-5

Chapter 8

DMM BASED APPROACH TO PRODUCE AEROENGINE DISCS

I. Balasundar*[*1], M. Sudhakara Rao[1], M. K. Rohatgi[1], T. Raghu[1] and B. P. Kashyap[2]
[1]Near Net Shape Group, Aeronautical Materials Division
Defence Metallurgical Research Laboratory, Hyderabad – 500 058, India
[2]Department of Metallurgical Engineering and Materials Science
Indian Institute of Technology Bombay, Mumbai - 400 076, India

ABSTRACT

Because of excellent creep properties near alpha titanium alloys are used as aeroengine components. Near alpha titanium alloy IMI 834 is one such material that is used as high pressure compressor discs, shafts and blades in aeroengines. This article deals with how dynamic materials modeling (DMM) has been used to understand the high temperature deformation behaviour of the material and to identify the optimum thermomechanical processing parameters i.e., temperature, strain rate and strain. Based on DMM, a two stage – two step process was designed to near isothermally forge 434 mm diameter discs to thier near net shape. Finite element analysis was used to evaluate the process and die design. The results obtained are presented and discussed in the paper.

Keywords: IMI 834; DMM; Processing Maps; Near isothermal forging; Compressor discs.

1. INTRODUCTION

Titanium alloys are used extensively in aerospace industries both as structural and engine component materials because of their high specific strength, high thermal stability and

---

[*] Corresponding author, E-mail: i_balasundar@yahoo.com

excellent corrosion resistance. Majority of titanium alloy aeroengine components are produced by thermomechanical processing. The ever increasing demand for higher performance of titanium alloys advocates for the development of new thermomechanical processing techniques or sequences to tailor the microstructure in order to achieve the desired combination of static and dynamic mechanical properties in these materials (Gangshu shen, 2000; Kuhlman, 2005). Therefore, to produce components with desired properties out of titanium alloys, it requires a genuine understanding of processing - structure - property correlations and also to have close control over both the processing and heat treatment variables which influence the microstructure evolution and hence the final properties that can be achieved. Although various material modeling techniques are in vogue today (Kutumbarao, 1996) to understand the high temperature deformation behaviour of materials, the technique of processing maps based on dynamic materials modeling (DMM) has gained wide acceptance. DMM can be used to characterize the high temperature deformation behaviour of materials and identify an optimum processing window in terms of temperature and strain rate which can be used in a manufacturing environment to produce various components (Prasad et al., 1984, 1990, 1997).

In this article, the thermomechanical processing parameters required to process the near alpha titanium alloy IMI834 were optimized using dynamic materials modeling (DMM). Based on DMM, a two stage – two step process was identified to forge the material in to an aeroengine disc of 434 mm in diameter. Finite element analysis was used to evaluate the process design and die design. The high temperature deformation behaviour of the material, deformation mechanisms operating in the material and the approach practiced to forge the airworthy discs to their near net shape with desired mechanical properties on a reliable and repeatable basis is presented and discussed.

2. Material

Near alpha titanium alloy IMI 834 with a nominal composition of Ti - 5.8%Al - 4.1%Sn - 3.5%Zr - 0.7%Nb - 0.55%Mo - 0.06%C was procured from M/s TIMET, U.K in the form of 172 mm diameter bars in the as-processed and machined condition. The as-received microstructure of the material is shown in Figure 1 and was found to have 38 - 41% primary α (Pα) in a transformed β (Tβ) matrix. The Pα in the as received material was found to vary from elongated to equiaxed in shape.

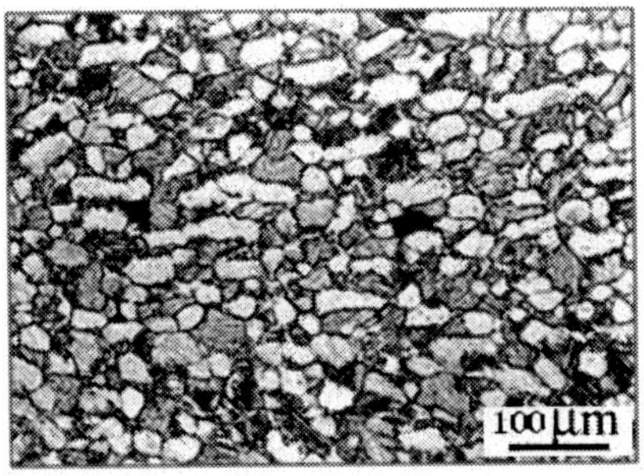

Figure 1. As-Received Microstructure of IMI 834.

This can be attributed to the different amount of deformation experienced by the material during the cogging operation in the alpha-beta region. The prior beta grain size was also found to be varying from 40μm to 65μm. The average prior beta grain size was estimated to be 50μm. The beta transus of the material was estimated to be 1060°C.

3. ISOTHERMAL COMPRESSION TESTING

To generate the high temperature flow behaviour data of titanium alloy IMI834, Isothermal compression tests were conducted over a range of temperatures and strain rates as shown in Table 1. For compression testing, cylindrical samples with a constant height to diameter ratio of 1.5 were electro-discharge machined from the as-received bar. The edges of the machined samples were chamfered to avoid fold over during the initial stages of testing. A small hole of 0.8 mm in diameter, reaching to the center of the sample was drilled at its mid-height and a K-type thermocouple was inserted through it to monitor and record the temperature (Prasad et al., 1997). Deltaglaze 347M® coated IMI 834 samples were then heated to the deformation temperature at a rate of 2°C/sec using a split type furnace, held for 30 minutes and compressed under constant true strain rates on a computer controlled servo hydraulic testing machine, custom built by M/s. DARTEC, UK. The adiabatic temperature rise during deformation was recorded. After testing, the samples were water quenched to freeze the high temperature microstructure. The quenched samples were then cut parallel to compression axis. The cut faces were then polished, etched with Kroll's reagent and were subjected to microstructural evaluation using an optical microscope and scanning electron microscope (SEM). The microstructure of the water quenched deformed samples were then used to evaluate the high temperature deformation mechanism in the material.

Table 1. Temperature and strain rate domain for processing map generation

Material: Near alpha titanium alloy IMI 834	
β Transus	1060°C
Temperature	850, 900, 950, 1000, 1030, 1060°C
Strain rate	$3x10^{-4}$, 10^{-3}, 10^{-2}, 10^{-1}, 10^{0}/sec
% Reduction	50% = True Strain (ε): 0.694

The load stroke curves obtained after compression testing were converted into true stress – true strain curves. They were corrected for elastic strain in order to obtain the true plastic strain. The flow stress at each true plastic strain was then corrected for adiabatic temperature rise (ATR) by applying linear interpolation (and extrapolation when required) between log flow stress and temperature inverse in absolute scale (Prasad, 1997). The ATR corrected flow stress values were used for evaluating the high temperature flow behaviour and processing map generation.

3.1. Effect of Temperature

The effect of temperature on the flow stress of IMI834 for a true plastic strain of 0.65 is shown in Figure 2. It can be seen that at deformation temperatures above 1030°C, the flow stress is insensitive to temperature. Below 1030°C the flow stress is very sensitive to temperature changes. This clearly indicates that the transition in the temperature sensitivity of flow stress occurs around 1030°C which is well below the β transus (1060°C) of the material. The change in flow behaviour can be attributed to the volume fraction of α and β phases present at the deformation temperature and the deformation characteristics of the dominant phase (Wiess, 1999; Wanjara et al., 2005, 2006; Vo, 2007; Tricot, 1998; Seetharaman, 1991). At temperatures well below 1030°C hexagonal close packed α being the dominant phase controls the deformation behaviour whereas above 1030°C body centered β is the dominant continuous phase and the deformation characteristic is controlled by this phase as it has been observed in this material and various other titanium alloys *(Wiess, 1999; Wanjara et al., 2005, 2006; Vo, 2007; Tricot, 1998; Seetharaman, 1991).*

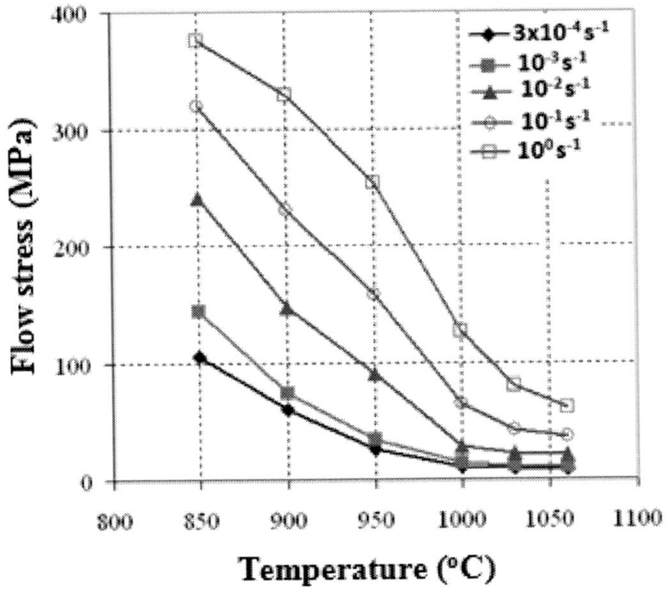

Figure 2. Variation of flow stress with temperature.

3.2. Effect of Strain Rate

The effect of strain rate on the flow stress for a true plastic strain of 0.65 is shown in Figure 3a. A progressive increase in the flow stress with increasing strain rate can be observed for the entire temperature range which indicates that the rate of dislocation generation increases with increasing strain rate. As the rate of dislocation annihilation depends on the dislocation density, at higher strain rates such as 0.1 and 1/sec recovery mechanisms such as climb, cross slip would be increasingly effective than at lower strain rates (Wanjara et al., 2005).

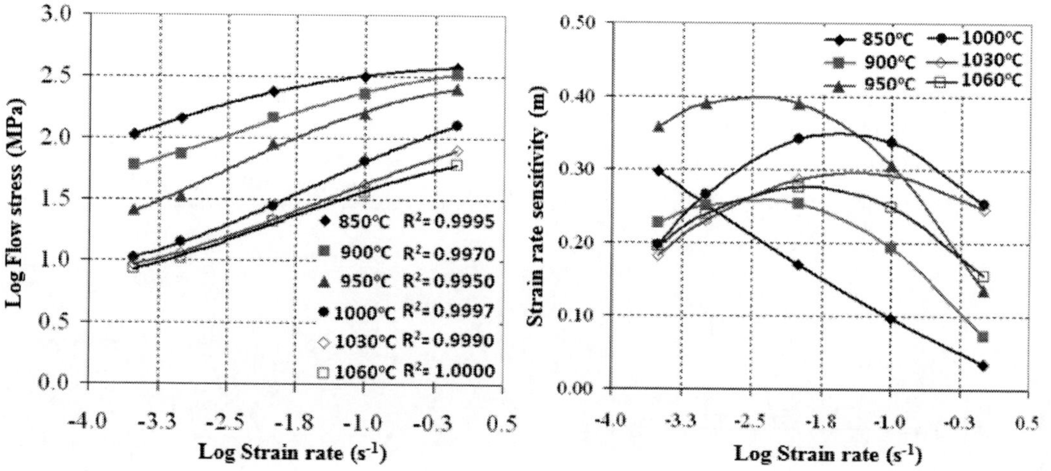

Figure 3. (a) Variation in flow stress as a function of strain rate on log-log scale. (b) Variation in strain rate sensitivity as a function of strain rate.

The variation of strain rate sensitivity with respect to temperature and strain rate is shown in Figure 3b. The plot exhibits two strain rate sensitivity peaks, one at 950°C and at a strain rate of $3\times10^{-3}s^{-1}$ and the other at 1000°C and at the strain rate of 3×10^{-2} s^{-1}. The peak values of m under these conditions are found to be 0.40 and 0.33 respectively. The variation in strain rate sensitivity can be attributed to different deformation mechanisms that are in operation under the processing conditions which shall be discussed latter.

4. Dynamic Materials Modeling

DMM is based on systems engineering concepts (Prasad et al., 1984, 1990, 1997). The hot working process is modeled as a system, where the system consists of source of power, stores of power and dissipator of power. The work piece is considered to dissipate power and is assumed to be nonlinear, dynamic, isolated and irreversible. The dynamic constitutive equation, which describes the manner in which the input energy is converted at any instant, is assumed to follow a power law relation [$\sigma = k\dot{\varepsilon}^{m}$], where k is a constant and m is strain rate sensitivity index. Large amount of input power is dissipated by the work piece as viscoplastic heat called the G Content and rest of the power is dissipated through microstructural changes, which is known as J Co-content (Prasad et al., 1984, 1990, 1997). The physical interpretations of G and J are based on thermodynamic principles described by Malvern (1969). The power partitioning between G and J is controlled by strain rate sensitivity of flow stress. The metallurgical power dissipation processes may be characterized by the variation of J co-content with temperature and strain rate, but normalization with the input power sharpens the variation (Prasad et al., 1984, 1990, 1997). This normalization leads to the definition of a dimensionless parameter called the efficiency (η) of power dissipation given as

$$\eta = \frac{J}{J_{max}} = \frac{2m}{m+1} \qquad (1)$$

The variation of η with temperature and strain rate constitutes the power dissipation map. Various domains in the map may be correlated with specific microstructural processes, which in turn can be applied for controlling the same. The stability condition described in the dynamic materials model by Kalyan Kumar (1987) is based on the extremum principles of irreversible thermodynamics as applied to large plastic flow proposed by Ziegler (1965). The condition for microstructural stability at constant temperature, in terms of a dimensionless instability parameter ξ, is given by

$$\xi(\dot{\varepsilon}) = \frac{\partial \ln\left[\dfrac{m}{m+1}\right]}{\partial \ln \dot{\varepsilon}} + m \tag{2}$$

Parameter ξ may be evaluated and plotted as a function of temperature and strain rate to obtain an instability map. Instabilities like adiabatic shear banding, flow localization, dynamic strain aging, kinking, mechanical twinning etc occur during plastic deformation in regimes where ξ is negative. Superimposition of the instability map over the power dissipation map gives the processing map.

4.1. Processing Map for IMI 834

The Processing map generated using PROMAP (Program for generating processing maps) software developed in-house (Balasundar, 2005) for a true plastic strain of 0.65 is shown in Figure 4. The shaded and unshaded regions in the processing map represent unstable and stable flow regimes respectively. In the unstable region, the material exhibits instability in the form of shear bands. For certain combinations of temperature and strain rates, the temperature of the sample was observed to increase beyond the deformation temperature during compression testing. At 850°C, 57°C increase in temperature was observed when IMI834 titanium alloy was deformed at a strain rate of 1 s^{-1}. The temperature rise measured here is within the range estimated and reported by Miaoquan Li (2007) and Young Niu (2008). The shear band instability observed in the region can be attributed to this adiabatic temperature rise. IMI 834 should not be processed under this temperature - strain rate regime to avoid inhomogeneous deformation of the material.

In the stable region, the map exhibits two domains. As per DMM, domains are the regions in which deterministic deformation mechanisms are operating (Prasad et al., 1984, 1990, 1997). The first domain is observed in the temperature range of 925 – 975°C and strain rate range of 10^{-3}-10^{-2} s^{-1} with a maximum efficiency of 62%. The efficiency values in the domain were found to increase steeply. The material also exhibits high strain rate sensitivity (Figure 3b) in the domain. The deformed material was found to exhibit a fine microstructure as compared to the as-received microstructure and to those microstructures obtained when processed outside the domain. The microstructure was found to have coarse primary alpha particles in a transformed beta matrix along with fine dynamically recrytallised alpha particles from the lath structure.

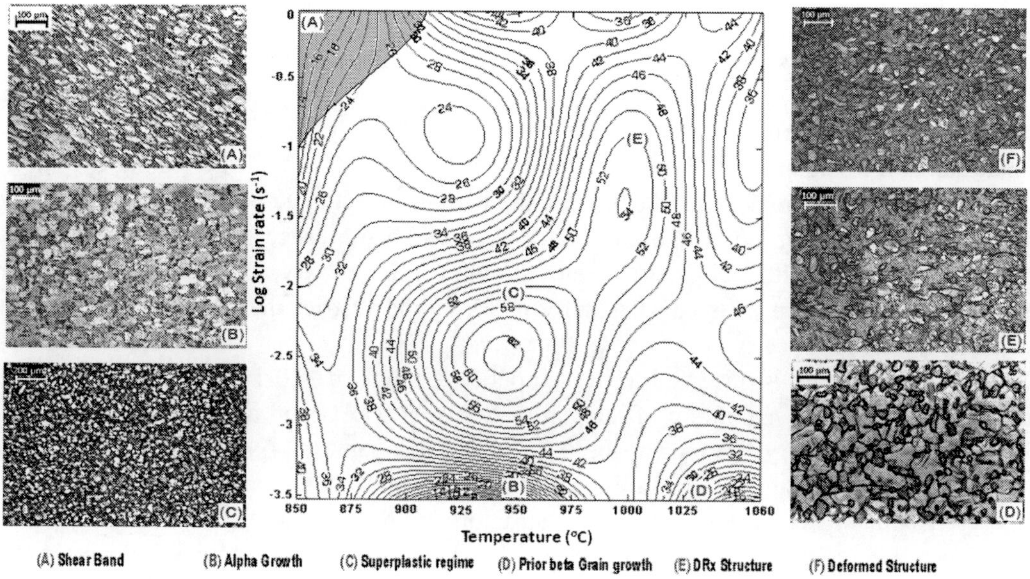

Figure 4. Processing Map – Microstructure correlation.

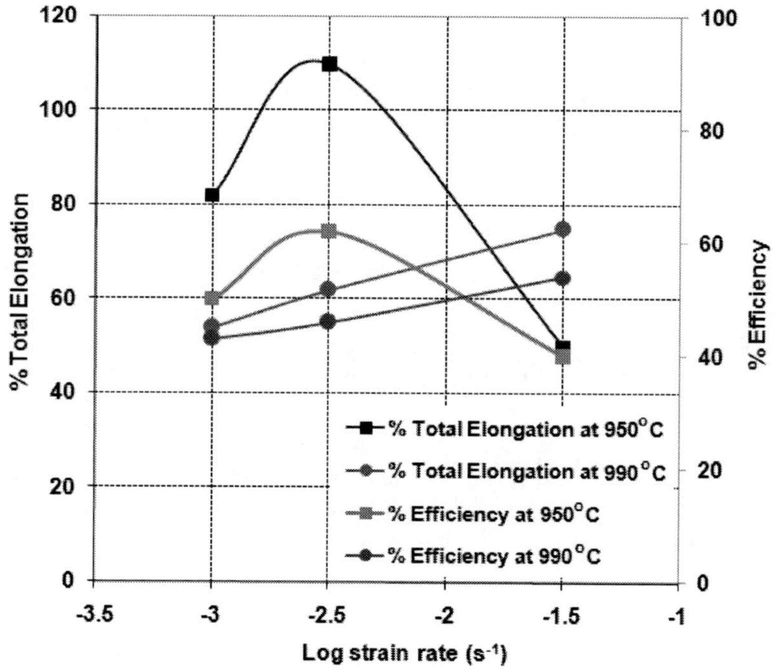

Figure 5. Variation of power dissipation efficiency and total elongation against strain rate.

Based on the microstructure obtained, the sharp increase in the efficiency and the high strain rate sensitivity observed in the domain, the material can be expected to exhibit super plastic characteristics (Prasad, 1997). For instance, a total elongation of 110% was obtained by tensile test at 950°C and with an average strain rate of $3.16x10^{-3}s^{-1}$(Figure 5). To some extent this substantiates the fact that the material exhibits superplastic characteristics. However, the maximum ductility attained here is less than that can be achieved in this class of materials. This can attributed to the bimodal (coarse and fine alpha) microstructure present in the material and to the growth of the primary alpha phase under the tensile testing condition. As the velocity of the cross head is constant, the true strain rate decreases during the tensile test. From the processing map (Figure 4, region B) it can be observed that, when the material is deformed at 950°C with a slow strain rate, growth of the primary alpha phase takes place. This growth in the alpha phase during deformation could restrict the maximum elongation that can be achieved.

The second domain is observed in the temperature range of 980-1020°C and 10^{-2} -$10^{-1}s^{-1}$ strain rate range with a maximum efficiency of 54%. At 975°C and a strain rate of 10^{-2} s^{-1}. The strain rate sensitivity is found to be 0.35 in this temperature - strain rate regime. The material undergoes dynamic recrystallisation in this domain which can be inferred from the partially dynamic recrystallised prior beta structure with wavy grain boundaries. A total elongation of 80% (Figure 5) was exhibited by the material when tested under tension in this temperature – strain rate regime. Further, a fine prior beta grain size (~10 to 20 μm) was obtained when the partially dynamic recrystallised compressed tested sample was subjected to static recrystallisation at the same temperature for an hour, which supports the argument that the material undergoes dynamic recrystallisation in this domain.

In the processing map, at temperatures above 1000°C, the efficiency values were found to decrease. The strain rate sensitivity is less in this temperature range when compared to the peak values. The microstructure of the samples processed under these conditions exhibit larger prior beta grains. Therefore, final processing of IMI 834 titanium alloy above 1020°C at slow strain rate is not recommended. If processed, undesirable prior beta grain growth could take place, which would then be difficult to refine during the post processing heat treatments. The manifestation of different microstructural and flow characteristics with respect to temperature and strain rate during high temperature deformation of IMI 834 titanium alloy are summarized in Table 2.

Table 2. Different processing domains for Ti – 600

Manifestation	Temperature	Strain rate
Shear banding.	850 - 925°C	10^{-1} – 10^{0}/sec
Super plastic Regime	920 - 975°C	10^{-3} to 10^{-2}/sec
Dynamic Recrystallisation	980 – 1020°C	10^{-2} to 10^{-1}/sec

5. NEAR ISOTHERMAL FORGING PROCESS DESIGN

The aeroengine disc to be produced out of IMI 834 is shown in Figure 6. It is an axisymmetric component with varying thickness along the radial direction.

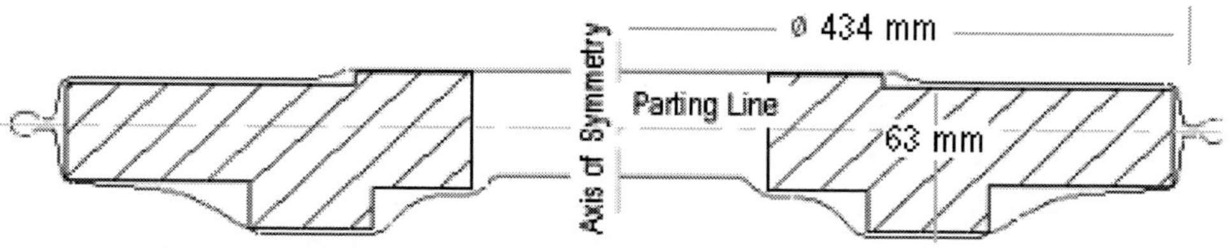

Figure 6. Sonic and Forge profile of aeroengine HPC disc.

The hatched and the outer line encompassing the hatched region correspond to sonic and forge profile respectively. The forge profile was established based on conventional die design principles (Sudhakara Rao et al., 2007). From the forge profile volume, the dimension of the billet (172 mm diameter and 377 mm height) required for forging was calculated. The height to diameter (h/d) ratio of the billet was found to be 2.20. It has been well established that when the h/d ratio exceeds beyond 2.0, the material undergoes inhomogeneous deformation (Dieter, 1988).

To avoid inhomogeneous deformation, it was decided to forge the material near isothermally in two stages. In the first stage the billet will be upset forged to 50% of its height under the superplastic regime. The preformed billet will then be forged into discs using the DRX domain to obtain a fine microstructure. Based on the processing map, the temperature for preforming and disc forging were decided to be 975°C and 1010°C respectively. In order to maintain the strain rate within the range identified from the processing map a two step process was designed i.e., during preforming and disc forging, the first 25% height reduction would be carried out using a certain cross head speed (step #1) and the rest of the reduction with a different cross head speed (step#2).

6. Finite Element Analysis (FEA)

Using the stress - strain data generated for the material over a range of temperature and strain rate, axisymmetric rigid plastic thermomechanical finite element analysis was carried out using MSC.Marc2007r1 software in order to ascertain the material flow pattern and load required for disc forging.

6.1. Preforming

For finite element analysis, IMI834 work piece material with an initial dimension of 172 mm in diameter and 377 mm height was assumed to be deformable and meshed with 4-node linear quadrilateral isoparametric elements. Though the initial mesh was coarse with only 360 elements, during deformation the mesh was refined by automatic remeshing in order to accommodate large plastic strains. The work piece was assumed to be at 975°C. The top and bottom flat dies were modeled as rigid bodies and were assumed to be at 910°C. The bottom die was assumed to be stationary. The boundary condition is represented mathematically as Ux=Uy=0, which implies that the displacement along the X and Y directions are zero. The top die was assigned with a downward displacement, the velocity of which was controlled by a control file that contains the information on the speed or the strain rate, time required, number of sub-steps to be taken to achieve the specified deformation or displacement. During the first 25% of height reduction (step#1), the top die was assumed to move with a velocity of 2mm/sec and for the second 25% it was assumed to move with a velocity of 1mm/sec. Axisymmetric boundary condition, represented mathematically as Ux=0 was imposed on the work piece along the symmetry axis. Adiabatic thermal boundary (dT/dx = dT/dy =0) condition was also imposed on the deformable work piece along the symmetry axis. Appropriate conductive and convective heat transfer coefficients between the work piece and dies, between the work piece and environment were also considered. In order to account for friction between the work piece and dies, a shear friction factor of 0.2 was assumed. It was also assumed that 90% of the total power input to the system is converted in to heat. This was specified in order to evaluate the increase in the temperature of the work piece due to deformation. The entire simulation of forging was carried out in a number of small steps as specified in the control file. The material flow pattern. equivalent plastic strain and temperature distribution during the preforming operation are shown in Figure 7a, 7b and 7c respectively. It can be observed that during preforming the material undergoes non-uniform deformation giving rise to a dumbbell shaped preform as shown in Figure 7a.

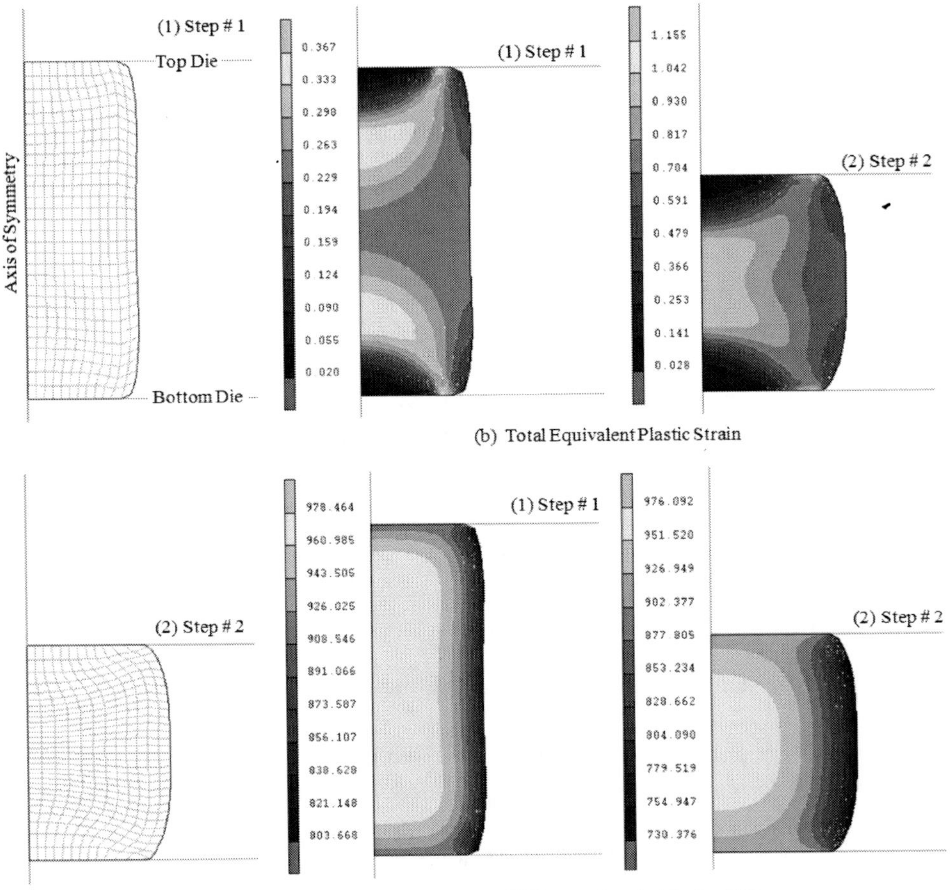

(a) Deformation Pattern (b) Total Equivalent Plastic Strain (c) Temperature distribution (°C)

Figure 7. FEA results on (a) Deformation pattern (b) Total equivalent plastic strain distribution and (c) Temperature distribution during preforming operation.

This dumbbell formation can be attributed to the effect of h/d ratio (Balasundar et al., 2004). During the initial stages of deformation, there are two strain concentration regions which are present just above and below the friction hill near the bottom and top die (Figure 7b-1). Friction hill is a region of very less strain that is formed due to friction between the die and work piece. The material above and below the friction hill only experiences deformation and there is very less strain at the mid height of the work piece. Due to this non-uniform deformation, dumbbell shape forms during early stages of deformation. As the deformation progresses, the height of the work piece is reduced and the two strain concentrated regions move towards each other and get united at 50% of height reduction as shown in Figure 7b-2. Though the strain concentrations unite, the dumbbell shape can still be observed in the work piece at the end of deformation as shown in Figure 7a-2. It can also be seen from Figure 7c, that the temperature rise in the material due to deformation is only marginal. This is because, the heat generated due to deformation has been offset by the heat lost by the material to the die and surrounding environment.

6.2. Disc Forging

In order to carry out the simulation of disc forging operation, the deformed geometry of the work piece at the end of preforming operation was taken as the initial work piece geometry for disc forging. The work piece was assumed to be at 1010°C. The top and bottom dies were modeled as rigid bodies and assumed to be at 910°C. Appropriate boundary conditions were applied as explained in the previous section. For the first 25% height reduction (step#1) the top die was assumed to move with a velocity of 2mm/sec and at 0.5mm/sec till a die gap of 1mm is reached. The entire analysis was carried out in a number of small steps as specified in the control file. The material flow pattern, equivalent plastic strain and temperature distribution during the disc forging operation are shown in Figure 8a, 8b and 8c respectively. The flow pattern of the material inside the die cavity during disc forging was found to be smooth and stream lined as shown in Figure 8a. It was observed that the dumbbell shape observed in the preform was eliminated during disc forging. The die corners were found to get filled only after the material reaches the flash region (Figure 8a-5). It is observed that similar to the preforming operation, the temperature rise (Figure 8c) in the material due to deformation during disc forging has been offset by the heat lost by the material to the die and surrounding environment. The load required for disc forging was found to be 1800MT.

7. Processing of IMI 834

As the virtual forging indicated a smooth streamlined die filling, the dies for disc forging were fabricated out of Ni alloy PER100. Forging experiments were carried out in a 2000MT oil hydraulic forge press custom built by M/s Pahnke, Germany. The dies were heated to the required temperature using a line frequency induction heater. The high pressure compressor discs were forged out of IMI 834 using the processing parameters listed in Table 3.

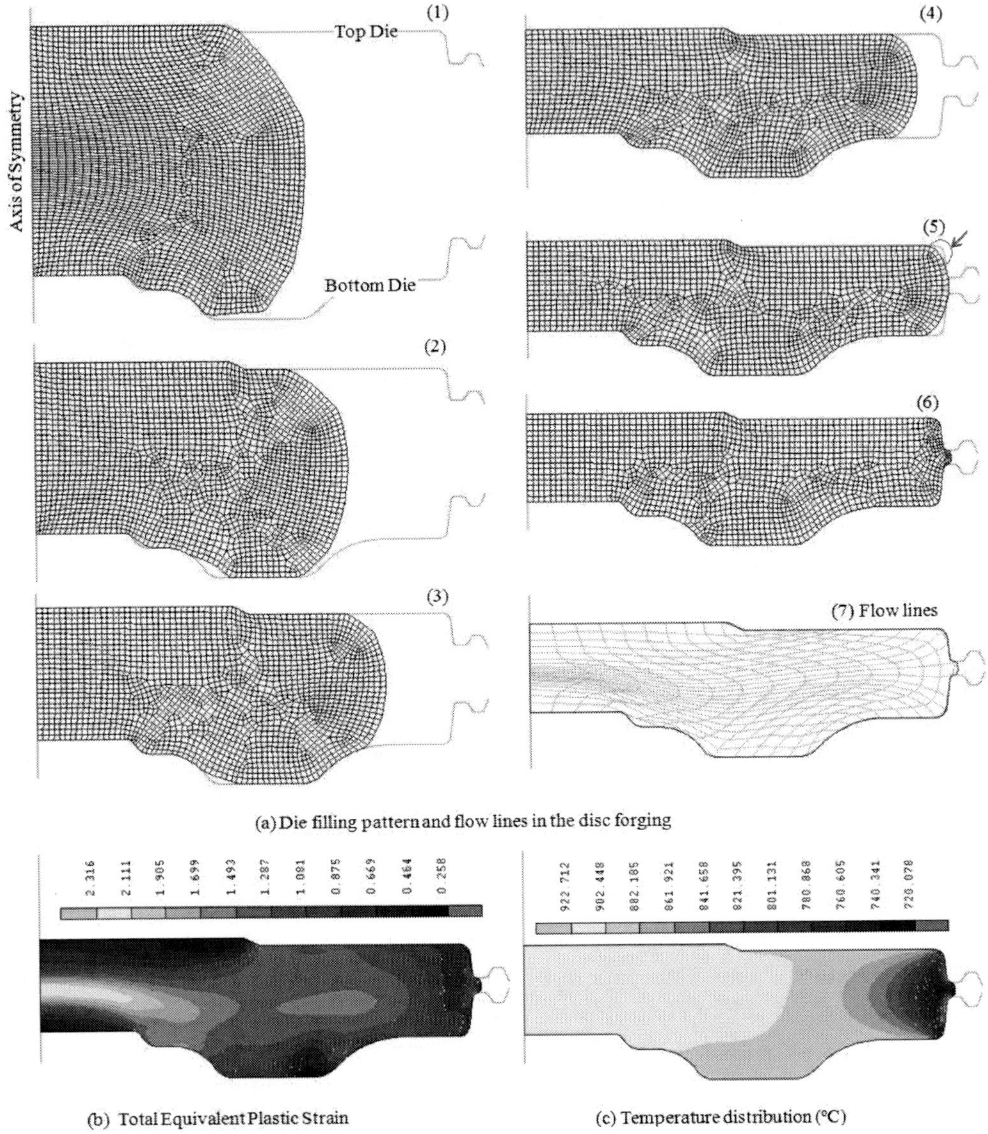

Figure 8a. FEA results on (a) Die filling pattern during disc forging and flow line in the disc forging (b) Totaal equivalent plastic strain distribution (c) Temperature distribution at the end of disc forging operation.

The load required for disc forging was found to be 1875 MT. The initial billet, preform and the disc forged out of IMI 834 are shown in Figure 9a and b.

It has been reported by various authors that a microstructure of 10-20% primary alpha in a transformed beta matrix gives optimum combination of creep and fatigue properties (Neal, 1988, 1995; Farthing, 1988; Daeubler,1995; Cope, 1988). Therefore, in order to achieve the desired microstructural features, the discs were first solution treated at appropriate temperature and then subjected to ageing for stress relieving. The solution treatment temperature to obtain 10-20% primary alpha in IMI 834 disc forging was estimated from the beta approach curve developed for the material. The disc forgings were heat treated as per the parameters listed in Table 4.

Table 3. Processing parameters for near isothermal forging of Ti-600

Parameters	Preforming	Disc Forging
Initial Dimension	172 ϕ X 380 mm	242 ϕ X 190 mm
Billet Temperature	975°C	1010°C
Die Temperature	910°C	910°C
Reduction	50%	Till 5 mm flash
Ram Speed	First 25% : 2mm/sec Second 25% : 1mm/sec	First 25%: 2 mm/sec Rest : 0.5 mm/sec
Final Dimension	242ϕ X 190 mm	436 ϕ x 44 mm

Table 4. Post processing heat treatment parameters

Treatment	Temperature	Soaking time	Quenchant
Solution Treatment	1040°C	2 Hours	Oil
Ageing	700°C	2 Hours	Air

Figure 9. (a) Initial billet and preform (b) As-forged disc and disc machined to sonic shape.

The flow lines (obtained by macro-etching) in the disc forging and the microstructure at various locations in the disc forging obtained after heat treatment are shown in Figure 10. It can be seen that the flow line pattern predicted by finite element analysis (Figure 8a) matches with the pattern obtained in the actual forging (Figure 10). The microstructure was found to be uniform throughout the disc forging. The average volume fraction of primary alpha present and the prior beta grain size were estimated to be 15% and 110μm respectively.

8. Mechanical Property Evaluation

The mechanical property requirement of a disc material is quite complex. The disc is expected to have good fatigue properties in order to withstand the cyclic stress encountered during normal flight and over speeding apart from high specific yield and tensile properties. As the material is exposed to high temperature, it is required to have good creep properties. The disc forging was then evaluated for various room and high temperature mechanical propertied as listed in Table 5. The disc forging was found to have the required airworthiness.

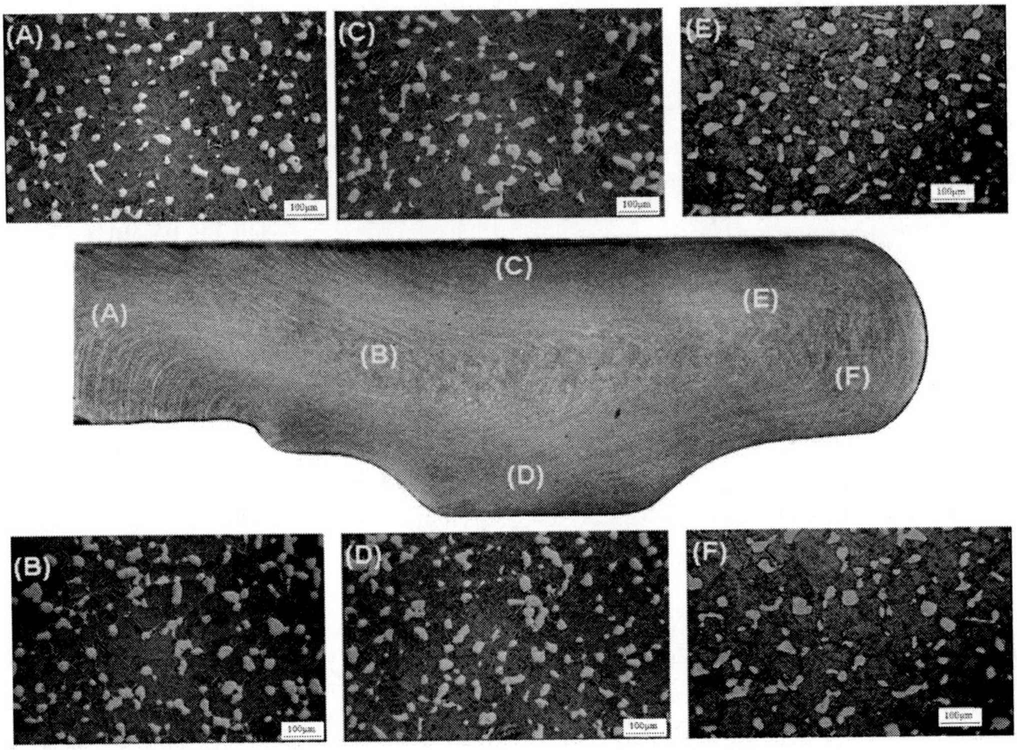

Figure 10. Flow lines and microstructure at various regions in disc forging.

Table 5. Room and high temperature mechanical properties

Property	Obtained
Room Temperature Tensile	
Tensile Strength (MPa)	936 - 979
Ultimate tensile strength (MPa)	1033 - 1079
% Elongation	11 - 16
% Reduction in Area	17 - 23
High Temperature Tensile (600°C)	
Tensile Strength (MPa)	485 - 521
Ultimate tensile strength (MPa)	616 - 657
% Elongation	17 - 21
% Reduction in Area	45 - 52
Notch Tensile - N/T Ratio	1.36-1.46
Creep	
600°C/150MPa/100Hr	0.04 - 0.14.
500°C/220MPa/100Hr	0.06 - 0.08
Low Cycle Fatigue	
17- 900 MPa / Trapezoidal Cycle (1-1-1-1sec)	2631 - 10091
High Cycle Fatigue (Plain)	
0-550 MPa	$1x10^5 – 1.07x10^7$
Fracture Toughness (MPa $m^{1/2}$).	41.47 – 42.52

SUMMARY

The beta transus of IMI 834 was estimated to be 1060°C and isothermal compression tests were carried out over a predetermined temperature and strain rate regime to evaluate the high temperature flow behaviour of the material. Based on dynamic materials modeling, the processing map for the material was generated using the data generated from the isothermal compression test. The material exhibits two potential domains for processing – superplastic domain (SPf) and dynamic recrystallisation (DRx) domain. A two stage - two step process was designed to forge the material in to compressor disc. Finite element analyses were carried out in order to ascertain the flow of the material during the performing and disc forging operations. The preforming and disc forging operations were carried out near isothermally using the superplastic and dynamic recrystallisation domain identified from the processing map. The forged discs were solution treated and aged at appropriate temperatures to obtain the desired microstructure comprising of 15% primary alpha in a transformed beta matrix. The heat treated discs were evaluated for room and high temperature mechanical properties. The results obtained indicate that the disc forged near isothermally can be used in the high pressure compressor region of the aeroengine. Thus following an DMM based approach, critical class- I components such as aeroengine discs can be produced without resorting to any shop floor trials on a reliable and repeatable basis.

ACKNOWLEDGMENT

The authors express their gratitude to Dr. G. Malakondaiah, Director, Defence Metallurgical Research Laboratory (DMRL) for encouraging us to publish his work. The authors thank Dr. P. K. Sagar and Dr. T. K. Nandy, Titanium alloy group of DMRL for providing IMI 834 material. The support rendered by Dr. Abhijit Dutta, Rolling and Formability Group of DMRL and his team for carrying out the hot compression test is acknowledged. The support rendered by Aeronautical Materials Testing Laboratory (AMTL), Hyderabad in carrying out mechanical testing is acknowledged. Authors acknowledge the funding provided by Defence Research and Development Organisation (DRDO) to carry out the work.

REFERENCES

Balasundar. I and Raghu. T, (2005), PROMAP - Program for generating processing Map, *DMRL Technical Report*, DMR TR 382 2005.

Balasundar. I, Sudhakara Rao. M, Raghu. T and Saha. G. G., (2004), Poster presented on Effect of height/diameter ratio on the deformation behaviour of titanium alloy IMI834, National Metallurgist Day - Annual Technical Meeting - 2004, 14-17 November, Trivandrum, Kerala, India.

Cope. M. T. and Hill. M. J, (1988), *Sixth world conference on titanium*, 153 - 158.

Daeubler. M. A, Helm D., Lutjering. G., (1995), Titanium'95*: Science and Technology*, 709 - 716.

Dieter E. George, (1988), *Mechanical Metallurgy*, Third Edition, McGraw Hill book Company.

Farthing. T.W, (1988), *Sixth world conference on titanium*, 37- 48.

Gangshu shen, David Furrer, (2000) Manufacturing of aerospace forgings, *J. of Mater. Process. Technol* 98, 189 - 195.

Gegel. H. L., Malas. J. C., Doraivelu. S. M. and. Shende. V. A, (1998), *ASM Metals Handbook*, Vol. 14, 417-438.

Kalyan Kumar. A. K. S., (1987) Criteria for predicting metallurgical instabilities in processing, M.S Thesis, *Department of Metallurgy*, IISc Bangalore, India.

Kuhlman. G. W., (2005) *Forging of titanium alloys*, ASM hand book Vol14A, Metal working: Bulk forming, 331- 353.

Kutumba Rao. V. V., and Rajagopalachary. T, (1996), *Bull. Mater. Sci.,* Vol. 19, pp. 677-698.

Malvern. L.E., (1969), Introduction to the Mechanics of Continuum Medium, Eaglewood Cliffs, Prentice-Hall, NJ, 197-271.

Miaoquan Li, Hongsi Pan, Yingying Lin, Jia Luo, (2007), High temperature deformation behaviour of near alpha Ti-5.6Al-4.8Sn-2.0Zr alloy, *J.Mar.Process. Tech.* 183, 71-76.

Neal. D. F., Alloy development, (1995) Titanium ' 95: Science and Technology, 2195 -2204.

Neal. D.F, (1988), *Development and Evaluation of high temperature titanium alloy* IMI 834, VI world conference on titanium, 253-258.

Prasad. Y. V. R. K., Gegel. H. L., Doraivelu S. M., Malas J. C., Morgan. J. T., Lark K. A., and Barker D.R., (1984), *Metall. Trans. A,* Vol. 15A, 1883 - 1892.

Prasad Y. V. R. K, (1990), *Indian J. of Technol.*, Vol. 28, 435 - 451.

Prasad Y. V. R. K., Sasidhara S. (1997), Hot Working Guide – A Compendium of Processing maps, *ASM International.*, Ohio.

Seetharaman. V and Lombard. C.M., (1991), Microstructure/property relationships in titanium aluminides and alloys, *Conference proceedings*, Eds. Young won Kim and Rodney Boyer, 237-251.

Sudhakara Rao. M, Balasundar .I and Raghu. T, (2007), Design of isothermal forging dies for high pressure compressor disc forgings. *DMRL Technical Report*, DMR 2007 428.

Tricot. R, (1998), *Thermomechanical Processing of Ti alloys*, VI world conference on titanium, 23-35.

Vo.P., Jahazi. M., Yue. S. and Bocher. P., (2007), Flow stress prediction during hot working of near alpha titanium alloys, *Mat.Sci and Engg* A447, 99-110.

Wanjara. P, Jahazi .M., Monajati .H., Yue. S., Immarigeon. J.-P., (2005), Hot working behaviour of near alpha alloy IMI 834, *Mat. Sci and engg* A 396, 50-60.

Wanjara .P., Jahazi. M., Monajati. H., Yue. S., (2006), Influence of thermomechanical processing on microstructural evolution in near alpha alloy IMI 834, *Mat. Sci and Eng* A416, 300-311.

Wiess. I. and Semiatin S. L., (1999), Thermomechanical processing of alpha titanium alloys - an overview, *Mat. Sci and Eng* A263, 243 - 256.

Young Niu, Hongliang Hou, Miaoquan Li, Zhiqiang Li, (2008), High temperature deformation behaviour of near alpha Ti600 titanium alloy, *Mat. Sci and Eng* A492, 24-28.

Ziegler. H. (1965), *Progress in Solid Mechanics* (ed. I.N. Sneddon and R. Hill), John Wiley and Sons, NY, Vol. 4, 91-193.

In: Machining and Forming Technologies, Volume 2
Editor: J. Paulo Davim

ISBN: 978-1-61122-820-5

Chapter 9

EBSD STUDIES ON THE INFLUENCE OF TEXTURE ON THE MECHANICAL PROPERTIES DEVELOPED IN WARM ROLLED DUPLEX STAINLESS STEELS

P. Chandramohan*[*,1], K. Manikandasubramanian[1] and P.Chandrasekar[2]

[1]Sri Krishna College of Engineering and Technology,
Coimbatore-641008, INDIA

[2]Faculty, School of Eng. and Science,
Swinburne University of Tech (Sarawak Campus),
Sarawak, Malaysia

ABSTRACT

This paper reports an investigation of micro texture in the cast condition and in the warm rolled conditions (50% size reduced) of nitrogen alloyed duplex stainless steel. The micro texture measurements were carried out using electron backscattered diffraction (EBSD) system. Grain orientations, Grain size distributions and Misorientation angles were also reported for the dual phases, ferrite (alpha) and austenite (gamma) under both the conditions of before heat treatment and after heat treatment. The results reveal that there is a definite improvement in the mechanical properties after reducing the specimen size to 50% by warm rolling, followed by heat treatment (solutionizing). This improvement in the mechanical property is correlated with the volume of low angle grain boundaries (EBSD analysis) and with the cleavage facets (SEM analysis).

Keywords: Rolling, Texture, Orientation, Gamma, Goss.

[*] Corresponding author, E-mail: pcmohu@yahoo.co.in

1. INTRODUCTION

Duplex stainless steels contain both phases, i.e. ferrite (Alpha-BCC) and austenite (Gamma-FCC) in almost equal proportions. Therefore, to make a study of texture in this rolled alloy, it is necessary to report few literatures related to texturing and orientations that exist in rolled FCC and in BCC alloys.

The major BCC and FCC texture components are listed by name, Euler angle and Miller indices {h k l}<u v w> in the table 1. The first indices {h k l} describes the crystallographic plane parallel to the sheet surface whereas the second indices <u v w> indicate the direction parallel to the rolling direction [Bruckner, 2001]. In FCC metals like aluminium, cube orientation (001) <100> is the recrystallisation texture component. A study carried out in an FCC AA6062 Al alloy reveals that, with an increase in the deformation temperatures , the dominance in nucleation appeared to have changed from cube (001) <100> to particle stimulated nucleation (PSN) to cube again and finally, to no preference. A deformation texture component called R-component (123) <412> is also seen in deformation textures of aluminium alloy which is misoriented from the cube component by 40° '<111>. S-orientation (123) <634> is not misoriented from the cube (100) <001> by 40° '<111> [Kashyap, 2001].

With respect to microstructural and crystallographic changes during the rolling process of DSS, it has been found that with decreasing plate thickness, i.e. .increasing rolling reduction, the grain size becomes progressively finer along with the preferred crystallographic orientations in the two separate phases [Hutchinson, 1985], [Akdut, 1996]. A study on the flat rolled DSS reveals that the increase in strength along the transverse direction is more than that of along rolling direction. Highest strength is found in the transversal direction which is due the mechanical anisotropy. This mechanical anisotropy observed for DSS is a direct consequence of crystallographic texture strengthening which means that there is a definite correlation between crystallographic and mechanical characteristics. [Mateo, 2003]. According to Hutchinson [1986], the ferrite texture in DSS consists mainly of a sharp fiber spread from (100) [011] to (211) [011] and minor components towards (311) [232] and (322) [241].

Table 1. Major BCC and FCC texture components

CRYSTAL LATTICE	NAME	φ_1	Φ	φ_2	{hkl} <uvw>
BCC	Rotated Cube	0°	0°	45°	{001} <110>
		0°	15°	45°	{115} <110>
	Inverse Brass	0°	35°	45°	{112} <110>
		0°	55°	45°	{111} <110>
	Inverse Copper	30°	55°	45°	{112} <110>
FCC	Cube	0°	0°	0°	{001} <100>
	Goss	0°	45°	0°	{011} <100>
	Brass	35°	45°	0°	{011} <211>

Table 1. (Continued)

CRYSTAL LATTICE	NAME	φ_1	Φ	φ_2	{hkl} <uvw>
FCC	Rotated Goss	90°	45°	0°	{011} <011>
	S	59°	37°	63°	{123} <634>
	C (Copper)	90°	35°	45°	{112} <111>
	R- Component				{123} <412>

They noted in their studies the absence of (111) [uvw] components which usually dominate bcc rolling. The main component in austenite texture is (110) [223] but less sharp than in ferrite phase.

Cold rolling induces austenite to martensite transformation in duplex stainless steels. 3% of the austenite initially present was transformed into martensite after 60% reduction, and 20% was transformed with 90% reduction at room temperature [Taveres, 2006]. It is observed that the sigma phase in the cold rolled DSS does not present a sharp texture component.

On the other hand a macro texture study by XRD diffraction in the samples of cold rolled and aged shows that the increase of the sigma phase precipitation enhances the amount of brass texture (110) [112] in austenite phase [Souza, 2008]. Hot rolling of X2CrNiMoN22-6-3 steel with solution annealing followed by cold-rolling in the direction perpendicular to the direction of hot-deformation strongly influence the formation of the final rolling textures.[Rys, 2008]

Texture inhomogeneity is one term that has to be given due consideration during this study. Characterizing the texture inhomogeneity is done using two parameters i.e. its intensity (a quantitative measure) and set of the dominating components (type of texture) [Bonarski 2005].As for as the determination of the texture of the near-surface layer the orientation distribution function (ODF) is a well known and applied procedure. [Bonarski, 1994, 1998[1], 1998[2], 2001]. The determining component of the texture is the orientation which is of greatest importance from the point of view of the expected properties dependent on the privileged orientation. For example, in the transformer plates such determined component is Goss orientation {110}<001>, in the car body sheets(for deep drawing)these are the two favourable orientations{111}<110> and {554}<225> [Bonarski 2005].

It is clear from the above stated literatures that, every metal forming process leads to a specific type of orientation which determines the mechanical property of the formed alloy. Therefore in this study the crystallographic texture and micro texture measurements were employed to bring out developments in the deformed (Rolled) and recrystallized microstructure of nitrogen alloyed duplex stainless steels. Correlation between the resulting mechanical properties over rolling and texture component is also reported.

2. Experimental Work

The starting material for the present work was nitrogen alloyed duplex stainless steels. Table 2 gives the chemical composition of the studied material. The high content of alloying elements, especially nitrogen, leads to improvements in mechanical properties and corrosion resistance as compared to the standard DSS grades. The steel was a cast material developed in-house (4A as per ASTM A 890).

The casting was subsequently warm rolled at 450°C, to 50% size thickness through various steps as shown in figure 1. Immediately after warm rolling, specimens were taken from one half of the casting (BHT-Before Heat Treatment) and the other half of the casting was subjected to solution annealing i.e. heated to 1130^0 C at the rate of 150^0 C / hr soaked for 2 hrs furnace cooled to 1040^0 C and then water quenched (AHT-After Heat Treatment).

30mm
↓
29.0
↓
28.0
↓
27.0
↓
26.0

Loaded in furnace for 400° C
25.0 mm
↓
24.0
↓
23.0
↓
22.0

Loaded in furnace for 400° C
21.0 mm
↓
20.0
↓
19.0
↓
18.0

Loaded in furnace for 400° C

17.0 mm
↓
16.0
↓
15.0
↓
14.0

Figure 1. Flow chart of steps in reduction.

Table 2. Chemical composition in wt%

Alloy	%C	%Si	%Mn	%S	%P	%Cr	%Ni	%MO	%Cu	%V	%W	%N
DSS	0.028	0.94	0.86	0.009	0.035	22.79	5.59	3.42	0.19	-	-	0.18

Samples were subjected to measurements of bulk crystallographic texture and micro texture. The measurements were obtained at the midthickness sections of the rolling plane (containing rolling directions (RD) and transverse directions (TD)) and of the long transverse plane (containing RD and normal directions (ND)) for the bulk and micro texture, respectively. All samples were electro polished using the standard [Samajdar, 1998] technique. For the bulk texture measurement, a PANalytical materials research diffraction (PANalytical, Almelo, The Netherlands) system was used. The orientation distribution functions (ODFs) were obtained through the inversion of four incomplete X-ray pole figures by the software MTM-FHM. [Van-Houtte, 1995] The software uses the standard series-expansion technique. [Bunge, 1982] The ODFs were further analyzed to obtain the volume fractions of the ideal FCC and BCC orientations.

The micro texture measurements were obtained in a Tex SEM Laboratories (TSL) energy- dispersive X-ray orientation-imaging microscopy or electron backscattered diffraction (EBSD) system on an FEI Company Quanta 200 high- vacuum scanning electron microscope (SEM). The following phenomena, which are directly linked to microstructure (i.e. individual grain orientation), were studied.

- Local property effects
- Interfacial parameters and properties (Grain misorientation)
- Morphological/geometrical grain parameters i.e. size, shape, location
- Orientation variations within individual grains
- Phase relationships
- Direct ODF measurement.

3. Results and Discussion

Electron back scattered image of the cast nitrogen alloyed duplex stainless steel alloy is shown in figure 2. The electron back scattered image of the specimen subjected to warm rolling and size reduced to 50% (BHT) is represented in figure 3 and the image of the specimen subjected to warm rolling and size reduced to 50%, followed by heat treatment (AHT) is represented in figure 4. Rolling direction and transverse direction are shown in all the three figures for clear understanding. It is observed from the colour coding of the figure 2 that the ferrite grains are predominantly oriented between <111> and <001> directions. Remaining ferrite grains are oriented in direction <001>. Austenite grains are also oriented between <111> and <001> directions. It can be observed from the figure 2 that the volume fraction of ferrite (alpha) is 46 percent and that of austenite (gamma) is 54 percent. It can also be observed from the figures 3 and 4 that the volume fraction of ferrite (alpha) is 71 percent and that of austenite (gamma) is 29 percent which means, the rolling process has increased the amount of ferrite by sacrificing austenite, irrespective of the heat treatment process. It is inferred from the colour coding of the figure 3 that the ferrite grains are predominantly oriented between <111> and <101>. Few ferrite grains are oriented between <001> and <101> directions and between <001> and <111> directions. Remaining ferrite grains are oriented in all the three directions <111>, <101> and <001>. Austenite grains are oriented in almost all the three directions namely <001>, <111> and <101>, but it follows a chain type network pattern. It is inferred from the colour coding of the figure 4 that the ferrite grains are predominantly oriented in <001>, <111> directions and in between <001> and <111> directions. Austenite grains are oriented in almost all the <001>, <111> and <101> directions but unlike the specimen before heat treatment, it does not follow a chain type network pattern. This shows that the formation of chain like arrangement of the austenite phase immediately after the rolling process is due to the load partitioning between ferrite and austenite phases which is due to the difference in the mechanical properties of both the phases. After heat treatment of the same rolled specimen, this chain type arrangement is disturbed.

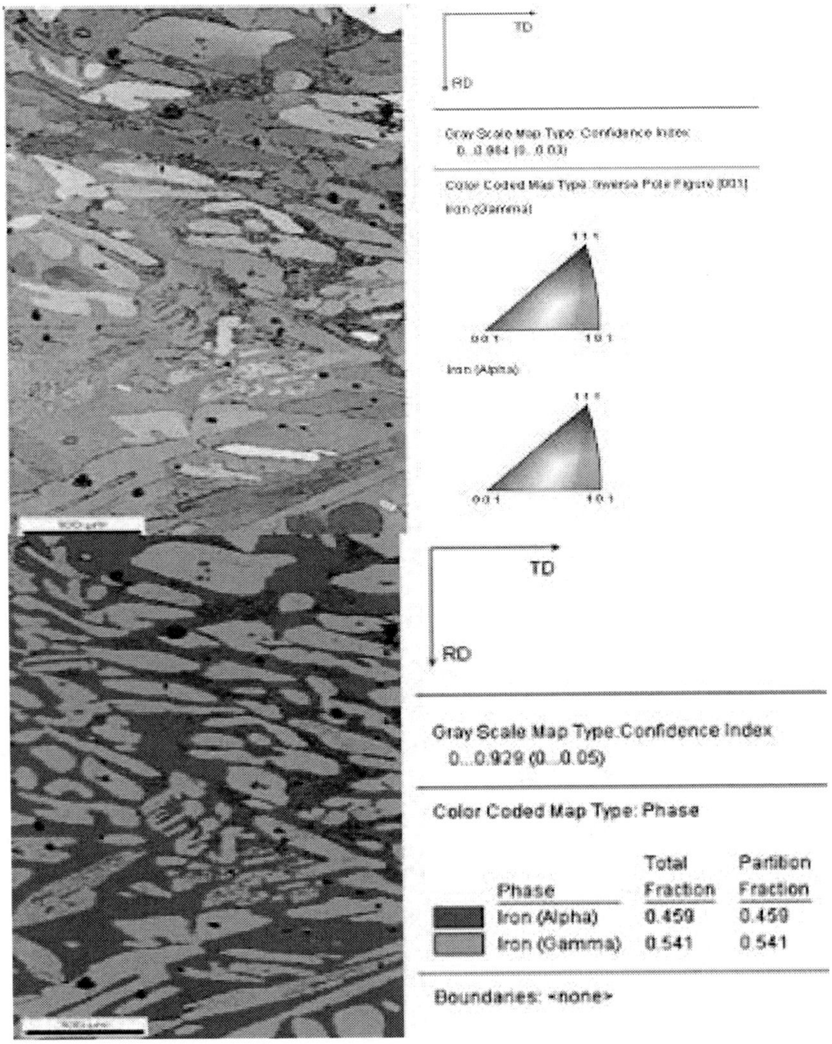

Figure 2. EBSD image of Cast and HT DSS alloy.

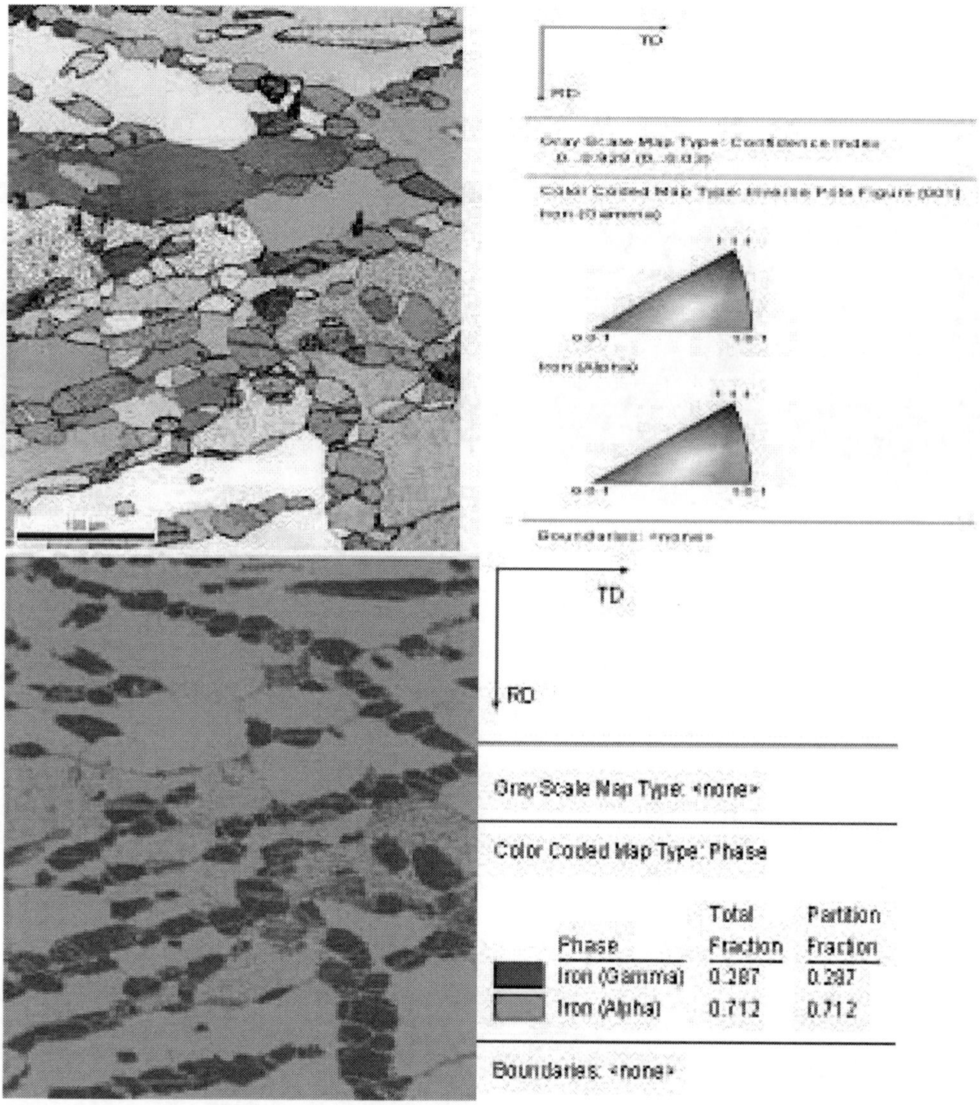

Figure 3. EBSD image of 50% Warm rolled –BHT.

Figures 5-8 represents the grain size distribution and misorientation angle of 50% warm rolled –BHT and 50% warm rolled –AHT specimens. The average grain size before heat treatment is 24 micron whereas, after heat treatment it is 44 micron, which means grain growth has occurred during heat treatment. The misorientation angle is widely distributed over a range, in both the specimens, upto $68.9\ ^0$ and 63^0 in BHT and AHT respectively. But the volume fraction of low angle grain boundary in AHT ($\leq 30.5^0$ - 0.23), is higher than in BHT ($\leq 30.5^0$ - 0.057) which is considered to be a positive sign as far as mechanical properties are concerned. Bulk texture measurement carried out using pole figure and ODF are shown in figures 9-10. Random orientation is observed. The volume fraction of individual texture components are given in table 3.

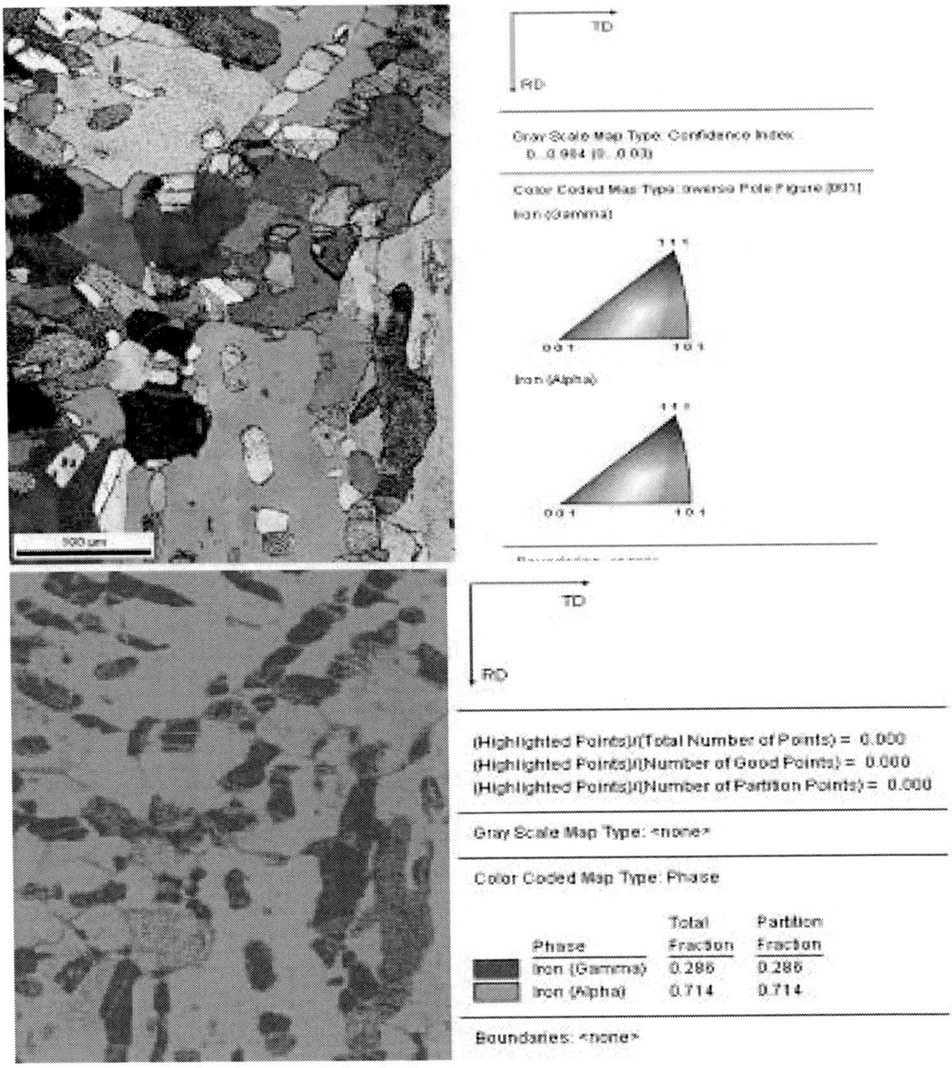

Figure 4. EBSD image of 50% Warm rolled –AHT.

Table 3. Volume fraction of texture components

Texture components	Volume (BHT)	Volume (AHT)
Cube	0.05	0.04
Taylor	0.09	0.09
Brass	0.11	0.08
S	0.20	0.15
Cu	0.09	0.08
CG	0.09	0.09
H	0.04	0.06
CH	0.09	0.11
Goss	0.05	0.05

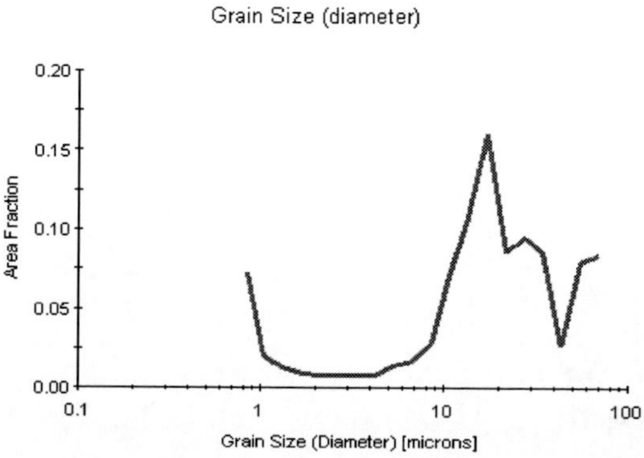

Chart: Grain Size (diameter)

Edge grains excluded from analysis

Diameter [microns]	Area Fraction
0.833978	0.072016
1.05209	0.0199082
1.32725	0.0128211
1.67437	0.00851773
2.11227	0.0084056
2.6647	0.00745833
3.36161	0.00745833
4.24078	0.00837854
5.34989	0.0137722
6.74906	0.0169233
8.51417	0.0289788
10.7409	0.070338
13.55	0.108128
17.0938	0.160043
21.5644	0.0863798
27.2042	0.0947428
34.319	0.0860704
43.2946	0.026079
54.6175	0.0795787
68.9018	0.0840019
Average	
Number	1.19266
Area	24.7402

Figure 5. Grain size distribution of 50% Warm rolled –BHT.

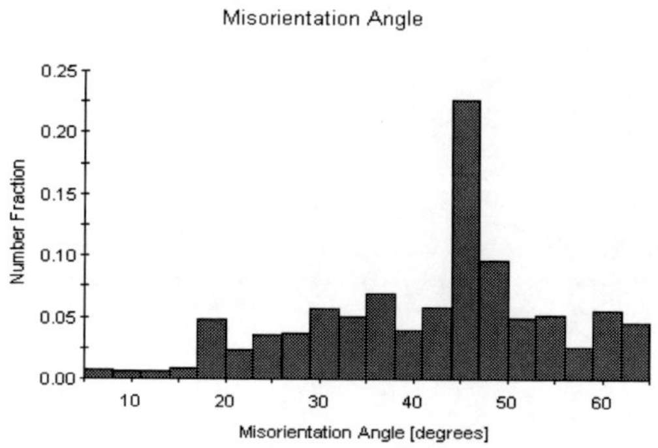

Chart: Misorientation Angle

Angle [degrees]	Number Fraction
6.5	0.00736333
9.5	0.00636821
12.5	0.00607785
15.5	0.00877566
18.5	0.0482788
21.5	0.0229131
24.5	0.0364671
27.5	0.0365657
30.5	0.0569868
33.5	0.0509654
36.5	0.0694667
39.5	0.0399768
42.5	0.0578382
45.5	0.225923
48.5	0.0964872
51.5	0.0496433
54.5	0.0516081
57.5	0.0256645
60.5	0.0562511
63.5	0.0463788
Average	
Number	41.5969

Figure 6. Misorientation angle of 50% Warm rolled –BHT.

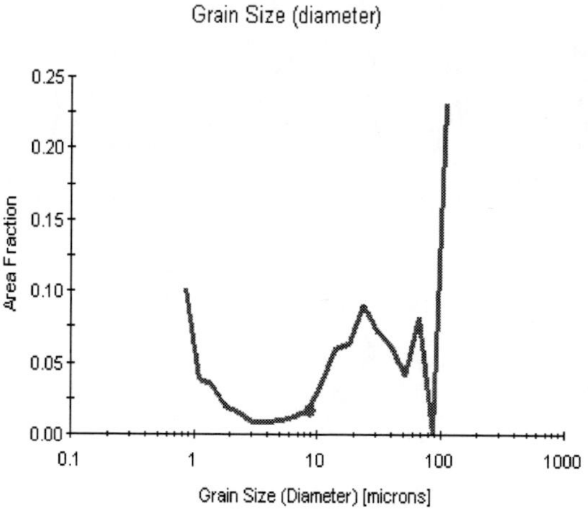

Chart: Grain Size (diameter)

Edge grains excluded from analysis

Diameter [microns]	Area Fraction
0.844426	0.101204
1.09213	0.0387339
1.41249	0.034176
1.82684	0.0193365
2.36272	0.0144347
3.0558	0.00873544
3.95218	0.00832573
5.11151	0.0095207
6.61092	0.0123008
8.55017	0.0169856
11.0583	0.0359392
14.3021	0.0597677
18.4975	0.0630112
23.9235	0.0908319
30.9412	0.0713443
40.0175	0.0616724
51.7562	0.041258
66.9384	0.0812088
86.574	0
111.97	0.231213
Average	
Number	1.05837
Area	44.1907

Figure 7. Grain size distribution of 50% Warm rolled –AHT.

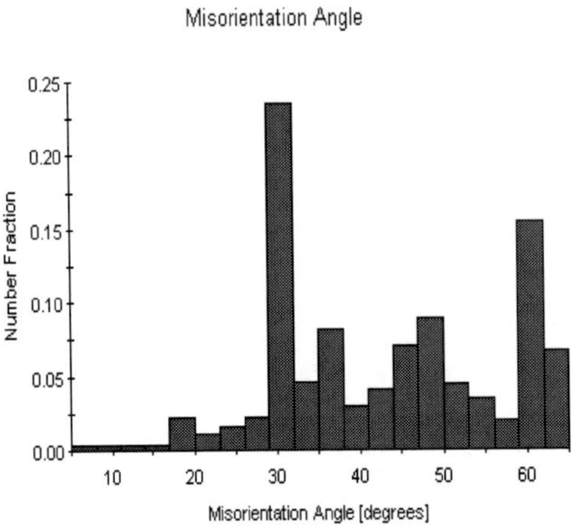

Chart: Misorientation Angle

Angle [degrees]	Number Fraction
6.5	0.0035863
9.5	0.00338084
12.5	0.00415912
15.5	0.00408129
18.5	0.0223288
21.5	0.0113488
24.5	0.0160325
27.5	0.0218556
30.5	0.234638
33.5	0.0455542
36.5	0.0820212
39.5	0.0292477
42.5	0.0405779
45.5	0.0702489
48.5	0.0894926
51.5	0.0443541
54.5	0.0350272
57.5	0.0195737
60.5	0.155184
63.5	0.067307
Average Number	42.6016

Figure 8. Misorientation angle of 50% Warm rolled –AHT.

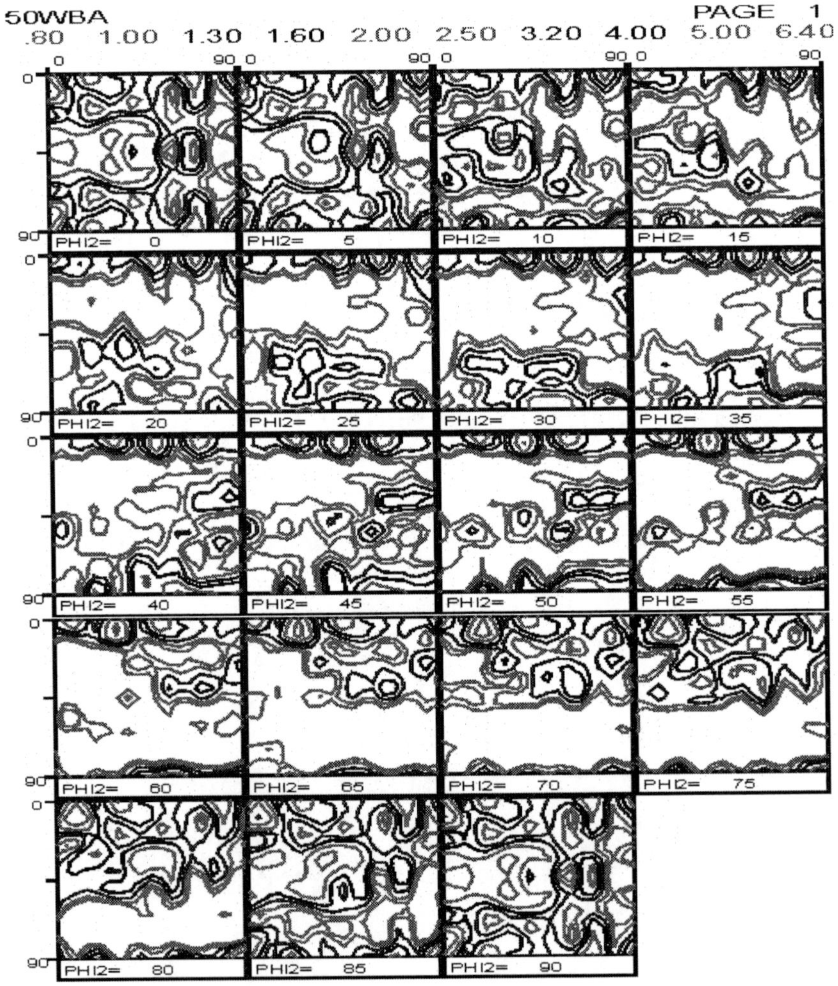

Figure 9. ODF of austenite phase (Bulk texture).

It can be observed that the cube orientation is the least and S orientation is the maximum. This confirms the standard that cube orientation is the recrystallised texture component and it gets turned to S, Brass and Cu orientations after rolling. The same is supported with improved mechanical properties, like tensile strength, yield strength and hardness in the AHT specimen as discussed below.

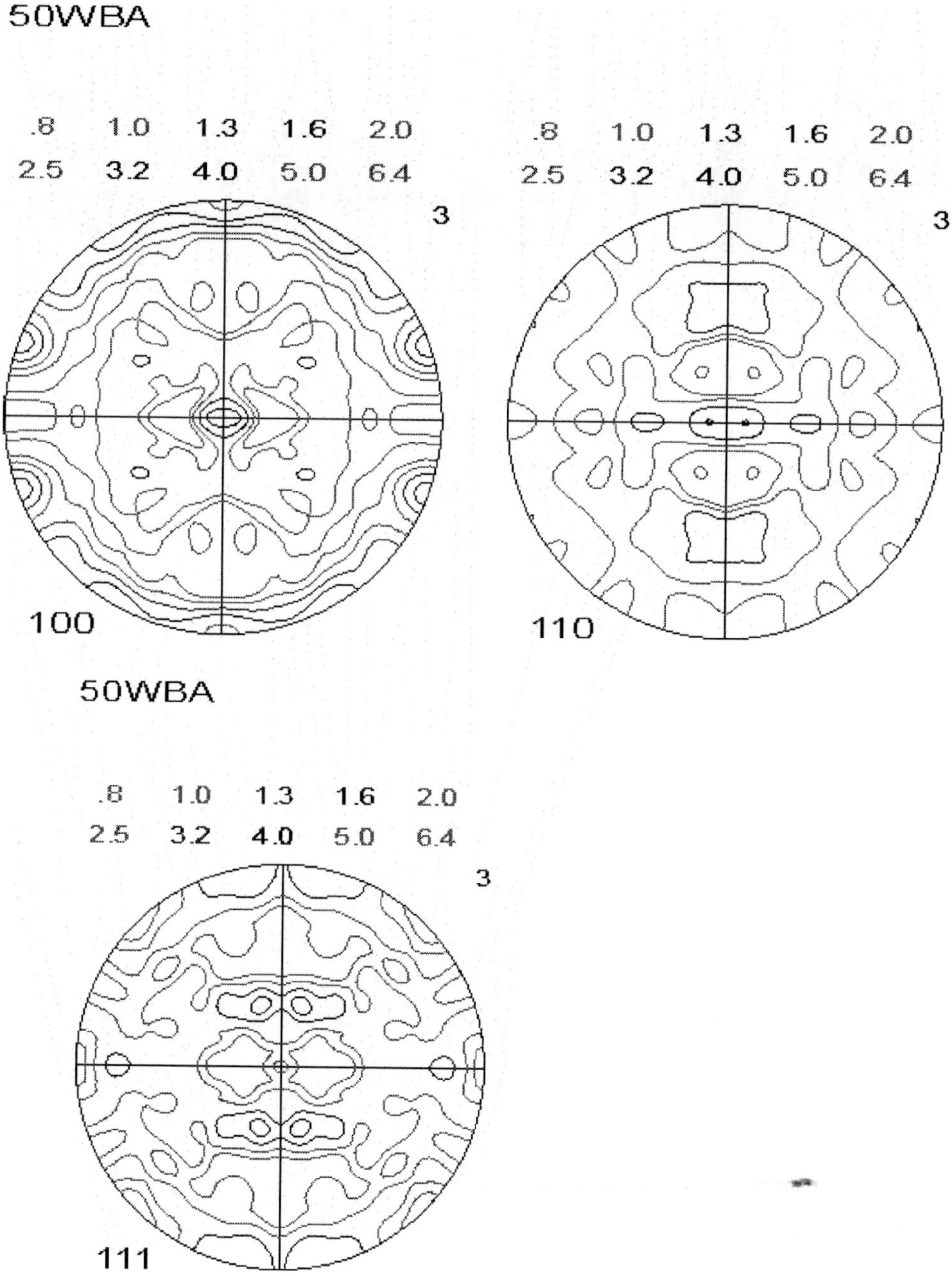

Figure 10. Pole figure of austenite phase (Bulk texture).

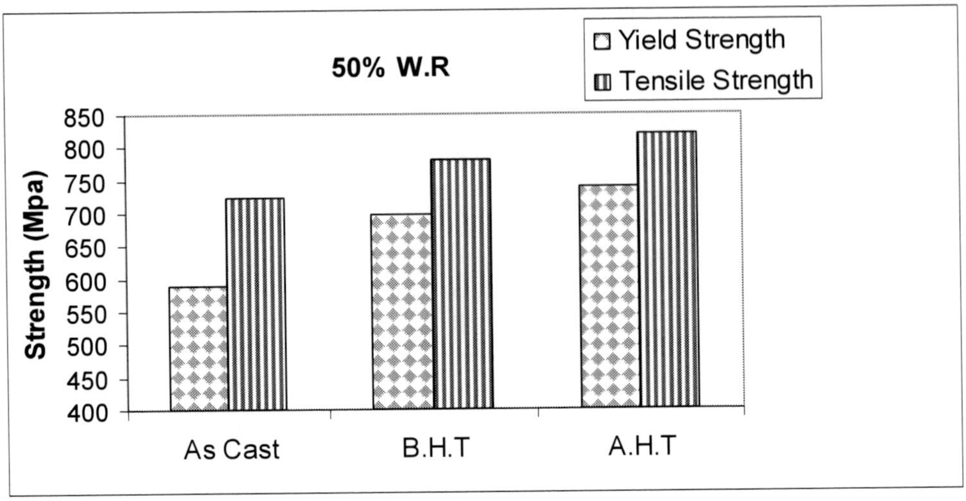

Figure 11. UTS and Yield strength.

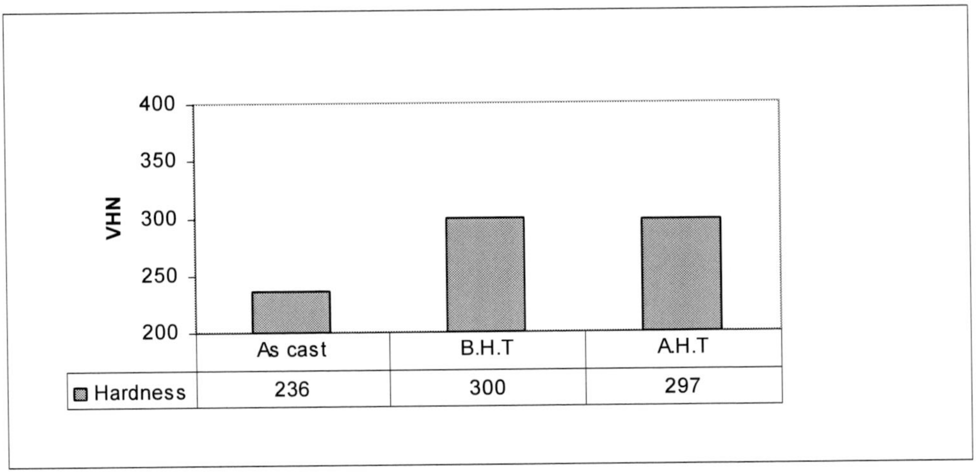

Figure 12. Hardness values of non rolled and rolled specimen.

The UTS and yield strength of the non rolled specimens of the alloy are 722 MPa and 588 MPa respectively. The effect of 50% size reduction by rolling, on the alloy, is shown in the figure 11. It is clear that there is a remarkable improvement in the UTS and yield strength of the 50% warm rolled specimen (820 MPa and 730 MPa respectively). The hardness values of the non rolled and rolled specimens are shown in figure 12. It can be observed that there is an improvement in hardness after the rolling process. The fracture surface analysis of this specimen is shown in the figure 13. Fracture surface of the rolled specimen reveals very fine dimples with cleavage facets. As per scale of fractures for estimation of fracture of test pieces and castings made of steel СЕП-2 (Russian standard), characteristics of warm rolled specimen structures are fibrous with fine grained aggregates in centre which occupies not more than 15% with shear lips from 8.1 to 10 mm. This observation reveals that the material becomes more ductile and tough with increase in % reduction by warm rolling.

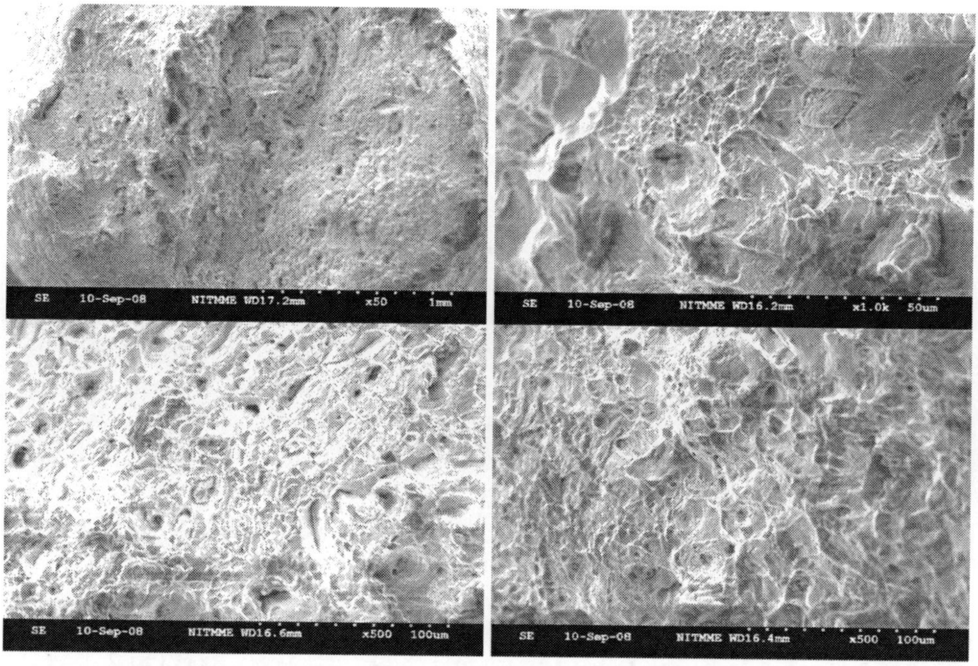

Figure 13. Fracture surface analysis of 50% WR alloy.

CONCLUSIONS

The directions of ferrite and austenite vary with the metal forming process and with the heat treatment procedure practiced. In the cast and heat treated specimen, both the ferrite and austenite grains are predominantly oriented between <111> and <001> directions. In the warm rolled (BHT) specimens, the ferrite grains are predominantly oriented between <111> and <101>. Austenite follows a chain type network pattern in BHT specimen which is not the case with the AHT specimen. In both the warm rolled (BHT and AHT) specimens, austenite grains are oriented in almost all the three directions namely <001>, <111> and <101>. In the warm rolled (AHT) specimens, the ferrite grains are predominantly oriented in <001>, <111> directions and in between <001> and <111> directions. Bulk texture measurement reveals random orientation of ferrite and austenite phases.

Rolling process increases the amount of ferrite (46 to 71) by sacrificing austenite (54 to 29), irrespective of the heat treatment process. Warm rolled (AHT) specimens reveal relatively large volume of low angle grain boundaries (EBSD analysis) and the cleavage facets (SEM analysis).

ACKNOWLEDGMENT

National Facility of Texture and Orientation Imaging Microscopy - Department of Science and Technology, India at IIT-Bombay, supported this work. The authors are grateful to m/s Auto shell co, m/s Commando engineers, Coimbatore for supplying and processing the DSS material.

REFERENCES

Akdut, N., Foct, J. and Gottstein, G. (1996), "Cold rolling texture development of α/γ duplex stainless steel", *Steel Research*, 67, 450-455.

Bonarski, J.T. (2001), "*Rentgenowska Tomografia Teksturowa*", IMIM PAN, Krakow (in polish).

Bonarski, J.T. (2005), "Quantification of texture inhomogeneity", *Materials Science Forum*, 495-497, 149-156.

Bonarski, J.T., Wcislak, L. and Bunge, H. (1994), "Investigation of inhomogeneous textures of coatings and near-surface layers". In *Proceedings of l0th International Conference on the Textures of Materials (ICOTOMIO)*, , ed. Bunge, H., Material Science,157-162, 111-118.

Bonarski, J.T[1]. and Pawlik, K. (1998), "Development of texture determination method for the near-surface-layers", *Archives of Metallurgy*, 43/2, 141-148.

Bonarski, J.T[2]., Bunge, H.J., Wcislak, L. and Pawlik, K. (1998), "Investigation of Inhomogeneous Surface Textures With Constant Information Depth: Part 1: Fundamentals", *Textures and Microstructures*. 31(1-2), 21- 41.

Bruckner,G. and Gottstein,G.(2001),"Transformation Textures during Diffusional α→γ→α Phase Transformations in Ferric steels", *ISIJ International*, 41(5), 468-477.

Bunge, H.J. (1982), "Texture analysis in material Science", *Butterwoths*, London.

Hutchinson, W.B., Schlippenbach, U.V. and Jonson, J. (1986), "Texture and anisotropy in duplex stainless steel SS2377". *Proceedings of Duplex Stainless Steel*, 1-2, 7-13.

Hutchinson,W.B., Ushioda,K and Runnsjo,G. (1985), "Anisotropy of tensile behaviour in a duplex stainless steel sheet", *Material science and technology*, 1,728-731.

Kashyap,K.T. (2001),"On the origin of recrystallisation textures", *Bulletin of Material Science*, Indian academy of sciences, 24(1), 2001, 23-26.

Mateo,A., Llanes,L., Akdut,N., Stolarz,J. and Anglada,M. (2003), "Anisotropy effects on the fatigue behaviour of rolled duplex stainless steels", *International journal of fatigue*, 25,481-488.

Rys, J. and Witkowska, M. (2008), "Rolling Texture Development in Nitrogen Alloyed Ferritic-Austentic Steel of Duplex Type", *Archives of Metallurgy and Materials*, 53(1),221-228.

Samajdar, I. and Doherty, R.D. (1998), "Cube recrystallisation texture in warm deformed aluminum: understanding and prediction", *Acta Materilia*, 46, 3145-3158.

Souza Jr, C.M., Abreu, H.F.G., Tavares, S.S.M. and Rebello, J.M.A (2008), "The sigma Phase formation in annealed UNS S31803 duplex stainless: Texture aspects", *Materials Characterization*, 59, 1301-1306.

Tavares, S.S.M., Silva da, M.R., Pardal., J.M. Abreu,H.F.G. and Gomes, A.M. (2006), "Microstructural changes produced by plastic deformation in the UNS S31803 duplex stainless steel", *Journal of Materials processing Technology*, 180(1-3), 318-322.

Van-Houtte, P. (1995), "*MTM-FHM software system*", Version 2, MTM KULeuven, Belgium.

In: Machining and Forming Technologies, Volume 2
Editor: J. Paulo Davim

ISBN: 978-1-61122-820-5

Chapter 10

COMPREHENSIVE REVIEW ON DRILLING OF MULTIMATERIAL STACKS

Vijayan Krishnaraj*, Redouane Zitoune and Francis Collombet
Université de Toulouse ; INSA ; UPS ; Mines Albi ; ISAE ; ICA
(Institut Clément Ader) 133c, avenue de Rangueil, F-31077 Toulouse, France

ABSTRACT

This paper reviews the challenges during drilling of composite, titanium, and aluminium stacks. The machinability aspects of these three material stacks commonly used in aerospace applications are reviewed individually based on influence of drill material, drill geometries, and drilling parameters. Composite material requires high speed and low feed rate, drilling titanium requires low speed and high feed rate whereas drilling aluminium requires a balance between speed to feed rate. The review reports critical results and gap in the literature for composite, titanium and aluminium. Drilling of multimaterial stacks require a compromise between many parameters. The problems and solutions related to drilling of multimaterial stacks are discussed and the directions in which the research on drilling of multimaterials may be carried out are suggested in this paper.

Keywords: Composite Laminates, Titanium, Aluminium, Multi-material stacks, Drilling, Machinability

1. INTRODUCTION

Composite materials are becoming more important to aerospace, naval, space, and automotive industries. Skin temperatures in next-generation military and commercial aircraft may approach 177°C (Kim and Ramulu, 2004). Because of their light weight and stability at elevated temperature, various titanium, carbon fibre reinforce plastic (CFRP), aluminium

* Corresponding author: E-mail: vkrishnaraj@hotmail.com

stacks have been used to solve the problem. In these composite structures, cut-outs and holes are to be produced in large numbers. For example, in an aircraft fuselage structure, millions of holes are required for joining purposes (Richardson, 1987). Drilling of such materials is a challenging task to manufacturing engineers because of differential machining properties.

The main aspect of machinabilityis to discuss in detail the force, torque, tool life, and surface finish. Though there are a number of interrelated factors which affect the machinabilityof the material, the most important factors are the cutting parameters, the properties of the work and tool materials, geometry of the cutting tool, clamping of work and the type of machine tool. Since material removal is a workshop related activity where very strong economic or production rate constraints prevail, it is important that the selection of cutting tool and work design parameter are based upon a rational optimization procedure. The cutting parameters in drilling are the spindle speed, and feed rate. The tool parameter involves the tool material and tool geometry. In this case, the work design parameter is concerned with the PMC material, titanium and aluminium stacks. Though the machining technique on PMC material is similar to that of metals, the fibre proportion and their orientation in PMC (fibres in continuous or chopped strands, fibres in unidirectional or bi-directional) play an important role in deciding the machinabilityof the material; moreover, machining of titanium and/or aluminium has other complications. This paper provides a detailed review of drilling of polymer matrix composite in chapter 2. Chapter 3 presents literature related to drilling of titanium alloy. In chapter 4 drilling of aluminium is discussed. In chapter 5 drilling of multimaterials is presented and in Chapter 6 conclusions based on these reviews and directions in which drilling of multimaterial stacks may be carried out are presented.

2. Drilling of Polymer Matrix Composites

2.1. Introduction

Polymer matrix composite materials are increasingly used in high performance applications because of superior strength to weight ratio and stiffness to weight ratio. Cut-outs and holes exist in most of the composite structures. Due to their laminated constructions several types of damages like matrix cracking and thermal alterations, fibre pullout and fuzzing, are introduced during drilling in addition to geometrical defects similar to those found in metal drilling (Abrate, 1997). About 60% of the rejections are due to the defects in the holes. These defects would create reduction in structural stiffness, leading to variation in the dynamic performance of the whole structure. Many of these problems are due to the use of non-optimal cutting tool designs, rapid tool wear, and machining conditions (Konig et al., 1985, 1989, 1990; Komanduri, 1993, 1997; Kohkonen and Potdar, 1998). Bhatnagar et al. (1995) have modelled the mechanism of chip formation of UD-CFRP and showed that cutting forces are dependent on the fibre angle as well as the direction of cutting.

According to the Polymer technology (Lubin, 1969) series data the cutting speeds from 77.36 to 154.72 m/mm, point angle 60° to 120°, helix angle 10° to 50°, clearance angle 9° to 20° can be selected for drilling PMC material. Apart from the drill diameter and feed employed, the torque and thrust force in drilling are mainly influenced by work material, drill geometry, drill wear and related features. The total drill torque is only slightly affected by the chisel edge length. By thinning the web, it is possible to reduce the axial thrust by 30 to 35%. Both thrust and torque rise sharply if the drill is allowed to dull too much. Figure 1.1 shows the thrust force and torque acting on a standard twist drill (Arshinov and Alekseev, 1976). In PMC, delamination near the entry and exit portion of the hole, fibre pull out, excessive tool wear are the main problems during drilling (Malhotra, 1990; Di Ilio, 1991).

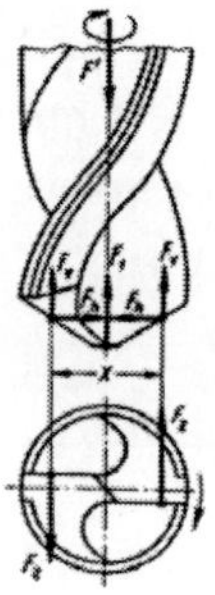

Figure 1.1. Forces acting on a drill (Arshinov 1976).

In drilling, the drill always exerts a compressive thrust force on the work piece. The laminate under the drill thus tends to be drawn away from the interlaminar bond around the hole. As the drill approaches the end, the uncut thickness becomes smaller and the resistance to deformation decreases. At some point the loading exceeds the interlaminar bond strength and delamination occurs. This happens before the laminate is completely penetrated by the drill as shown in Figure 1.2 (a). The chisel edge of the drill will first abrade the laminate initially. It, then by moving forward, tends to pull the abraded material away along the flute. The material spirals up before it is machined completely as shown in Figure 1.2 (b). This action introduces a peeling force upwards to separate the upper laminate from the uncut portion held by the downward acting thrust force. The cutting force acting in the peripheral direction is the driving force for delamination. It generates a peeling force in the axial direction through the slope of the drill flute and is a function of tool geometry and friction between the tool and workpiece. Delamination caused by peel-up becomes progressively more difficult as drilling proceeds, since the thickness resisting the lamina bending becomes greater.

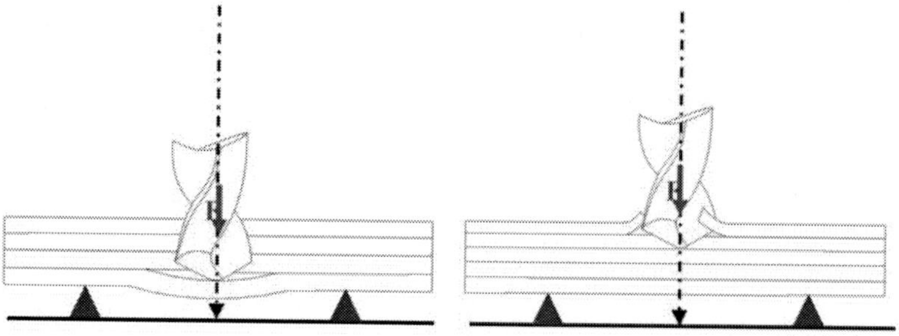

(a) Push – Out at Exit

(b) Peel – Up at Entrance

Figure 1.2. Drill bit showing delamination at exit and at entrance (Zitoune 2006).

2.2. Effect of Tool Material

In general, the desirable properties for a cutting tool material are, small grain size to be able to produce a sharp cutting edge, high hot hardness to provide excellent abrasive wear resistance, good toughness to maintain a sharp cutting edge without chipping or deformation under a cutting force's dynamic action, good thermal conductivity to remove heat from cutting zone, and thermal stability to maintain integrity at cutting temperatures and low chemical affinity to the workpiece material. The degree to which each of these properties is needed depends on the workpiece material. The micro constituents in the workmaterial reduce the tool life. In case of fibre reinforced composite materials, the fibre size, shape, position (orientation) and proportion influence the cutting forces and tool life (Sakuma and Seto, 1978, 1981a, 1981b, 1983, Sakuma et al., 1985).

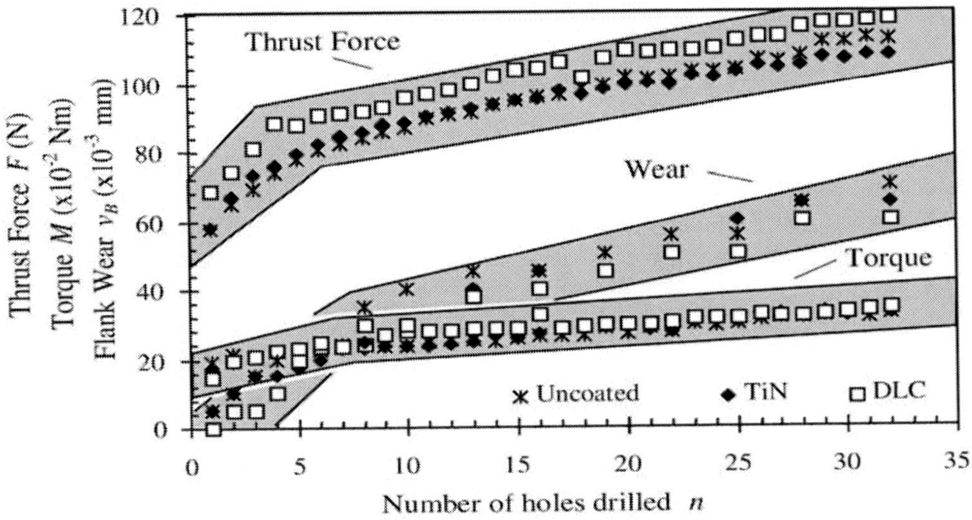

Figure 1.3. Variations in the maximum thrust force, torque and Flank wear with number of drilled holes [Murphy 2002].

As high speed steel is less resistant to wear, its tool life is less when used on fibres like glass and graphite which are highly abrasive in nature (Santhanakrishnan et al. 1989). Sakuma et al., (1984) drilled holes using four drill materials and investigated drill wear pattern, thermal conductivity of tool material, flank wear width and cutting forces and reported that K01 grade carbide drill has the highest wear resistance when compared to HSS, P10 grade carbide and ultra fine grain drills. It is found that when cutting speed is increased, the rate of wear on the nose in every tool material starts to increase remarkably at a certain speed (Sakuma and Seto, 1981a, 1983). Davim and Reis (2003a) investigated drilling of CFRP using HSS, cemented carbide (K10) helical flute and cemented carbide four flute drills and concluded that helical flute carbide drill is better because of the hot hardness when compared to HSS drill and positive rake angle when compared to four flute cemented carbide drills. The various types of wear were classified into four groups which may depend upon the cutting speed, feed rate, geometry of contact, coolant, tool material and work material. The types of wear include adhesive, abrasive, fatigue, and corrosive wear. One of the best methods to improve the productivity or tool life of a drill is to add a coating or surface treatment. Coatings and surface treatments build a barrier between the drill and the workpiece but coatings do not seem to have influence while drilling of PMC materials (Figure 1.3) (Murphy et al., 2002; Arul et al., 2006b). Ramulu, et al., (1999) have showed that better quality holes can be achieved while drilling Graphite/Bismaleimide composites using PCD four-faceted drill. Table 1.1 gives the typical tool materials used for drilling and its critical machining parameters. For drilling PMC, carbide drill of grade K10 or K20 and PCD is found suitable. While drilling with carbide drills, thrust force was found to be smaller than that of HSS drills, whereas the thrust force of the polycrystalline diamond drill was one third of that of the HSS drill. The PCD drills produced the highest quality holes and suffered the least amount of wear, but the number of literature related to the life or economy seems to be less. Apart from tool materials, tool geometry has direct influence on forces, and quality of drilled holes.

Table 1.1. Typical tool materials and critical drilling parameters on composites

Cutting parameters Machinability Characteristics									
Reference	**Work material**	**Cutting Tool**	**Vc m/min**	**F (mm/r)**	**Th N**	**Tq Nm**	**Wear mm**	**No holes**	**Ra (μm)**
Murphy (2002)	CFRP Vf0.62	WC One shot drill, Φ6.35 mm	60	0.048	60	0.2	0.07	32	
Arul (2006)	GFRP Vf 0.4	Tipped WC Φ6 mm	18.85	0.02	30	-	-	50	-
		TiN HSS Φ6 mm	18.85	0.02	35	-	-	30	-
		HSS Φ6 mm	18.85	0.02	40	-	-	30	-
Varatharajan (2006)	GF/Polyester	Carbide tipped Φ6 mm	47	0.1	74	-	0.13-0.21	30	-
Ramulu (1999)	Gr/Bi	HSS Φ6.35 mm	54	0.028	150	-	0.45	18	5.24
		PCD Φ6.35 mm	54	0.028	50	-	-	40	1.13
Bhattacharyaa (1998)	KFRP Vf0.55	HSS Candle stick drill	30	0.1	410	-	-	250>	2-3

Vc= Cutting speed m/min, f= Feed rate mm/rev, Fz= Thrust N, Tq = Torque Nm, Ra = Surface roughness (Ra in μm)

2.3. Effect of Tool Geometry

Many modifications are done on cutting tools to improve the quality of machining. Jain and Yang (1994) have showed that it is possible to reduce the thrust force significantly by reducing the chisel edge width. Won and Dharan (2002a), Tsao and Hocheng (2003) conducted drilling experiments on CFRP to determine quantitatively, the effect of chisel edge and pre drilling on thrust force. In a standard twist drill point angle and chisel edge have higher influence on thrust force, by reducing the point angle and chisel edge the thrust force during drilling can be reduced (Chen, 1997). Point angle of 75° to 118°, cutting speed of 45-

65 m/min and feed rate of 0.03 mm/rev to 0.04 mm/rev were recommended so as to drill CFRP with less thrust force and delamination (Enemuoh et al., 2001). Lesser point angles were also recommended by Gindy (1998), Singh, Bhatnagar, and Viswanath (2008).

Bhattacharyya and Horrigan (1998) showed that by introducing a negative point angle and high clearance angle (Modified drill with -20° point angle), Kevlar fibre reinforced plastic (KFRP) can be successfully drilled by applying liquid nitrogen at the drilling site. Mathew, Ramakrishnan and Naik (1999) have proved that trepanning tools are giving reduced thrust while making holes in thin laminated composites. Piquet et al. (2000) used a drill similar to dagger drill in comparison with twist drill and showed improvement.

Table 1.2. Typical tool geometries and critical drilling parameters on composites

Cutting parameters Machinability Characteristics							
Reference	**Work material**	**Tool geometry**	**Vc (m/min)**	**Feed (mm/rev)**	**Thrust (N)**	**Torque (Nm)**	**Df**
Bhattacharyya (1997)	KFPR Vf 0.36	Candle stick drill HSS Φ 10	30	0.01	430	-	-
Piquet (2000)	CFRP	Special drill Φ10 Carbide K20	18	0.05	120	0.3	-
Mathew (1999)	GFRP Vf 0.4	Trepanning tool HSS Φ 10	50	0.2	40	0.5	-
Enemuoh (2001)	CFRP	Φ 6.35 Point angle of 75° to 160°	45-65	0.03-.04	40	-	-
Davim (2003)	CFRP Vf 0.55	Helical flute Φ 5 Carbide K10	16	0.04	-	-	1.042
	CFRP Vf 0.55	Brad & Spur Φ 5 Carbide K10	30	0.05	-	-	1.01
Davim (2004)	GFRP Vf 0.65	Brad & Spur Φ 5 Carbide K10	55	0.05	6.27	-	1.024
Tsao (2004)	CFRP Vf 0.55	Candle stick Saw drill	31	0.02	-	-	1.30
Tsao (2006)	CFRP Vf 0.55	Twist drill Step drill Saw drill Candle stick Core drill	28-31	0.007 0.049 0.0051 0.069 0.0047	35 29.8 28.6 25.1 42.6	-	-
Tsao (2008)	CFRP Vf 0.55	Step core drill	1200 rpm	0.006	-	-	-
Ramkumar (2004)	GFRP Vf 0.4	Φ 6 3 flute solid carbide	12	0.04	20	-	-
Singh (2008)	UD GFRP Vf 0.56	Φ 6 HSS 90° point angle	28	0.075	60	-	1.03

Vc= Cutting speed m/min, f= Feed rate mm/rev, Fz= Thrust N, Tq = Torque Nm, Df : Delamination factor.

Davim and Reis (2003 a) showed that helical flute K10 carbide drill performs better than HSS and four flute K10 carbide drills. Ramkumar et al. (2004b) concluded that the wear land of 3 flutes solid carbide drill is less when compared to tipped WC, and 2 flute solid carbide drills while drilling glass fibre reinforced plastics. This could be because of the less thrust force and temperature at the cutting edges when compared to the 2 flute solid carbide and WC tipped drills.

Several geometries of drill tools have been investigated, aiming at the reduction of thrust force and delamination (Davim and Reis 2003b, 2004; Hocheng and Tsao, 2003, 20105, 2006, 2004; Tsao 2007, 2008, Abrao et al., 2008, Krishnaraj et al., 2008). Table 1.2 presents typical tool geometries used and its critical drilling parameters on composites. Apart from standard twist drill, the most investigated geometries are the candlestick drill, saw drill, core drill and step drill. Drilling thrust of the twist drill is the highest, followed by saw drill and core drill, while the candle stick drill and step drill are the lowest (Figure 1.4). When compared to candle stick drill, step drill seems to be more promising in terms of life, where as candle stick drill seems to be a better choice in terms of delamination. Selecting right process parameters is critical for drilling of polymer matrix composites; since the parameters not only affect the life of the tool, also damage the composite product.

2.4. Effect of Speed and Feed on Drilling

Konig et al. (1989, 1990) concluded that apart from physical properties of the materials, cutting energy and time of contact affect the drill temperature; these depend on cutting conditions, workpiece, and tool material. While drilling of CFRP thermoset and thermoplastic materials by varying feed rates, continuous and curly chips were observed in thermoplastics. However, a fracture with discontinuous chips and delamination were observed at the hole exit in thermoset fibre reinforcement (Hocheng and Puw 1993, 1992).

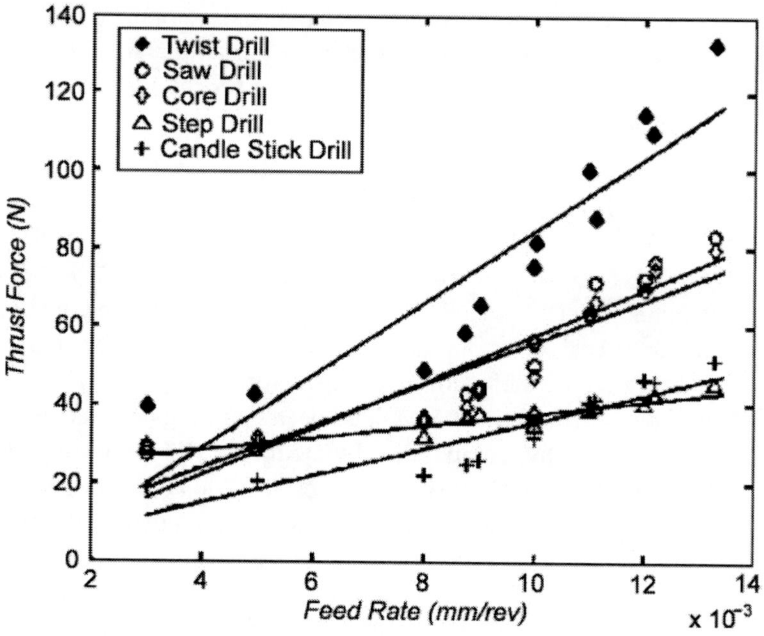

Figure 1.4. Correlation between thrust force and feed rate for special drills (Hocheng 2006).

While drilling PMC, feed rate has direct influence on thrust force where as cutting speed does not have higher influence on thrust force (Gindy, 1998; Chen, 1997; Won and Dharan, 2002b). Davim et al. (2003, 2004) conducted experiments in drilling in polymer matrix composites materials and reported that the delamination factor increases with both cutting parameters, which means that the composite damage is bigger for higher cutting speed and for higher feed. Influence of process parameters, tool diameter, laminate thickness, on cutting force and torque, and delamination during drilling of glass fibre polyester reinforced composites was reported by Mohan et al. (2005, 2007). Zitoune et al. (2005) showed that, depth of cut, direction of fibre angle and cutting speed have significant influence on the shape of chips and the damage extent.

Lin et al. (1996, 1999) concluded that drill wear is the main problem to be overcome in high speed drilling of fibre reinforced plastics. Drilling of composites at high speeds have been studied by Krishnaraj et al. (2005), Karnik et al., (2008) & Rubio et al., (2008). At high spindle speeds (40000 rpm) the increase in feed rate did not result in high damage, this could be because of the rise in temperature at high speeds.

The damage in the drilled hole and the instantaneous cutting forces has been correlated with the fibre orientation (Dipaola, 1996). Drilled hole quality is dependent on the tool wear under the same drilling conditions (Aoyame et al. 1995). Chen (1997) introduced the concept of delamination factor in his experimental investigation on CFRP material and showed that delamination free drilling is possible by proper selection of drilling parameters and tool geometry. The quality of the hole is normally represented as delamination factor. The delamination factor is achieved by diameter of the damaged zone by the diameter of the hole (Eq.1).

$$\mathrm{Fd} = \mathrm{Dmax}/\mathrm{D} \tag{1}$$

(Figure 1.5 (a)). Figure 1.5 (b) shows the delaminated composite plate. Davim et al. (2007) further developed an equation (2) for the inclusion of damage area in the adjusted delamination factor (Fda) obtained through digital image processing, which allows a more realistic assessment of drilling induced damage.

$$Fda = Fd + \frac{Ad}{(A\max - Ao)}(Fd^2 - Fd) \tag{2}$$

where Fd is the delamination factor, Ad is the damage area, Amax is the area corresponds to the damage diameter, and Ao is the nominal hole area. Zhang et al. (2001 b) have reported that delamination and fuzzing are the major mechanism at exit defects.

Quality of the holes drilled in PMC has been investigated using Tool maker's microscope, enhanced radiography, computerized tomography (CT), ultrasonic C scan, digital analysis (Davim 2007, Tsao 2005 c). Ravishankar et al. (2000a, 2000b) showed that, it is possible to evaluate the drilling induced damages in composites through AE signal characterization. A simple and cost effective technique to evaluate delamination in drilling composite laminates using digital scanning has been proposed by Khashaba (2004). Seif et al. (2007) used

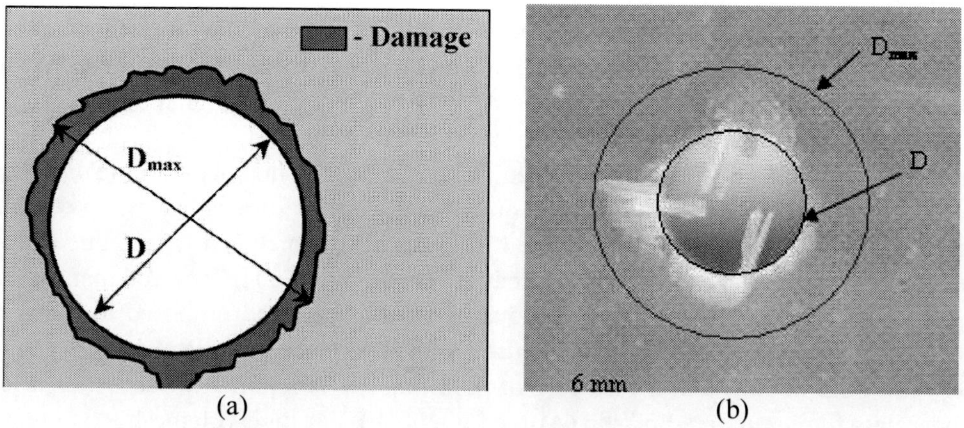

Figure 1.5. Evaluation of Delamination Factor ((a) Davim 2004 (b) Zitoune 2006).

Shadow Moiré laser based imaging technique for dark composite parts where visual inspection is difficult. Zitoune et al. (2007, 2008) have studied further delamination at entry and exit while drilling of long fibre composite, and reported that there is influence of manufacturing processes of the composite plate materials and concluded that the delamination of the laminates manufactured in oven is larger compared to the one of the drilled plates manufactured in autoclave. While drilling T700-M21 (3rd generation resin) in comparison with T2H-EH2 (2nd generation resin) presented enhanced machining conditions i.e, better surface finish and minimal defect at hole exit (Zitoune et al. 2009) because of the presence of thermoplastic nodulus.

Table 1.3. Critical drilling parameters on composites

Cutting parameters Machinability Characteristics						
Reference	**Work material**	**Tool geometry**	**Speed (m/min)**	**Feed (mm/rev)**	**Thrust N**	**DF**
Ramkumar (2004)	GFRP Vf 0.4	HSS Φ6 mm	12	0.04	30	-
Arul (2006)	GFRP Vf 0.4	Tipped WC Φ6 mm	18.5	0.02	30	1.005
Velayudham (2007)	GFRP Vf 0.63	Tripod, Carbide K10, Φ6.5 mm	80	0.1	50	1.02
Velayudham (2005)	GFRP Vf 0.66	Tipped carbide K10, Φ6.5 mm Point angle 85°	80	0.08	50	-
Mohan (2007)	GFRP Vf 0.63	Coated carbide	19	0.08	-	-
Zitoune (2008)	GFRP Vf 0.65	Carbide K20 Φ6 mm	10	0.02	-	-

Fz= Thrust N, Df : Delamination factor

Khashaba (2004) showed that by increasing the cutting speed in drilling cross-winding, woven and chopped composites reduced the push-out delamination as a result of decreasing the thrust force. The depth of the affected zone and the severity of the damage decreases with an increase in cutting speed (Sadat, 1988). At lower feed rates, delamination occurred at the sub-laminate, whereas at higher feed rates it occurred at the early stage of the drilling. The relationship between spindle speed and feed rate are important in terms tool life and quality of hole. The effect of spindle speed on hole quality and cutting force is less when compared to feed rate. Increasing cutting speed resulted in lower thrust force and torque due to the high temperatures produced which softens the matrix. Ramkumar (2004a) showed that maintaining critical thrust below 70 N improves the quality of the hole, and proposed to use a cutting speed of 12 m/min, feed rate of 0.04 mm/rev with workpiece vibration. A spindle speed of 1000 rpm (18.5 m/min) and a feed rate of 0.02 mm/rev is reported to be better in case of drilling glass fibre reinforced plastic (Arul et al. 2006b). For high volume fraction glass fibre reinforced plastics, Velayudham (2005a, 2007) suggested to use a cutting speed of 80 m/min, feed rate of 0.1 mm/rev and maintain thrust force during drilling below 100N. Spindle speed of 490 rpm (10 m/min), and feed rate of 0.02 mm/rev is reported to be the critical process parameters for glass fibre reinforced plastics (Zitoune 2008). Table 1.3 presents preferred cutting speeds and feeds for various combinations.

The variations in the preferred process parameters (Table 1.3) are based on manufacturing process, volume fraction of the composite, tool material and geometry. In general high spindle speed and low feed rate is preferable for drilling polymer matrix composites (Enemuoh 2001). But at very high speeds, tool wear is a major problem; still there is a gap in the literature in the fields of optimum parameters at high spindle speed, drill life and economy of using high spindle speed. It is presumed that there is a critical thrust force beyond which delamination is initiated. The following chapter discusses the mathematical models derived by various researchers to predict critical thrust force for various drills. A brief review of drilling of composite is presented by abrao et al (2007).

2.5. Analytical and Numerical Models to Predict Critical Thrust Force

Mathematical models are helpful in deciding the process parameters and could be used as an online process control tool while drilling. Everstine et al. (1971) derived the forces required for continuous machining by considering both elastic and plastic stress responses of polymer matrix composites. Hocheng and Dharan (1990) developed the first analytical model (Eq.3) by considering the lamina of composite as an isotropic material, to determine the critical thrust force (F_Z in N) during the drilling of composite laminates based on linear elastic fracture mechanics (LEFM).

$$F_A = \pi \left[\frac{8 G_{IC} E\, h^3}{3(1-\nu^2)} \right]^{1/2} \tag{3}$$

where, G_{IC} is the critical energy release rate in J/m^2, E is the modulus of elasticity in Pa, h is the thickness of the uncut laminate in m, and ν is the Poisson's ratio. Jain and Yang (1994) developed an anisotropic model for unidirectional lamina and found to have a satisfactory agreement with experimental results (Caprino and Tagliaferri, 1995). Sadat (1996) developed first orthotropic model to predict the load that causes delamination at a given ply locations. Several mathematical models inspired by Hocheng's equation have been developed to predict drilling force of composites (Chandrasekarn, 1995; Stone, 1996, Tsao and Chen, 1997; Lachaud et al. 2001; Langella, 2005; Kim, 2005; Hocheng 2005, Tsao, 2005). Dharan and Won (2000) proposed an intelligent machining scheme to drill holes below critical thrust force to avoid delamination using the Eqn. 3.

The 2 dimensional numerical model reported by Sadat et al., (1992) to predict critical thrust force, used isotropic material properties and distributed load on the last plies. Later Bhattacharyya et al., (1998), Singh et al., (2008) used 3 dimensional isoparametric brick elements by distributing loads on the last plies and analysed stress levels at the intersection between the uncut plies and the wall of the drilled hole. Durào et al., (2006, 2008) used 3 dimensional models to study the delamination. Zitoune et al., (2007) used a much realistic 3 dimensional model by distributing the load predicted by experiments on chisel edge, and the main cutting edges so as simulate the real cutting effect and also by incorporating shear force effect in the model. These finite element models are helpful in understanding the stress levels at various locations (especially at the interface between drill wall and the bottom plies) as well as predicting the limiting thrust force so as reduce the defects, but still there is no FE model developed to match with the realistic situations to analyse effect of cutting parameters.

The analytical and numerical models developed for various tool geometries and cutting conditions could be effectively used to select spindle speed and feed rate and monitor quality of hole.

2.6. Drilled Hole Quality on Residual Mechanical Properties

The quality of the holes is important since it has direct effect on its mechanical strengths. The most commonly used method for the evaluation of the residual mechanical properties is the tensile test, the static, cyclic bearing load tests and the fatigue test. Drilling on GFRP laminates was performed and reported that the damage-extent strictly depends upon the spindle speed to feed ratio (Tagliaferri et al. 1990; Capello and Tagliaferri, 2001). The lower the speed-feed ratio the poorer will be the quality. However, an optimum speed-feed ratio seems to exist resulting in maximum bearing strength. Mohamed et al. (1993) evaluated 3D orthogonal woven carbon/epoxy composites with holes for joint locations. Based on the testing results, they concluded that 3D woven composites are not sensitive to machining because the structure is well bound by Z yarns. Persson et al. (1997) compared the performance of KTH (Kungl Tekniska Hogskolan) method, PCD (Poly Crystalline Diamond) and dagger type drill tools. Holes drilled by KTH method yielded the highest strength and fatigue life with fine hole quality compared to the holes done by PCD and dagger tools. Kishore et al. (2009) reported that the drill point geometry, spindle speed and feed have effect on residual mechanical strength of the laminate. Mariatti et al. (2004) and Rao et al. (2008) have shown that tensile properties of composite decrease with increasing hole sizes. Influence of the drilling quality of a composite material was tested using cyclic compression (Hamdoun et al., 2006) and concluded that the drilled hole quality has less influence on compressive strength. Krishnaraj et al. (2007) correlated the effect drilling parameters on mechanical strength of GFRP, and suggested to use a spindle speed of 3000 rpm (56 m/min), 0.02 mm/rev feed rate in order to drill holes so that maximum strength could be achieved.

2.7. Results and Discussion

Since high degrees of abrasive wear are experienced by the tools, wear-resistant, harder tools are required while drilling PMC. For drilling PMC, carbide drill of grade K10/K20 or PCD is found suitable. Several geometries of drill tools have been investigated, aiming at the reduction of thrust force and delamination. Drilling thrust of the twist drill is the highest, followed by saw drill and core drill, while the candle stick drill and step drill are the lowest. The effect of spindle speed on hole quality and cutting force is less when compared to feed rate. Increasing cutting speed resulted in lower thrust force and torque due to the high temperatures produced which softens the matrix. In general high spindle speed and low feed rate are preferable for drilling polymer matrix composites. But at very high speeds, tool wear seems to be a major problem; still there is a gap in the literature in the fields of optimum parameters at high spindle speed, drill life and economy of using high spindle speed. During vibration assisted drilling of composites, the thrust force and torque can be reduced by 20 to 30% when compared to conventional drilling. The holes in millions are drilled on composites in order to assemble with mating parts in aircraft industry. From the industrial point of view large number holes are to be drilled with out defects in the composites at high speed and with low cost. Use of titanium in the aircraft structures is increasing, and in the following chapter the critical aspects of drilling of titanium alloy is discussed.

3. Drilling of Titanium Alloy

3.1. Introduction

The most commonly used titanium alloy in the aerospace industries is Ti–6Al-4V. Like stainless steel, the low thermal conductivity of titanium inhibits dissipation of heat within the workplace itself, thus requiring proper tools and application of coolants.

3.2. Effect of Tool Material

Use of sharp tools is vital, because dull tools will accentuate heat build-up to cause undue galling and seizing, leading to premature tool failure. Dry drilling of alloy Ti–6Al–4V was carried out by Cantero et al. (2005) and studied tool wear evolution, quality of machined holes and surface integrity after machining. Cutting time up to drill failure is specially increased when holes were machined with a pause between each hole to cool the drill. SEM–EDS analysis showed progressive loss of TiN coating of drill and workpiece material adhesion in rake surface. Likewise diffusion of Ti alloy in rake surface and drill helical flute was observed by them in both conditions tested. In some holes combustion of the chip was observed, the frequency of this phenomenon increased as the cutting time increased (Ezugwu, 1997). Sharif et al. (2007) have reported that non uniform flank wear, chipping and catastrophic failure were observed when drilling using TiAlN coated and un-coated drills. Tool wear was observed to be very high at high cutting speeds where as, at lower speeds surface roughness was found to be higher. Uncoated drills were found to be inferior when compared to coated drills.

Straight grade cemented carbides (K20) are regarded as the most suitable tool material available commercially for the machining of titanium alloys as a continuous operation, as a basic rule, cutting fluid must be applied when machining titanium alloys. The correct use of coolants during machining operations greatly extends the life of the cutting tool. The super hard cutting tool materials (CBN and PCD) have also shown a good performance in terms of wear rate in the machining of titanium (Ezugwu, 1997).

3.3. Effect of Tool Geometry

In general drills for machining titanium based alloys require a very high lip relief angle (10°) to provide a sharp cutting edge and to minimize the contact length between the tool and the workpiece. The point angle is usually in the range of 130°-135° and reported that higher point angle and helical point style reduce the burr height and thickness (Dornfeld, 1999). Li et al. (2007a, b) further showed that WC-Co spiral point drill has lower negative rake angle which can generate low thrust force and further has advantages of low thrust force, torque, energy, and burr size. Figure 2 shows drills used by Li et al. (2007 b).

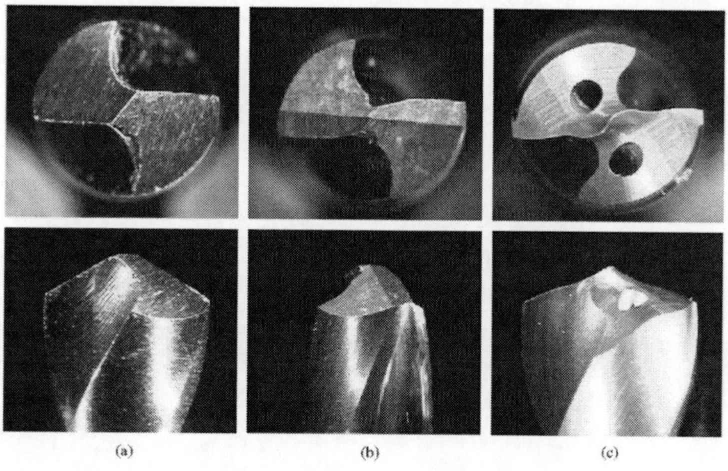

Figure 2 (a). HSS Twist, (b) WC–Co Twist, and (c) WC–Co Spiral (Li 2007).

3.4. Effect of Speed and Feed on Drilling

Dornfeld et al. (1999) studied the burr formation in titanium alloy, and reported that roll back burr due to thermal effects has been observed in dry cutting with relatively high feed rates and high cutting speed and ring formation burr in wet cutting. Lower burr height was found at high feed rate. The feasibility of high-throughput drilling of Ti–6Al–4V and the importance of feed and internal cutting fluid supply to improve the drill life has been reported by Li et al. (2007a). They could able to achieve material removal rate (MRR) of 156 mm^3/s with peripheral cutting speed up to 183 m/min using a commercially available 4 mm diameter

WC–Co spiral point drill. In comparison, a conventional HSS twist drill could only work below 27.4 m/min cutting speed and achieve 23.3 mm^3/s MRR. Sharif (2007) showed that during drilling with TiAlN coated drill flank wear was observed to increase gradually when cutting speeds of 25 m/min – 35 m/min were employed. The highest tool life was obtained with TiAlN coated drills when drilling at a cutting speed of 25 m/min.

Table 2. Overview of drilling parameters and the machinability characteristics of Titanium

Cutting parameters Machinability Characteristics										
Reference	**Work material**	**Tool geometry**	**Vc m/min**	**Feed mm/rev**	**Th N**	**TqNm**	**Sf µm**	**T°C**	**Hd mm**	**Burr height mm**
Cantero (2005)	Ti-6Al-4V	Carbide K40 Φ 6 Split point TiN coated	50	0.07	–	–	0.7	–	0.045	0.1
Dornfeld (1999)	Ti-6Al-4V	Carbide 6.35 118° point angle and 2 flute 25° helix angle With coolant	–	–	–	–	–	–	–	0.12-0.16
		Carbide 6.35 150° point angle and 3 flute 25° helix angle With coolant	–	–	–	–	–	–	–	0.085-0.1
Sharif (2007)	Ti-6Al-4V	Carbide Φ6 (TiAlN) Point angle 125° Helix angle 25° With coolant	25	0.06	–	–	0.095	–	–	–
Li (2007)	Ti-6Al-4V	Carbide Φ9.92 Spiral point With coolant	61	0.152	–	–	–	–	–	–
		Carbide Φ4 Spiral point Internal coolant	91	0.10	275	1.3	0.6	–	–	–
		Carbide Φ4 Spiral point Point angle 130° 30° helix angle Internal coolant	61	0.152	290	1.4	0.8	–	–	–
Zeilmann (2006)	Ti6Al-4V	Carbide Φ8.5 K10, Through coolant	30-40	0.1	–	–	–	160	–	–
	Ti-6Al-4V	Carbide Φ8.5 K10, MQL	30-40	0.1	–	–	–	300	–	–

Vc = Cutting speed (m/min), Fz = Thrust force N, Tq = Torque (Nm), Sf= Surface finish Ra (µm), T = Temperature °C, Hd= Hole deviation(mm).

Li et al. (2007b) further, investigated the temperature reached during drilling of titanium alloy, with K type thermocouple, and reported that 50% reduction in temperature is possible with application of minimum quantity lubrication (MQL) internally through the tool. The results of Li (2007a) suggest to going for lower peripheral speed and higher feed, to get less drill temperature.

3.5. Results and Discussion

Use of straight grade cemented carbides (K20) with coating are regarded as the suitable tool material for the drilling of titanium alloys. The correct use of coolants during machining operations greatly extends the life of the cutting tool. In general drills for machining titanium based alloys require a very high lip relief angle (10°) to provide a sharp cutting edge and to minimize the contact length between the tool and the workpiece. The point angle can be in the range of 130°-135° and reported that higher point angle and helical point style reduce the burr height and thickness. Highest tool life was obtained with TiAlN coated drills when drilling at a cutting speed of 25 m/min. While drilling titanium, slower speeds and heavier feeds are required. According to our survey the literature available for drilling of Ti alloy seems to be less. Use of aluminium in aerospace is high and drilling aluminium has its own difficulties. In the following chapter the problems related to drilling of aluminium are discussed.

4. DRILLING OF ALUMINIUM ALLOY

4.1. Introduction

Applications that involve drilling large number of holes encompass industries such as the aerospace, automotive, and potentially many other industries. Although the speed at which aluminium gets drilled has been climbing through the years, there appears to be yet a large margin to grow. Drilling at many times the conventional rates undoubtedly translates into faster productivity, the savings of which typically would dwarf the cost associated with excessive tool wear (Hamade, 2005).

The most important wear mechanisms in machining of aluminium alloy are: built-up edge (BUE), adherent layer and diffusion. At low cutting conditions, the built-up edges are formed on the tool rake face and take over the function of the cutting edge. At severe cutting conditions, temperature is higher; mechanisms of tool wear involve chemical action and diffusion. On the one hand, aluminium elements diffuse into the tool through the cobalt binder phase.

4.2. Effect of Tool Materials

Aluminium is commonly machined with high speed steel, diamond and carbide tooling; Silicon nitride based ceramic tools are generally not used with aluminium because of high solubility of silicon in aluminium (Kelly, 2002). Nouari et al. (2005) show that HSS tool is not suitable for dry machining of AA2024 aluminium alloy. Although high speed steel drills remain the main choice today for drilling aluminium at conventional speeds, carbide drills constitute the workhorse at medium/high speeds (List, 2005). These carbide drill varieties are commonly treated with even more durable coatings and are able to operate at previously impossible high cutting speeds (even under dry or minimal coolant conditions). Thin, hard, single or multi-layer coatings translate into extending the usefulness of the drill especially at high speed drilling. Grzesik (2000), Rivero (2006), and Nouari (2005) demonstrated the effect of multilayer coatings on lowering the coefficient of sliding friction (and consequently the friction heat flux) during machining. In general, the coating on the flutes of the drill is more important than at the cutting edge, that it both remains adherent to the drill and allows smooth chip evacuation (non-adhesion of aluminium), even as the chips become hot. Dasch (2006), Wain (2005) studied (Carbon based, Graphite and MoS2 based) tool coatings and reported that there is an enormous potential for carbon and graphite based coatings for dry machining. PCD tools have higher lives of carbides while machining aluminium-silicon alloy compositions but of course at a high cost to match.

4.3. Effect of Tool Geometry

Standard twist drills are generally quite satisfactory for the drilling of aluminium alloys but for deep holes, high spiral drills with a helix angle of 40°- 48° instead of the normal 24°- 28° are recommended. Higher helix angle increases the rake angle. The point angle may be increased with advantage from the normal *Figure* of 118° to 130°-140°. This produces narrow chip, which is more easily expelled through the flutes of the drill (Kelly, 2002). Rivero (2006) reported that less torque, temperature, and burr height were observed at 200 m/min, 0.3 mm/rev while drilling with TiAlN coated 3 flute carbide drills. Bhowmick et al. (2008) have showed that while drilling diamond like carbon coated HSS, with minimum quantity lubrication at 50 m/min, 0.25 mm/rev feed rate gave lower thrust force and torque.

4.4. Effect of Speed and Feed on Drilling

High feed rate drilling of aluminium alloy A1050, A2017 and A6061 has been attempted with TiN coated SKH56 drills under cutting conditions of 1500 rpm and 1.0 mm/rev (Sakurai et al. 2000).

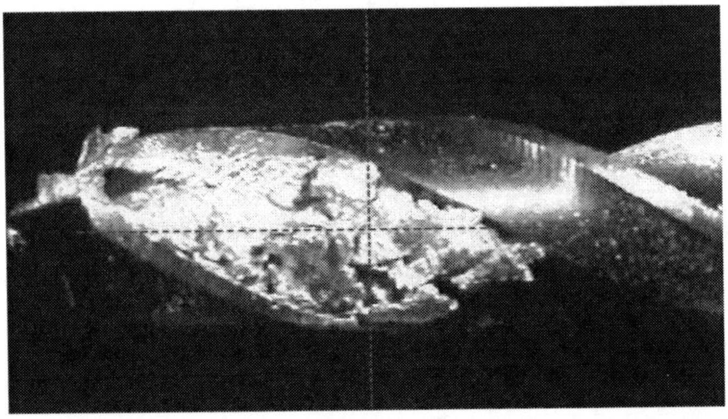

Figure 3. Clogging of the drill flutes with hot aluminium chips (Dasch 2006).

It is found that the high feed drilling of aluminium alloy A1050 is difficult practically, while high feed drilling in A2017 and A6061 alloys is very feasible. In fact, aluminium alloy A1050 (almost pure aluminium at composition of Al 99.50%) is conventionally classified as a soft, gummy alloy with poor machinability(worst rating, E), while A2017-T6 and A6061-T6 are both wrought precipitation-hardened aluminium alloys that are assigned a rating of B (second best) and C (third best), respectively.

According to Metals Handbook (ASM Intl), many of the chipping and surface finish difficulties encountered in machining aluminium alloys can be eliminated by increasing cutting speed. For some applications, speeds of 35–50 m/s are used to obtain smooth surfaces. Rivero et al. (2006) reports that the smallest burrs occur when feed increases. As feed force increases, the last layer of the remaining material at the exit hole becomes thicker, thus reducing burr size. Even now in industries, especially dealing with drilling of aluminium alloy, controlling burr is a major task. In general, flood lubrication is superior for drilling aluminium alloy at lower cutting speeds and feed rates, whilst mist application is superior for higher cutting speeds and feed rates (Kelly, 2002).

Table 3. Overview of drilling parameters and the machinability characteristics of aluminium

Cutting parameters Machinability Characteristics									
Ref-erence	**Work material**	**Tool geometry**	**Vc m/min**	**Feed mm/rev**	**Th N**	**Tq Nm**	**SF µm**	**T°C**	**Hd (mm)**
Nouari (2005)	AA2024	Carbide Φ 6 Point angle 130° (Dry drilling)	165 65 25	0.04	–	–	1- 3 1-3.5 1-2	180 125 90	0.09 0.12 0.18 (after 50 holes)
Kelly (2002)	ACP 5080	HSS Φ 10 (flood coolant)	105 25	0.15 0.30	450 690	2.4 2.8	5 2.9	90	0.03 0.02
		HSS Φ 10 (Mist coolant)	105 25	0.15 0.30	400 800	2.2 3.1	4 5	–	0.02 0.015
Rivero (2006)	Al. 7075-T6	Carbide Φ 10 Helix angle 30° Point angle 130° TiAlN coated dry drilling	250 200 150	0.3 0.3 0.3	–	11 9.5 10	–	- 246 -	–
Bhowmik (2008)	Al-6%Si	HSS Φ 6.35 DLC coated (MQL 30ml/h)	50	0.25	150	1.6	–	–	–
Hamade (2005)	A390	Carbide Φ 6.35 Helix angle 33° Point angle 119°	800 600	0.12	45 50	–	–	–	–
Kurt (2009)	Al 2024	Standard HSS drill	30	0.15			3.37		

Vc = Cutting speed (m/min), Tq = Torque (Nm), SF= Surface finish Ra (µm), T=Temperature Deg C, Hd= Hole deviation(mm).

Higher cutting speeds generate higher chip sliding velocities and consequently generate a significant temperature on the rake face of the tool (Nouari, 2005). The HSS drill is used at a lower speed (61 m/min) and the temperatures observed are much lower, whereas the higher speeds of the carbide drill (213 m/min) result in higher temperatures.

Failure then, was not due to wear at the cutting edge as with most machining operations, but rather to adhesion and build-up of material in the drill flutes so that chips could not be evacuated, resulting in catastrophic failure (Figure 3). The effect of chip size on hole surface finish is examined by Batzer et al. (1998). The results show that the significant variables affecting chip size include the feed, material, drill type, and to a lesser extent cutting fluid presence and drill speed. Bhowmick et al. (2008) showed that drilling Al-Si alloy using H_2O-MQL spray required lower thrust and torque compared to dry drilling. Similar conclusion was made by Braga (2002). Hamade (2005) showed that at high spindle speeds thrust force of 40–50N is observed. Drilling aluminium at lower cutting speeds, BUE is a major problem where as if the cutting speed is increased; there is a transition in wear mechanisms from abrasion and/or adhesion to diffusion. Controlling burr and temperature are other important aspects in drilling of aluminium. Coated carbides and MQL are preferable at high speeds in order to avoid diffusion. Table 3 presents some of the critical drilling parameters used for drilling aluminium.

4.5. Results and Discussion

Standard twist drills are generally quite satisfactory for drilling of aluminium alloys but for deep holes, high spiral drills and high point angle are preferred. In general, flood lubrication is superior for drilling aluminium alloy at lower cutting speeds and feed rates, whilst mist lubrication is superior at higher cutting speeds and feed rates.

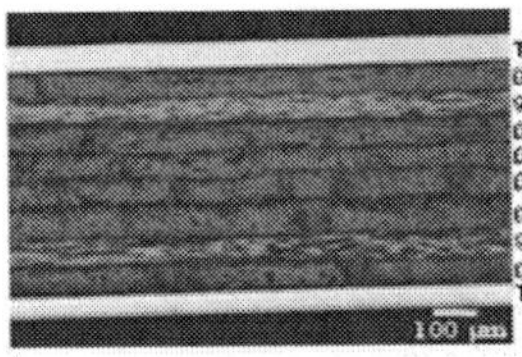

(a) Ti/Gr/Ti Stack

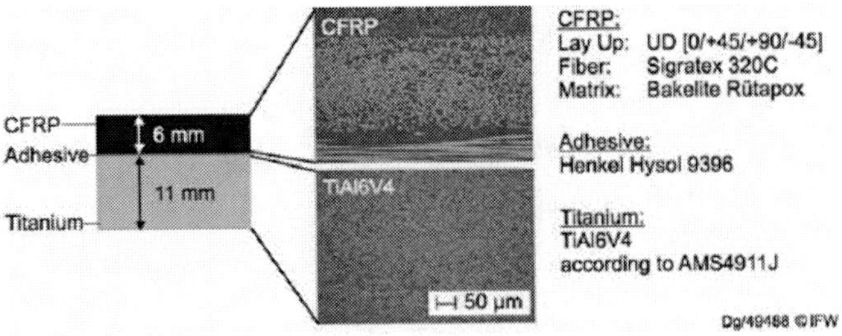

(b) CFRP/TI Stack

Figure 4. Various material combinations ((a) Ramulu 2001 (b) Denkena 2008.

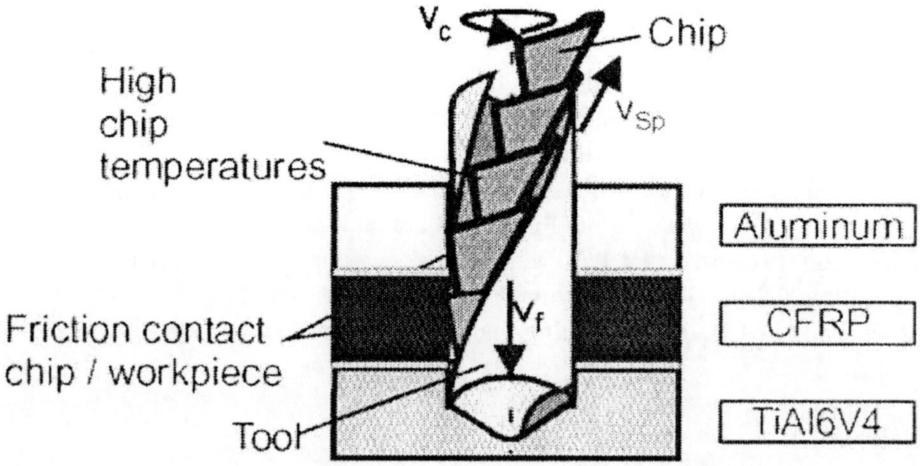

Figure 5. Chip removal problems when drilling multimaterial (Brinksmeier 2002).

Tungsten carbide with TiAlN coating is preferred to drill aluminium which eliminates the increase of torque at the end of hole, reduces burr size, power consumption and increases tool life. PCD tools have higher lives of carbides while machining aluminium-silicon alloy compositions but at a high cost to match. Failure, of drills during drilling of aluminium, was not due to wear at the cutting edge as with most machining operations, but rather to adhesion and build-up of material in the drill flutes so that chips could not be evacuated, resulting in catastrophic failure. At most care is required to drill aluminium successfully. For aerospace applications, drilling of dissimilar material like carbon composite and titanium stack or carbon composites and Aluminium stack or carbon, titanium and aluminium stack are required. The following chapter discusses the difficulties of drilling multimaterials.

5. Drilling of Multimaterials

5.1. Requirements of Cutting Tool

Drilling of multimaterials is a challenging task to manufacturing engineers because of differential machining properties. Different combinations of material stack up are shown in Figure 4. Due to different material properties, holes with small diameter tolerances are difficult to drill. The modulus of elasticity (E) of the materials causes different elastic deformations and therefore varying tolerances along the entire hole. Additionally, the chips transport through the hole as well as built up edges of aluminium at the primary cutting edges combined with increased tool wear affect the hole quality (Brinksmeier, 2002). Figure 5 shows the chip removal problems when drilling multimaterial stack.

For machining multi-materials, sharp and high hot hardness tool materials are required. The tool materials should not react with the multi-material stack. High speed steel and carbide were used to drill Gr/Bi Titanium stack by Kim and Ramulu (2004). They found that tool wear occurred rapidly when drilling Gr/Bi-Ti stacks. Crater wear, flank wear and more extreme wear on the helical cutting edges were observed during drilling. Carbide drills performed much better when drilling Gr/Bi–Ti stacks because of its hot hardness (Figure 6).

Multi-material theoretically calls for different tools, one that fits the attributes of a composite and a different one that fits the attributes of titanium, and etc. Composite are extremely abrasive, making PCD drills the preferred choice. However PCD does not last long when drilling titanium, which creates a great deal of heat. Machinists generally end up selecting carbide drills, as wearing out PCD drills in titanium is expensive.

Garrick (2007) reported that PCD veined drill developed at Precorp Inc–USA, may be applied to drilling CFRP/Ti stacks (Figure 7), which increases tool life and hole quality. Since temperature is a large factor in machining titanium, best results could be achieved when using liquid coolant.

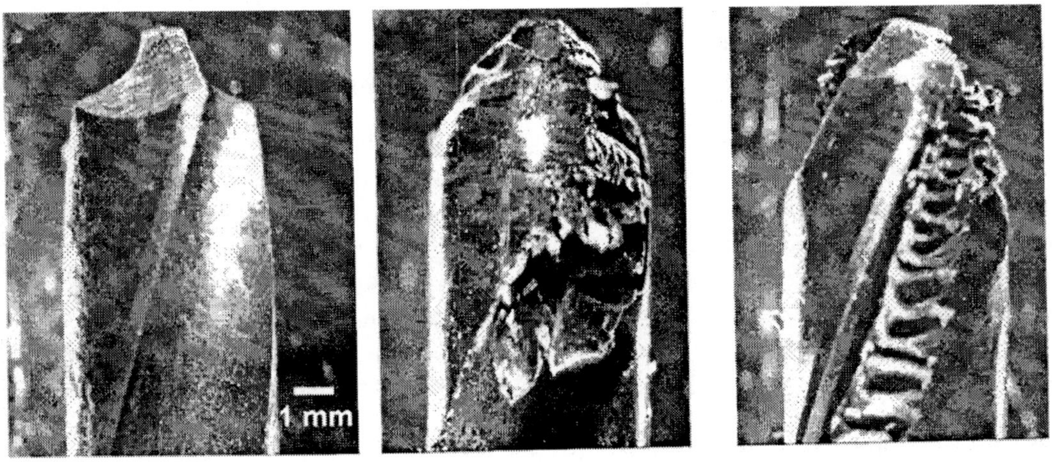

(a) Carbide (b) HSS (c) HSS-Co

Figure 6. Micrographs showing wear of drills (Ramulu 2001).

5.2. Quality of Machined Holes

Ramulu et al. (2001) examined the challenges involved in drilling Ti/Gr made of thermoplastic polymer-matrix composite (PMC) plies with titanium foils as the outer plies. They reported that the most significant issue arising when drilling Gr/Bi-Ti stacks is the low thermal conductivity of Ti alloys, resulting in heat generation around the region and increased tool temperatures. High temperature which is produced by Ti cutting was found to increase the radial distance of damage at the interface and tool wear, which leads to increased cutting forces, burr heights, heat induced delamination, and surface roughness. The combination of high speed and low feed is detrimental to the exit Ti burr height which is due to Ti's low thermal conductivity. They further recommended using carbide drills instead of HSS at a feed rate of 0.08 mm /rev. and spindling speed of 660 rpm (13 m/min) for drilling Ti/Gr stacks. Multi-layer materials drilled with adapted step drills improved diameter tolerances, surface quality, and reduced tool wear (Brinksmeier et al, 2002). Improvements are found as for as tool life is concerned when a coating of TiB2 is used with minimum quantity lubricant.

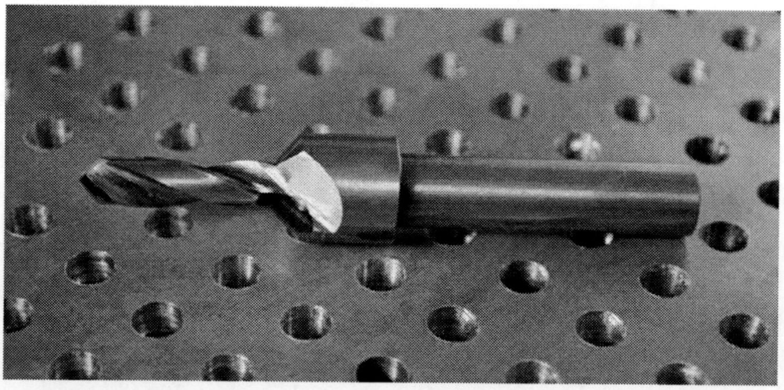

Figure 7. PCD Combination Drill (Precorp Inc, Utah, USA).

Kim and Ramulu (2004) have stated that the optimum process conditions for achieving desired hole quality and process cost are found to be a combination of low speed and low feed when using carbide drills and high feed and low speed in drilling with HSS-Co drills. Kim further (2005) reported that while drilling at low feed rates, the PMC was sheared continuously and smeared into the gaps among the fibres. At higher feeds, the cutting mechanism of PMC was characterised by compression induced fracture perpendicular to the fibres and interlaminar shear fracture along the fibre/matrix interface due to bending. It was also shown by them that at higher cutting speeds increased roundness and at increased feed rates reduced roundness. High spindle speed and low feed rate were detrimental to the exit Ti burr height. Drilling with low feed rate produced a smoother surface and shallow fibre pull out due to reduced cutting forces. Kim and Ramulu (2007) stated that consolidation processing, such as induction heating press and autoclave processes have effect on drilling characteristics on TiGr composites. Roudge et al, (2008) have developed a quantification method particularly for delamination on drilling of multi-layer materials. By helical milling of CFRP-Ti stacks, the quality of the holes was found to be higher (Denkena, 2008).

5.3. Results and Discussion

Drilling of multimaterial stacks has problems like chip disposal, change in dynamic cutting forces, tool temperature and tool wear. Titanium chips are long and continuous at low speeds, and became shorter and stiffer as feeds increased, whereas polymer composite chips are continuous at low feeds and became dust-like chips as feed increased. Aluminium chips will be continuous when the spindle speed is high and feed is less. At low spindle speeds BUE deteriorates drilling performance. When the aluminium or Ti is stacked at the bottom of CFRP, the continuous chips passing through the CFRP deteriorate the quality of the hole. If the drill passes through the Ti or Aluminium, increase in cutting temperature or burr will be a source of damage to the CFRP.

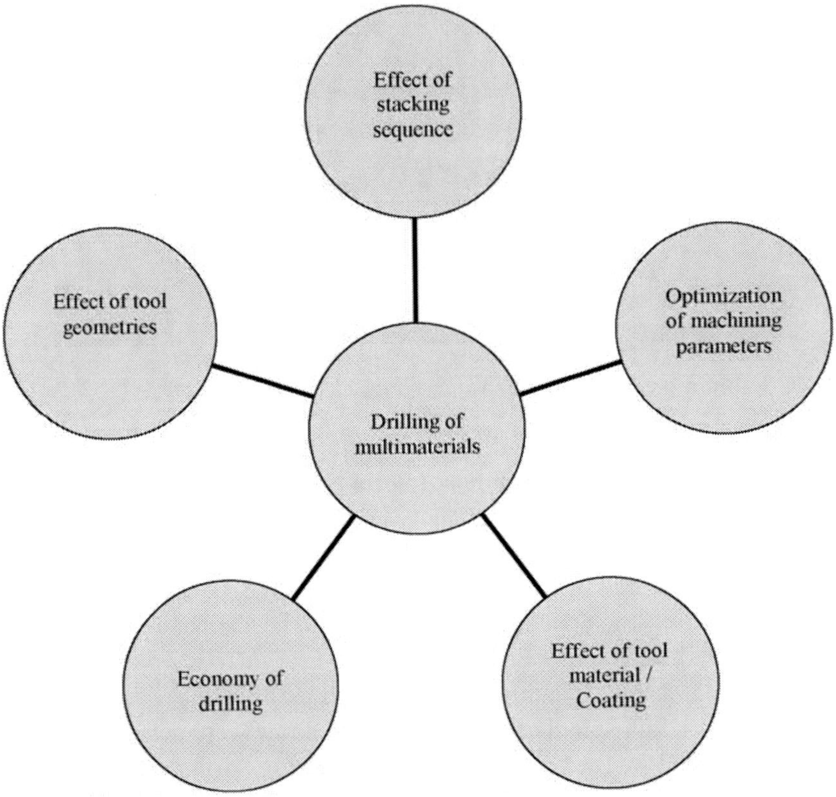

Figure 8. Proposed activities to study drilling of multimaterials.

The drill passing through titanium worn out at a faster rate may not cut the fibres properly. PCD is preferable for composite and aluminium whereas for titanium machining titanium is preferable. The choice of cutting tool should be a compromise, to withstand high temperature involved during titanium drilling, as well as drilling highly abrasive carbon fibre. Each material requires different cutting parameters like spindle speed, feed, and drill geometry. Composites require lower point angle whereas for aluminium and titanium higher point angle is preferred. Use of coolant is important while drilling titanium and aluminium where for composite use of coolant has adverse effects. Worthwhile research has been attempted in drilling of Gr.composite/Ti metal stacks. Major concerns with these studies include severe tool wear, heat induced damage, hole size, roundness, shape, surface texture and presence of metallic burrs. Delamination and exit hole burrs can be classified as exit hole damage which is very critical in structural durability of the multi-material structures. High speed and low feed rate is detrimental to the exit Ti burr height which is due to Ti's low thermal conductivity. For drilling CFRP composite high speed and low feed rate is preferable. Drilling aluminium requires a balance between spindle speed to feed rate and different tool geometry when compared to CFRP. Industries are looking for improvements to drill multimaterials successfully including orbital drilling. However open literature on study of drilling hybrid composites are very limited.

Figure 8 shows the directions in which the problem of multimaterial drilling may be studied. Lot of work have to be done in terms of optimising drilling process parameters for different stacking sequences (CFRP/AL, Ti/CFRP, Ti/CFRP/Al), tool geometry, improving tool life and hole quality.

6. CONCLUSIONS

Aircraft orders are expected to grow gradually over the next few years, but machining applications for composite parts may double by 2010 (Mike, 2007). For drilling PMC, carbide drill of grade K10/K20 or PCD is preferred. To drill with lower thrust force candle stick drill or step drill can be used. Spindle speed range of 10-20 m/min and feed rate in the range of 0.01-0.05 mm/rev may be used. Reducing feed rate during exit greatly reduces the delamination. During vibration assisted drilling of composites, it is shown that the thrust force and torque are reduced by 20 to 30% when compared with conventional drilling (Zhang et al., 2001a; Wang et al., 2004; Velayudham et al., 2005a, b). Drilling GFRP is feasible at 200 Hz frequency and 15 μm amplitude of vibration for minimum thrust (Ramkumar 2004a; Arul 2006a).

Use of straight grade cemented carbides (K20) with coating are regarded as the suitable tool material for the drilling of titanium alloys. The correct use of coolants during machining operations greatly extends the life of the cutting tool. In general drills for machining titanium based alloys require a very high lip relief angle (10°) to provide a sharp cutting edge and to minimize the contact length between the tool and the workpiece. The point angle is usually in the range of 130°-135° and reported that higher point angle and helical point style reduce the burr height and thickness. Highest tool life was obtained with TiAlN coated drills when drilling at a cutting speed of 25m/min. While drilling titanium, slower speeds and heavier feeds are required.

For drilling aluminium alloys, high spiral drills and high point angle are preferred. In general, flood lubrication is superior for drilling aluminium alloy at lower cutting speeds and feed rates, whilst mist lubrication is superior for higher cutting speeds and feed rates. Tungsten carbide with TiAlN coating is preferred to drill aluminium which eliminates the increase of torque at the end of hole, reduces burr size, power consumption and increases tool life. PCD tools have higher lives of carbides while machining aluminium-silicon alloy compositions but of course at a high cost to match.

For machining multimaterials, sharp and high hardness tool materials are required. Carbide drills are preferred while drilling Gr/Bi-Ti stacks. Carbide drills at a feed rate of 0.08 mm/rev, cutting speed of 13 m/min may be applied. PCD veined drills can be applied to drilling multimaterial stacks. Improvements are found as for as tool life is concerned with coatings (TiB2) and minimum quantity lubrication.

REFERENCES

Abrate, S. (1997) *Machining of composites,* In:Mallick P.K (ed.), Composites engineering hand book, (pp.777-807) New York: Marcel Deckker Inc.

Abrao, A. M. Faria, P. E. Campos Rubio, J. C. Reis, P and Davim, J.P. (2007) 'Drilling of fiber reinforced plastics: A review', *Jl. of Materials Processing Technology,* Vol.186,

No.1-3, pp.1-7.

Abrao, A. M. Campos Rubio, J. C. Faria, P. E. Davim, J. P. (2008) 'The effect of cutting tool', *Materials & Design*, Vol.29, No.2, pp.508-513.

Aoyame, E. Nobe, H. Hirogaki, T. (2001) 'Drilled hole damage of small diameter drilling in printed wiring board', *J. Mater Process Tech.* Vol.118,No.1-3, pp.436-441.

Arshinov, V. and Alekseev, G. (1976) *Metal cutting theory and cutting tool design,* Mir Publishers, Moscow.

Arul, S. Vijayaraghavan, L. Malhotra, S. K. Krishnamurthy, R. (2006a) 'The effect of vibratory drilling on hole quality in polymeric composites', *Int. J. of Machine Tools & Manuf.* Vol.46, pp. 252–259.

Arul, S. Vijayaraghavan, L. Malhotra, S. K. Krishnamurthy, R. (2006b) 'Influence of tool material on dynamics of drilling of GFRP composites', *Int. Jl. Adv Manuf Tech.* Vol.29, pp.655-662.

Batzer, S. A. Haan, D. M. Rao, P. D. Olson, W.W. Sutherland, J.W. (1998) 'Chip morphology and hole surface texture in the drilling of cast *Aluminum alloys', J. of Materials Processing Tech.* Vol.79, pp.72–78.

Bhatnagar, N. Ramakrishnan, N. Naik, N. K. Komanduri, R. (1995) On the machining of fiber reinforced plastic composite laminates, I*nt. J. Mach. Tools Manufact.* Vol.35, No.5, pp.701-716.

Bhattacharyya, D. and Horrigan, P. W. (1998) 'A Study of hole drilling on Kevlar composites', *Comp. Sc. and Tech.* Vol.58, pp.267-283.

Bhowmick, S. Alpas, A. T. (2008) 'Minimum quantity lubrication drilling of aluminium–silicon alloys in water using diamond-like carbon coated drills', *Int. J. of Machine Tools & Manuf.* Vol.48, pp.1429– 1443.

Braga, D. U. Diniz, A. E., Miranda, G. W. A. and Coppini, N. L. (2002) 'Using a minimum quantity of lubricant (MQL) and a diamond coated tool in the drilling of aluminum–silicon alloys', *Jl. of Mat. Proc. Tech.* Vol.122, pp.127–138.

Brinksmeier, E. and Janssen, R. (2002) 'Drilling of multi-layer composite materials consisting of carbon fiber reinforced plastics (CFRP), Titanium and aluminum alloys', *Annals of the CIRP.* Vol.51, No.1, pp.87-90.

Cantero, J. L. Tardio, M. M. Canteli, J. A. Marcos, M and Miguelez, M. H. (2005) 'Dry drilling of alloy Ti-6Al-4V', *Int. J. of Machine Tools & Mafr.* Vol.45, pp.1246-1255.

Capello, E. and Tagliaferri, V. (2001) 'Drilling damage of GFRP and residual mechanical behaviour-Part II: Static and cyclic bearing loads', *J. Comp. Tech. and Res.* pp.131-137.

Caprino, G. and Tagliaferri, V. (1995) 'Damage development in drilling GFRPs', *Int. Jl. Machine Tool and Mfr.*, Vol.35, No.6, pp.817-829.

Chamberlain, B. Machinability of Aluminum Alloys, *Metals Handbook,* Vol. 2, ASM International, Metals Park, OH 44073, pp.187–190.

Chandrasekharan, V. Kapoor, S. G. and DeVor, R. E. (1995) 'A mechanistic approach to predicting the cutting forces in drill with application to fiber reinforced composite materials', *J. Engg. Ind.* Vol.117, pp.559-570.

Chen, W. C. (1997) 'Some experimental investigations in the drilling of Carbon fibre reinforced composite laminations', *Int. J. Mach. Tools and Mfr.* Vol.37, No. 8, pp.1097 - 1108.

Dasch, J. M., Ang, C. C., Wong, C. A., Cheng, Y. T., Weiner, A. M., Lev, L. C. and Konca, E. (2006) 'A comparison of five categories of carbon-based tool coatings for dry drilling of aluminium', *Surf. & Coatings Tech.* Vol.200, pp.2970 – 2977.

Davim, J. P. Pedro Reis. (2003a) 'Study of delamination in drilling carbon fiber reinforced plastics (CFPR) using design experiments', *Comp. Str.* Vol.59, pp.481-487.

Davim, J. P. and Reis, P. (2003b) 'Drilling carbon finer reinforced plastics manufactured by autoclave-experimental and statistical study', *Mat. and Des.* Vol.24, pp.315-324.

Davim, J. P. Pedro Reis and Conceicao Antonia. (2004) 'Experimental study of drilling glass fiber reinforced (GFRP) manufactured by hand lay-up', *Comp. Science and Tech.* Vol.64, pp.289-297.

Davim, J. P. Campos Rubio, J. Abrao, A.M. (2007) 'A novel approach based on digital image analysis to evaluate the delamination factor after drilling composite laminates', *Comp. Sci. and Tec*h. Vol.67, pp.1939–1945.

Denkena, B. Boehnke, D. and Dege, J. H. (2008) 'Helical milling of CFRP–titanium layer compounds', *CIRP J. of Manuf. Sc. and Tech.* Vol.1, pp.64-69.

Dharan, C. K. H. and Won, M. S. (2000) 'Machining parameters for an intelligent machining system for composite laminates', *Int. J. Mach. Tools & Mfr,* Vol.40, pp.415-426.

Di Ilio, A. Tagliaferri, V. and Venialli, F. (1991) 'Cutting mechanism in drilling Aramid composite', *Int. J. of Machine Tool & Mfr.* Vol.31, No.2, pp.155-165.

Dipaola, G. Kapoor, S. G. and Devor, R. E. (1996) 'An experimental investigation of the crack growth phenomenon for drilling of fiber-reinforced composite material', *J. of Engg for Ind.* Vol.118, pp.104-11.

Dornfeld, D. A. Kim, J. S. Dechow, H. Hewson, J. Chen, L. J. (1999) 'Drilling burr formation in titanium alloy', Ti-6Al-4V, *Annals of the CIRP.* Vol.48, No.1, pp.73-76.

Drozda T. J. (1983) Tool and manufacturing engineer's handbook, Fourth Edition, *Machining*, Vol.1, SME. USA.

Durào L. M. P. Moura M.F.S.F. Marques A.T. (2006) 'Numerical simulation of the drilling process on carbon/epoxy composite laminates', *Composites Part A,* Vol.37, pp.1325–1333.

Durào L. M. P. Moura M.F.S.F. Marques A.T. (2008) 'Numerical prediction', *Engg. Frac. Mech.,* Vol.75, No.9, pp.2767-2778.

Enemuoh, U. E. Sherif, E. L. Gizawy, A. and Okafor, C. A. (2001) 'An approach for development of damage free drilling of carbon fibre reinforced thermosets', *Intl. J. Machine Tools and Mfr.* Vol.41, pp.1795-1814.

Everstine, G. C., and Rogers, T. G., (1971) 'A theory of machining of fiber reinforced materials', *J. Comp. Mat.* Vol.15, pp.94-106.

Ezugwu, E. O. and Wang, Z. M. (1997) 'Titanium alloys and their machinability - a review', *J. of Mat. Proc. Tech.* Vol.68, pp.262-274.

Fernandes, M. Cook, C. (2006) 'Drilling of carbon composites using a one shot drill bit. Part I: Five stage representation of drilling and factors affecting maximum force and torque', *Int. J. of Machine Tools & Manuf.* Vol.46, pp.70–75

Garrick, R. (2007) *'Drilling advanced aircraft structures with PCD drills'*, SAE, No.2007-01-3893.

Gindy, N. N. (1998) 'Selection of drilling conditions for glass fiber reinforced plastics', *Int. Jl. Production Research,* Vol.26, No.8, pp.1317-1327.

Grzesik, W. and Nieslony, P. (2000) 'The influence of thin hard coatings on friction in the orthogonal cutting process', *Transact.* NAMRI/SME, 28, pp.113–118.

Hamade, R. F. and Ismail, F. (2005) 'A case for aggressive drilling of aluminum', *Jl. of Mat.*

Proc. Tech. Vol.166, pp.86–97.

Hamdoun, Z. Guillaumat, L. and Lataillade, J. (2006) 'Influence of the drilling quality on the fatigue compression behaviour of carbon epoxy laminates', *Int. J. of Fat.* Vol.28, pp.1–8.

Hocheng, H. and Dharan, C. K. H. (1990) 'Delamination during drilling in composite laminates', *Trans. ASME,* Vol.112, pp.236-239.

Hocheng, H. and Puw, H. Y. (1992) 'On the drilling characteristic of fibre reinforced thermoset and thermoplastic', *Int. J. Machine Tool & Manuf.* Vol.32, No.4, pp.583-592.

Hocheng, H. and Puw, H. Y. (1993) 'Machinability of fibre reinforced thermo Plastics in drilling', *Trans.ASME,* Vol.114, pp.146-149.

Hocheng, H. and Tsao, C. C. (2003) 'Comprehensive analysis of delamination in drilling of composite materials with various drill bits', Jl. of Mat. Proc. Tech. 140, pp.335-339.

Hocheng, H. and Tsao, C. C. (2005) 'The path towards delamination-free drilling of composite materials', Jl of Mat. Proc. Tech. Vol.167, pp.251-264.

Hocheng, H. and Tsao, C. C. (2006) 'Effects of special drill bits on drilling-induced delamination of composite materials', *Int. J. Machine Tools & Manuf.* Vol.46, pp.1403-1416.

Jain, S. and Yang, D. C. H. (1994) 'Delamination free drilling of composite laminates'. Transaction of ASME, *Jl. Engg. Ind.* Vol.116, pp.475-461.

Karnik, S. R. Gaitonde, V. N. Rubio, C. J. Correia E. A. Abrao, A. M. and Davim, J. P, (2008) 'Delamination analysis in high speed', *Mat. & Des.* Vol.2, No. 9, pp.1768-1776.

Kelly, J. F., Cotterell, M. G. (2002) 'Minimal lubrication machining of aluminium alloys', *J. of Mat. Proc. Tech.* Vol.120, pp.327–334.

Khashaba, U. A. (2004) 'Delamination in drilling GFR thermoset composites', *Comp. Struct.* Vol.63, No.3, pp.313-327.

Kim, D. and Ramulu, M. (2004) 'Drilling process optimization for graphite/bismaleimide-titanium alloy stacks', *Comp. Stru.* Vol.63, No.1, pp.101- 114.

Kim, D., and Ramulu, M. (2005) 'Machinability of Titanium/Graphite Hybrid Composites in Drilling', *Trans. NAMRI/SME,* Vol.33, pp.445-452.

Kim, D., and Ramulu, M. (2007) 'Study on the Drilling of Titanium/Graphite Hybrid Composites', *J. Eng. Mater. Tech.* Vol.129, No.3, pp.390-397.

Kim, G. W. and Lee, K. Y. (2005) 'Critical thrust force at propagation of delamination zone due to drilling of FRP metallic strips', *Composite Structures,* Vol.69, pp.137-141.

Kishore, R.A. Tiwari, R. Dvivedi, A. Singh, I. (2009) 'Taguchi analysis of the residual tensile strength after drilling in glass fiber reinforced epoxy composites', *Mat. and Des.* Vol.30, No.6, pp.2186-2190.

Kohkonen, K. E. and Potdar, N. (1998) *Composite machining,* Hand book of Composites, Ed by Peters ST, Chapman & Hall.

Komanduri, R. (1993) 'Machining of fibre-reinforced composites', *Mech. Eng.* Vol.114, pp.58-64.

Komanduri, R. (1997) 'Machining of fibre reinforced composites', *Mach. Sc. and Tech.* Vol.1, No.1, pp.113-152.

Konig, W., Wulf, Ch., Graß, P. and Willercheid, H. (1985) 'Machining of fibre reinforced plastics', *Annals of the CIRP.* Vol.38, pp.119-24.

Konig, W. and Graβ, P. (1989) 'Quality definition and assessment in drilling of fiber reinforced thermosets', *Annals of the CIRP*. Vol.38(1), pp.119-124.

Konig, W., Cronjager, L., Spur, G. and Tonshoff, H. K. (1990) 'Machining of new materials',

Annals of the CIRP. Vol.39, No.2, pp.673-680.

Krishnaraj, V., Vijayarangan, S. and Rameshkumar, A. (2007) 'Effect of drilling parameters on Mechanical Strength in drilling GFRP', *Int. Jl. of Com. Appl. in tech.* Vol.28, No.1, pp.87-93.

Krishnaraj, V., Vijayarangan, S., and Suresh, G., (2005) 'An investigation on High speed drilling of GFRP', *Indian Jl. of Engg. and Mat. Sci.* Vol.12, pp.189-196.

Krishnaraj, V. Vijayarangan, S. and Paulo Davim, J. (2008) 'An experimental and statistical study on the effect of drill geometries on force and hole quality in drilling of glass fibre reinforced plastic', *Int. Jl. of Mat. and Product Tech.* Vol.32, No.2/3, pp.264-275.

Lachaud, F. Piquet, R. Collombet, F. and Surcin, L. (2001) 'Drilling of composite structures', *Comp. Str.* Vol.52, pp.511-516.

Langella, A. Nele, L. and Maio, A. (2005) 'A torque and thrust prediction model for drilling of composite materials', *Composites Part A: App. Sc. and Manuf.* Vol.36, pp.83-93.

Lubin, G. (1969). Hand book of fibre glass and advanced plastic composites, *Polymer Technology Series.*

Li, R. Albert, J. and Shih. (2007a) 'Spiral point drill temperature and stress in high-throughput drilling of titanium', *Int. J. of Mach. Tools and Manf.* Vol.47, pp.2005-2017.

Li, R. Hegde, P. Albert, J. and Shih. (2007b) 'High throughput drilling of titanium alloys', *Int. J. of Mach Tools and Manf.* Vol.47, pp.63-74.

Lin, S. C. and Chen, I. K. (1996) 'Drilling carbon fiber-reinforced composite material at high speed', *Wear,* Vol.194, pp.156-162.

Lin, S. C. and Shen, J. (1999) 'Drilling unidirectional glass fiber-reinforced composite materials at high speed', *J. Comp. Mat.* Vol.33, No.9, pp.827-851.

List, G. Nouari, M. G′ehin, D. Gomez, S. Manaud, P. Le Petitcorps, Y. Girot, F. (2005) 'Wear behaviour of cemented carbide tools in dry machining of aluminium alloy', *Wear,* Vol.259, pp.1177–1189.

Malhotra, S. K. (1990) 'Some studies on drilling fibrous composites', *J. Mat. Proc. Tech.* Vol.24, pp.291-300.

Mohamed, M. Carlson, Pu, Gu, and Sadler, R. (1993) 'Effect of machining on the performance of carbon/epoxy 3D woven composites', International conference on advanced composites, Ed. by Chandra T. and Dhingra A.K. *The Minerals Metals and Materials Society,* pp.723-728.

Mariatti, M. Nasir, M. Ismail, H. and Backlund, J. (2004) 'Effect of hole drilling techniques on tensile properties of continuous fiber impregnated thermoplastic (COFIT) plain weave composites', *J. Reinf. Plastics and Comp.* Vol.23, No.11, pp.1173-1186.

Mathew, J. Ramakrishnan, N. and Naik, N.K. (1999) 'Trepanning on unidirectional composites: delamination studies', *Composites Part A,* Vol.30, pp.951-959.

Mike, T. (2007) 'Composites Challenge Cutting Tools', *Manuf. Engg.* Vol.138, No.4.

Mohan, N. S. Ramachandra, A. Kulkarni, S. M. (2005) 'Influence of process parameters on cutting force and torque during drilling of glass–fiber polyester reinforced composites', *Comp. Str.* Vol.71, pp.407–413.

Mohan, N. S. Ramachandra, A. Kulkarni, S. M. (2007) 'Delamination analysis in drilling process of glass fiber reinforced plastic (GFRP) composite materials', *J. of Mat. Proc. Tech.* Vol.186, pp.265–271.

Murphy, C. Byrne, G. and Gilchrist, M. D. (2002) 'The performance of coated tungsten carbide drills when machining carbon fibre-reinforced epoxy composite materials', *Proc.*

I Mech Engrs., Vol.216, pp.143-152.

Nouari, M. List, G. Girot, F. and Ge'hin, G. (2005) 'Effect of machining parameters and coating on wear mechanisms in dry drilling of aluminium alloys', *Int. J. of Machine Tools & Manuf.* Vol.45, pp.1436–1442.

Persson, E. Eriksson, I. and Zackrisson, L. (1997) 'Effects of hole machining defects on strength and fatigue life of composite laminates'. *Composites Part A,* Vol.28(A), pp.141-151.

Piquet, R. Ferret, B. Lachaud, F. and Swider, P. (2000) 'Experimental analysis of drilling damage in thin carbon/epoxy plate using special drills', *Composites Part A,* Vol.31, pp.1107-1115.

Rao, B. S. Rudramoorthy, R. Srinivas, S. and Nageswara Rao, B. (2008) 'Effect of drilling induced damage on notched tensile and pin bearing strengths of woven GFR-epoxy composites', *Mat. Science and Engg. A,* Vol.472, pp.347–352.

Ramkumar, J. Aravindan, S. Malhotra, S. K. and Krishnamurthy, R. (2004a) 'An enhancement of the machining performance of GFRP by oscillatory assisted drilling', *Int. J. Ad. Manuf. Tech,* Vol.23, pp.240-244.

Ramkumar, J. Malhotra, S. K. and Krishnamurthy, R. (2004b) 'Effect of workpiece vibration on drilling of GFRP laminates', *J. Mat. Proc. Tech.* Vol.152, v329-332.

Ramulu, M. Young, P. and Kao, H. (1999) 'Drilling of Graphite/Bismaleimide composite material', *J. of Matl. Engg. & Perf.* Vol.8, No.3, pp.330-338.

Ramulu, M. Branson, T. and Kim, D. (2001) 'A study on the drilling of composite and titanium stacks', *Comp. Str.* Vol.54, pp.67-77.

Ravishankar, S. R. and Murthy, C. R. L. (2000a) 'Application of acoustic emission in drilling of composite laminates', *NDT&E Int.* Vol.33, pp.429-235.

Ravishankar, S. R. and Murthy, C. R. L. (2000b) 'Characteristics of AE signals obtained during drilling composite laminates', *NDT&E Int.* Vol.33, pp.341-348.

Richardson, T. (1987) *Composites - A design guide,* Industrial Press Inc., New York.

Rivero, A. Aramendi, G. Herranz, S. Lopez de Lacalle L. N. (2006) 'An experimental investigation of the effect of coatings and cutting parameters on the dry drilling performance of aluminium alloys', *Int J Adv Manuf Technol,* Vol.28, pp.1-11.

Roudge, M., Cherif, M., Cahuc, O., Darnis, P., and Danis, M. (2008) 'Multi-layers materials. Qualitative approach of the process', *Int. J. Materials Forming,* Vol.1, No.1, pp.949-952.

Rubio, C. J. Abrao, A. M. Faria, P. E. Esteves Correia, A. Davim J. P. (2008) 'Effects of high speed', *Int. Jl. of Machine Tools and Mafr,* Vol.48, No.6, pp.715-720.

Sadat, A. B., (1988) 'The effect of cutting speed and width of cut on interlaminar tensile strength of Graphite/Epoxy composite material', *Jl. of Comp. Tech. & Res.* Vol.10, pp.173-176.

Sadat, A. B. Chan, W. S. and Wang, B. P. (1992) 'Delamination of graphite/ epoxy laminate during drilling operation', *J. Energy Resources Tech.* Vol.114, pp.139-141.

Sadat, A. B. (1996) 'Prediction of delamination load in drilling of graphite/epoxy composites', *Engg systems design and analysis.* Vol.3, pp.21-26.

Sakuma, K. and Seto, M. (1978) 'Tool wear in cutting GFRP materials (Effect of physical properties of tool material)', *Bulletin of JSME,* Vol.22, No.163, pp.107-114.

Sakuma, K. and Seto, M. (1981a) 'Tool wear in cutting GFRP composites, (Relation between cutting temperature and tool wear)', *Bulletin of JSME,* Vol.24, No.190, pp.748-755.

Sakuma, K. and Seto, M. (1981b) 'Flank wear characteristics of cutting tools in different types of tests (Comparison between results of conventional turning, step turning and facing tools)', *Bulletin of the JSME,* Vol.24, No.196, pp.1871-1878.

Sakuma, K. and Seto, M. (1983) 'Tool wear in cutting glass fiber reinforced plastics (The relation between fiber orientation and tool wear)', *Bulletin of the JSME,* Vol.26, No.218, pp.1420-1427.

Sakuma, K. Yakoo, Y. and Seto, M. (1984) 'Study on drilling of reinforced plastics (GFRP & CFRP)-Relation between tool material and wear behavior', *Bulletin of JSME,* Vol.27, No.228, pp.1237-1244

Sakuma, K. Seto, M. Taniguchi, M. and Yokoo, Y. (1985) 'Tool wear in cutting carbon FRP materials (effect of physical properties of tool material)', *Bulletin of JSME,* Vol.28, No.245, pp.2781-2788.

Sakurai, K. Adachi, K., Kawai, G., Sawai, T. and Ogawa, K. (2000) 'High feed rate drilling of aluminium alloy', *Mater. Sci. Forum,* Vol.331–337, No.1, pp.625–630.

Santhanakrishnan, G. Krishnamurthy, R. and Malhotra, S. K. (1989) 'High speed steel tool wear studies in machining of glass fiber reinforced plastics', *Wear,* Vol.132, pp.327-336.

Seif, M. A. Khashaba,U. A. and Rojas-Oviedo,R. (2007) 'Measuring delamination in carbon/epoxy composites using a shadow moire laser based imaging technique', *Comp. Str.* Vol.79, pp.113–118.

Sharif, S. and Rahim, E. A. (2007) 'Performance of coated – and uncoated carbide tools when drilling titanium alloy- Ti-6Al4V', *J. of Matl. Proc. Tech.* Vol.185, pp.72-76.

Singh, I. Bhatnagar, N. and Viswanath, P. (2008) 'Drilling of uni-directional glass fiber reinforced plastics: Experimental and finite element study', *Mat. and Des.* Vol.29, pp.546–553.

Stone, R. and Krishnamurthy, K. (1996) 'A Neural network thrust force controller to minimise delamination during drilling of Graphite/Epoxy laminate', *Int. Jl. Machine Tool and Mfr.* Vol.36, No.9, pp.985-1003.

Tagliaferri, V. Caprino, G. and Diteriazzii, A. (1990) 'Effect of drilling parameters on the finish and mechanical properties of GFRP composites', *Int. Jl. Machine tool and Mfr.* Vol.10, No.1, pp.77-84.

Tsao, C. C. and Chen, W. C. (1997) 'Prediction of the location of delamination in the drilling of composite laminates', *Jl. of Matl. Proc. Tech.* Vol.70, pp.185-189.

Tsao, C. C. and Hocheng, H. (2003) 'The effect of chisel length and associated pilot hole on delamination when drilling composite materials', *Int. J. of Machine Tools & Manf.* Vol.43, pp.1087–1092.

Tsao, C. C. and Hocheng, H. (2004) 'Taguchi analysis of delamination associated with various drill bits in drilling of composite material', *Int. J. Machine tools & Mfr.*, Vol.44, pp.1085-1090.

Tsao, C. C. and Hocheng, H. (2005a) 'Effect of eccentricity of twist drill and candle stick drill on delamination in drilling composite materials', *Intl. J. Machine Tools & Manuf.* Vol.45, pp.125-130.

Tsao, C. C. and Hocheng, H. (2005b) 'Effects of exit back-up on delamination in drilling composite materials using a saw drill and core drilling', *Int. Jl. of Machine Tools & Mafr.* Vol.45, pp.1261-1270.

Tsao, C. C. and Hocheng, H. (2005c) 'Computerised tomography and C-scan for measuring delamination in the drilling of composite materials using various drills', *Int. J. Machine tools & Mfr.* Vol.45, pp.1282-1287.

Tsao, C. C. (2007) 'Taguchi analysis of drilling quality associated with core drill in drilling of composite material', *Int. Jl. Adv Manuf. Tech.* Vol.32, pp.877–884.

Tsao, C. C. (2008 a) 'Experimental study of drilling composite materials with step-core drill', *Materials and Design,* Vol. 29, pp.1740-1744.

Tsao, C. C. Hocheng, H. (2008b) 'Evaluation', *Jl. of Materials Processing Technology,* Vol.203, No.1-3, pp.342-348.

Tsao, C. C. and Hocheng, H. (2008c) 'Effects of peripheral drilling moment on delamination using special drill bits', *J. of Mat. Proc. Tech.* Vol.201, No.1-3, pp.471-476.

Velayudham, A. Krishnamurthy, R. and Soundarapandian, T. (2005a) 'Evaluation of drilling characteristics of high volume fraction fiber glass reinforced polymeric composites', *Int. Jl. Mach Tools and Mfr,* Vol.5, pp.339-406.

Velayudham, A. Krishnamurthy, R. Soundarapandian, T. (2005b). 'Acoustic emission based drill condition monitoring during drilling of glass/phenolic polymeric composite using wavelet packet transform', *Matls. Sci. and Engg A.* Vol.412, pp.141–145.

Velayudham, A. Krishnamurthy, R. (2007) 'Effect of point geometry and their influence on thrust and delamination in drilling of polymeric composites', *Jl. of Mat. Proc. Tech.* Vol.185, pp.204–209.

Wain, N. Thomas, N. R. Hickman, S. Wallbank, J. Teer, D. G. (2005) 'Performance of low-friction coatings in the dry drilling of automotive Al–Si alloys', *Surf. & Coatings Tech.* Vol.200, pp.1885 – 1892.

Wang, X. Wnag, L. J. and Tao, J. P. (2004) 'Investigation on thrust in vibration drilling of fiber-reinforced plastics', *Jl. Matl. Proc. Tech.* Vol.48, pp.239-244.

Won, M. S. and Dharan, C. K. H. (2002a) 'Chisel edge and pilot hole effects in drilling composite laminates', *Jl. of Manuf. Sc. and Engg.* Vol.124, pp.242-247.

Won, M. S. and Dharan, C. K. H. (2002b) 'Drilling of aramid and carbon fiber reinforced polymer composites', *J. Manuf. Sci. and Engg.* Vol.124, pp.778-783.

Zeilmann, P. R. and Lindolfo Weingaertner W. (2006) 'Analysis of temperature during drilling of Ti6Al4V with minimum quantity of lubricant', *Int. J. of Matl. Proc.* Tech. Vol.179, pp.124-127.

Zhang, L. B. Wang, L. J. Liu, X. Y. Zhao, H-W. Wang, X. and Luo, H. Y. (2001) 'Mechanical model for predicting thrust and torque in vibration drilling fibre reinforced composite materials', *Int. J. of Machine Tools and Mfr.* Vol.41, pp.641-657.

Zhang, H., Chew, W., Chen, D. and Zhang, L. (2001) 'Assessment of the exit defects in carbon fiber-reinforced plastic plates caused by drilling', *Key Engg. Mats*. Vol.196, pp.43-52.

Zitoune, R. Collombet, F. Lachaud, F. Piquet, R. and Pasquet, P. (2005) 'Experiment-calculation comparison of the cutting conditions representative of the long fibre composite drilling phase', *Comp. Sc. and Tech.* Vol.65, No.3-4, pp.455-466.

Zitoune, R. and Collombet, F. (2007) 'Numerical Prediction of the thrust force responsible of delamination during drilling of the long fibre composite structures', Comp. *Part A. Apld. Sc. and Manuf.* Vol.38, pp.858-86.

Zitoune, R., Collombet, F., and Hernaiz Lopez, G. (2008) 'Experimental and analytical study of the influence of HexFit® glass fibre composite manufacturing process on delamination during drilling', *Int. J. of Mach. and Machinabilityof Mails.* Vol.3, No.3-4, pp.326 - 342.

Zitoune, R. and Collombet, F. (2009) 'Influence of machining quality on composite part manufacturing, J. P. Davim(ed.), Drilling of Composite Materials', *Series: Materials and Manufacturing Technology,* ISBN: 978-1-60741-163-5.

In: Machining and Forming Technologies, Volume 2
Editor: J. Paulo Davim

ISBN: 978-1-61122-820-5

Chapter 11

3D MODELING OF DRILLING PROCESS OF AISI 1010 STEEL

***Riaz Muhammad*[*1], *Naseer Ahmed* [1], *Muhammad Abid* [1] and *V.V. Silberschmidt* [2]**

[1] Faculty of Mechanical Engg. Ghulam Ishaq Khan Institute of Engineering Sciences and Technology, Topi, Swabi, N.W.F.P, Pakistan

[2] Mechanics of Advanced Materials Research Group, Wolfson School of Mechanical and Manufacturing Engineering Loughborough University Loughborough Leicestershire LE11 3TU, U.K.

ABSTRACT

The paper outlines the development of 3D finite element model for drilling of AISI 1010 steel using commercial FE code MSC Marc Mentat. Thrust force, Torque and temperature distribution in the cutting region are calculated for selected cutting conditions. For experimental measurement of thrust force and torque, a strain gauge based three component dynamometer is used. Standard k-type thermo couple is used for the measurement of temperature at the tool-workpiece interaction zone during drilling process which is inserted in to the coolant hole of a drill bit. The effect of lubrication on thrust force and torque is also studied. To study the effect of lubrication on thrust force and torque two contact conditions, contact with friction and contact without friction are used in finite element analysis of drilling process.

Keywords: Drilling, AISI 1010, thrust force, Torque, temperature, lubrication, 3D Finite element analysis.

[*] Corresponding author: E-mail: eriazm99@yahoo.com

1. Introduction

Cutting is a very useful way to obtain industrially required components. The ordinary and broadly used manufacturing operation such as drilling, turning, boring, grinding and milling are used to manufacture a diversity of mechanical products in manufacturing industries. Drilling is very important and most complex machining process of originating holes in industrial products. The main characteristic that distinguishes drilling from other machining operations is the combined cutting operation at the lips and surfaces such as rake angle and inclination angle. The cutting velocity too is not uniform over the surface of the drill bit; it varies along the drill radius. Most of the drilling operations are normally performed at the final steps of fabricating a part therefore causes costly production waste. The cutting forces and thermal aspect of drilling process are the key issues to be studied in order to control the possible flaws of drilling in manufacturing. Many researchers have investigated drilling process analytical, numerically and experimentally to gain a better understanding of process. Watson A R (Watson, 1985) (Watson, 1985) (Watson, 1985) (Watson, 1985) developed a detailed machining model for the cutting lips and chisel edge of a drill bit to investigate the thrust force and torque during machining by assuming the cutting edge of the drill bit as a collection of few elements that were removing material from the workpiece. Watson A R approach was further utilized by (Strenkowski et al, 2004) to predict the thrust force and torque on the cutting lips and chisel edge by using analytical finite element technique. (Elhachimi, 1999) (Elhachimi, 1999) determined the continuous distribution of thrust forces and torque on the cutting lips and chisel edge of the drill bit by using oblique cutting model for the lips and orthogonal cutting model for the chisel edge. Based on mechanics and dynamics of drilling process Pirtini M et al (Pirtini, 2005) developed a force model to investigate the cutting forces and hole quality. Bono M et al (Bono, 2006) developed FE model to investigate the temperature distribution in the drill bit. Similarly FE approach is used to investigated the temperature distribution in spade drilling by Shen Q et al (Shen, 1997)

Lin S C et al (Lin, 1996) has experimentally studied the effect of high speed on thrust force and torque on fibre reinforced composites by using Kestler 9273 drilling dynamometer. Li R et al (Li, 2007) experimentally investigated the various cutting condition, tool geometry and tool material on tool life and performance in high throughput drilling of Ti-6Al-4V. The effect of cutting parameter on temperature distribution of the cutting tool is studied experimentally in a step by step and continuous dry drilling by Bagci E et al (Bagci, 2006). Ueda T et al (Ueda, 2007) investigated the temperature of the cutting edges of a drill bit experimentally by using a two-color pyrometer with an optical fiber.

A fully thermomechanically coupled orthogonal cutting process was simulated in Third Wave AdvantEdge FE commercial code to investigate the effect of cutting parameters on temperature of the drill bit by Ozcelik B et al (Ozcelik, 2006). Bono M et al (Bono, 2002) has developed a model to predict the heat flow into the work piece during dry drilling process which are used as an input during 2D FE simulation of drilling process.

In the present study a 3D thermomechanically coupled finite element simulation of drilling of AISI 1010 steel is performed. The thrust force, Torque and temperature distribution in the workpiece and in the cutting tool are calculated for selected cutting conditions.

2. Model Description

2.1. General Feature

Commercial finite element code MSC MARC/MENTAT (MSC-MARC-Users-Guide, 2005), has been used to perform 3D finite element analysis (FEA) of drilling process. The FE model is using the updated Lagrangian formulation. Due to updated Lagrangian formulation, the element shape is changing with the material flow and the calculation embeds a computational mesh in the material domain and solves the problem for the position of the mesh at discrete points in time. The updated Lagrangian formulation includes all kinematics nonlinear effects due to large displacements, large rotations and large strains.

During drilling large deformation occurs in the tool-workpiece interaction zone. Tool chip friction and plastic deformation transforms largely into heat and raises the temperature of the workpiece thus inducing thermal stresses and volumetric expansion as well as affects the material properties of the workpiece. The nature of the problem thus required a fully thermomechanically coupled analysis.

Strain, strain rate, temperature and strain hardening has a major influence on the flow stress or instantaneously yield strength at which the materials start to deform plastically or flow. Accurate and reliable flow stress models are highly necessary to represent work material constitutive behaviour under different cutting condition. Johnson Cook material model (Johnson, 1985) is used to describe the mechanical behaviour of the AISI 1010 steel at high strain, strain rate and elevated temperature in simulation of the drilling process in which strain, strain rate and strain hardening are combine in a multiplicative manner.

$$\sigma_y = \left[A + B\left(\varepsilon_p\right)^n\right]\left[1 + C\ln(\varepsilon^{\circ})\right]\left[1 - \left(T_H\right)^m\right],$$

Where σ_y is flow stress, ε_p is the effective plastic strain, $\varepsilon^{\circ} = \frac{\varepsilon_p^{\circ}}{\varepsilon_0^{\circ}}$, $\overset{\circ}{\varepsilon}_p$ is the plastic strain rate, $\overset{\circ}{\varepsilon}_o = 1\text{S}^{-1}$ is the reference strain rate, $T_H = \frac{T - T_R}{T_M - T_R}$ is the homologous temperature, T is the absolute temperature, T_M is the melting temperature of the mild steel, T_R is the room temperature, A=250 Mpa is the initial yield stress at the reference strain rate, B=275 Mpa is the hardening modulus, C=0.078 is strain rate sensitivity coefficient, n=0.36 is the hardening coefficient, m is the thermal softening coefficient. All these parameters are determined experientially using the Split Hopkinson Pressure Bar method and were adopted from (Umeda, 2005).

Friction plays a very important role in metal cutting. It not only determines the power requirement for removing a given volume of metal, but also controls the surface quality of a finished product and the rate of wear of a cutting tool. Various friction models are available to adequately formulate the friction at the tool chip interface. These includes the coulomb friction model in which the friction stress is proportional to the normal interface stress $\tau = \mu\sigma_n$ (Baker, 2002), (Fassi, 2003), (Ng, 2002) to modelled the friction at the tool tip interface, shear friction model $\tau = \mu k$ where k is the shear yield strength (Yen, 2004), modified coulomb friction model $\tau = \min(\mu p, \tau_{th})$ (Shi, 2002), stress-based polynomial model (Yang, 2002) and modified shear friction model (Liu, 2000) for friction modeling in metal cutting.

When the normal force or stress becomes large, the Coulomb friction model may not correlate well with experimental observations. This is caused by the fact that the Coulomb friction model predicts that the frictional shear stresses increase to a level that can exceed the flow stress or the failure stress of the material. As this is not physically possible, a different friction calculation should be applied. Oden, J. T. and Pires, E. B. (1983) The choices are either to have a nonlinear coefficient of friction, or to introduce the friction stress limit in the bilinear model or to use the modified shear friction model. Modified shear friction model is implemented using approximations of the theoretical step function. Because this step function causes numerical difficulties in FEA calculation of the process. The arctangent function used in modified shear friction model is used to smooth out the step function in order to avoid the numerical difficulties. As a result of the smoothing procedure, a node in contact always has some slipping. Besides the numerical reasons, this 'ever slipping node' model has a physical basis. (Oden et al. 1983) pointed out that for metals; there is an elasto-plastic deformation of asperities at the microscopic level (termed as 'cold weld') which leads to a nonlocal and nonlinear frictional contact behavior. The arctan representation of the friction model is a mathematical idealization of this nonlinear friction behavior. Therefore shear friction between the tool and workpiece was modelled with a modified constant shear friction model. Therefore a modified shear friction model (MSC-MARC-Users-Guide, 2005) is used in the FEA simulation of drilling process.

$$\sigma_{\mathrm{fr}} \leq -\mu \frac{\bar{\sigma}}{\sqrt{3}} \frac{2}{\pi} \operatorname{sgn}(V_r) \arctan\left(\frac{V_r}{V_{\mathrm{cr}}}\right),$$

Where $\bar{\sigma}$ is the equivalent stress, V_r is the relative sliding velocity, V_{cr} is the critical sliding velocity below which sticking is simulated, μ is the friction coefficient and $\operatorname{sgn}(x)$ is the signum function.

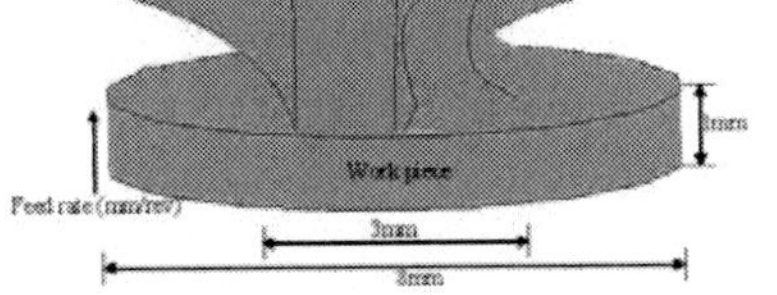

Figure 1. Scheme of relative movements of the workpiece and cutting tool in 3D simulations of drilling process.

2.2. FE Model

A 3D model of drilling process with relative moment of the workpiece and drill bit is shown in Figure 1. Various studies are available in literature for turning, milling and grinding using FEM. The drill bit geometry has a complex shape; therefore modeling of a drill bit is a difficult process in FEA packages. Therefore relatively high volume of memory and higher processing power is required for FEA simulation of drilling process. To reduce the computational cost and time, only one cutting lips of the drill bit having a code of MZS0980S-DIN is modelled in FE simulation. The dimensions of the workpiece used in simulation are 8mm external diameter, 3mm pre hole internal diameter and 1 mm thickness.

Table 1. Mechanical and Thermal dependent properties of workpiece and tool

AISI 1010 Steel	Elastic modulus (GPa)	205
	Poisson ratio	0.29
	Tensile strength (MPa), Ultimate	365
	Density (Kg/ m^3)	7870
	Tensile strength (MPa), Yield	305
	Thermal Conductivity K W/mK	12.5 (20<T≤132)
		14.5 (132<T≤244)
		15.676 (244<T≤300)
		18358 (300<T≤500)
		23.93 (500<T≤827)
	Specific Heat Capacity J/KgK	426.2 (20<T≤83)
		455 (83<T≤146)
		499 (146<T≤291)
		563 (291<T≤500)
		677 (500<T≤870)
	Coefficient of Thermal Expansion 10-6/K	12.5 (20<T≤82)
		13.2 (82<T≤175)
		14.0 (175<T≤376)
		14.8 (376<T≤621)
		16.9 (621<T≤866)
AISI M2 steel	Elastic modulus (GPa)	210
	Poisson ratio	0.3
	compressive strength (MPa), Yield	3250
	Density (Kg/ m^3)	8150
	Thermal Conductivity K W/mK	21.3 (20<T≤95)
		23.5 (95<T≤200)
		25.6 (200<T≤400)
		27.0 (400<T≤540)
		29.2 (540<T≤870)
	Specific Heat Capacity J/KgK	55
	Coefficient of Thermal Expansion 10-6/K	11.9

In the present FE simulation of drilling process, the cutting tool rotate with spindle speeds of (300Rpm, 650Rpm, 1000Rpm, 1300Rpm and 2000Rpm) and the workpiece is moved against the cutting tool with a feed rates of (0.12mm/rev, 0.175mm/rev, 0.24mm/rev and 0.35mm/rev). Modified shear friction model is used to incorporate the effect of lubrication in FE simulation of drilling process at the interaction zone of the workpiece and tool. The properties of the workpiece and cutting tool used in FE simulation are that of AISI 1010 steel and AISI M2 steel respectively which are shown in Table 1. (ASM-Hand-book, 1990).

The chip separation is achieved through re-meshing techniques based on a pre-defined strain limit. Due to the plastic deformation of the chip, the elements near the cutting lips of the drill bit distorts badly and negative value comes in the Jacobean matrix resulting in non convergence of the simulation. Hence a re-meshing technique is used to replace those distorted elements with ones of better shape. During re-meshing the current state of deformation, strains, stresses, etc. are transferred to the new mesh, and the analysis is continued. Figure 2 shows the chip formation after the successful implementation of the remeshing techniques.

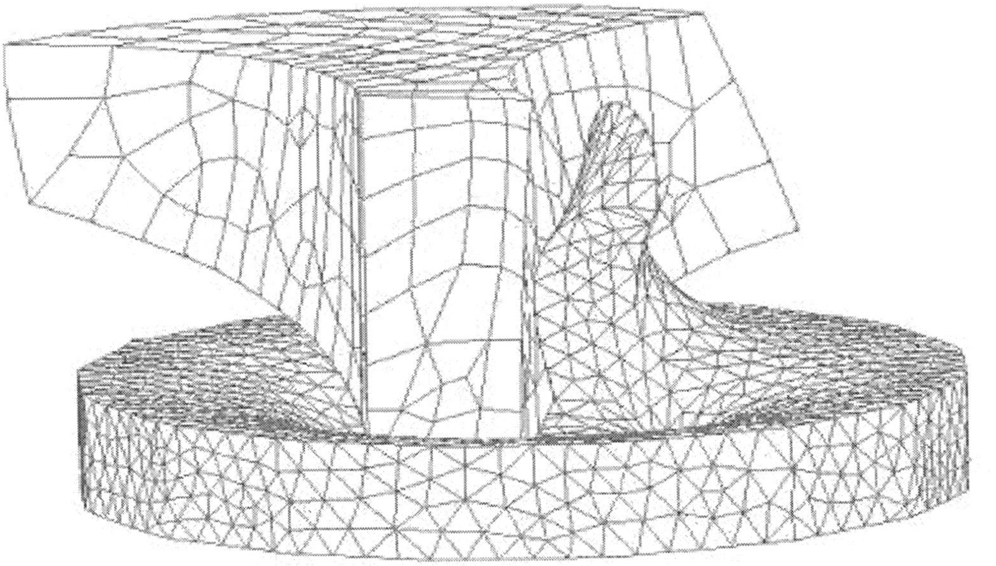

Figure 2. Formation of chip.

3. Experimental Setup

For experimental verification of thrust force, torque and temperature distribution in the cutting region during drilling process various experiments were conducted on CYL 6240 conventional lathe machine with a maximum spindle speed of 2000 rpm and 7.5 Hp drive motor.

3.1. Cutting Condition

Table 2 shows the detail of variables used in experimentation including feed rates, cutting speeds and lubrication. Thrust force, torque and temperature were experimentally measured in dry drilling conditions.

Table 2. Detail of variable used in experimental work

Parameter used in analysis	Magnitude of parameter
Lubrication	Off
Spindle Speed (Rpm)	300, 650, 1000, 1300, 2000
Feed Rate (mm/rev)	0.12, 0.175, 0.24, 0.35

3.2. Thrust Force and Torque Measurement

Thrust forces and torques were measured using an in-house build three component dynamometer. The experimental verification is carried out only for dry drilling condition. The FMS DL48 data logger was used to transfer data from dynamometer to computer where the data were analyzed and recorded. The experimental setup is shown in Figure 3. For the purpose of simplification only average values of thrust forces and torques were taken. The experimental and numerical results were in good agreement.

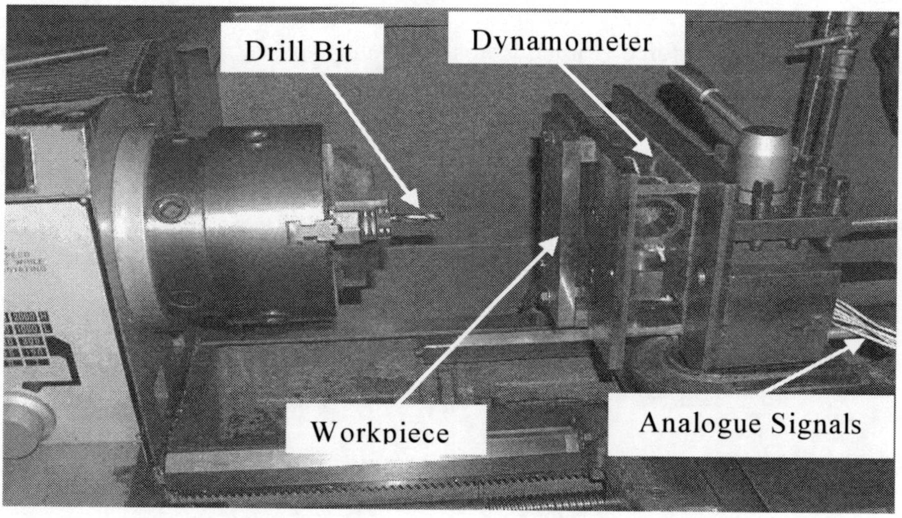

Figure 3. Experimental for Thrust force and Torque measurement.

3.3. Temperature Measurement

To measure the temperature of a drill bit during dry drilling process, a standard Teflon coated k type thermocouple is inserted into the coolant hole of a drill bit. The diameter of the thermocouple is 127μm and can take values up to 500Ċ with a response time of 10μs. HH501DK 4-channel K-type thermometer is used to record the measured temperature values;

Fourier's law is then used to calculate actual temperature at the cutting lips of drill bit. The setup used for the measurement of a drill bit temperature is shown in Figure 4. The numerical and experiment results are in good qualitative agreement.

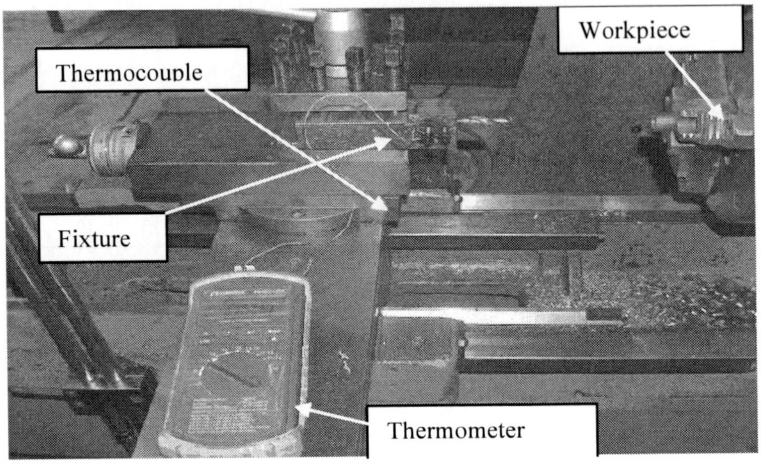

Figure 4. Experimental setup for Temperature Measurement.

4. Results and Discussion

In order to get a better understanding of the effect of various process parameters of drilling process on thrust force, torque and temperature in the cutting zone, a series of different experiments and simulations were performed.

Figure 4 shows the thrust force calculated both experimentally and through FE simulation at various cutting conditions. In all the experiments the cutting speed was kept constant and the feed rates were varying. Good agreements were found between the measured and FE predicted thrust forces. Average errors of about 7-8% were observed between the experimentally measured and FE Predicted thrust forces.

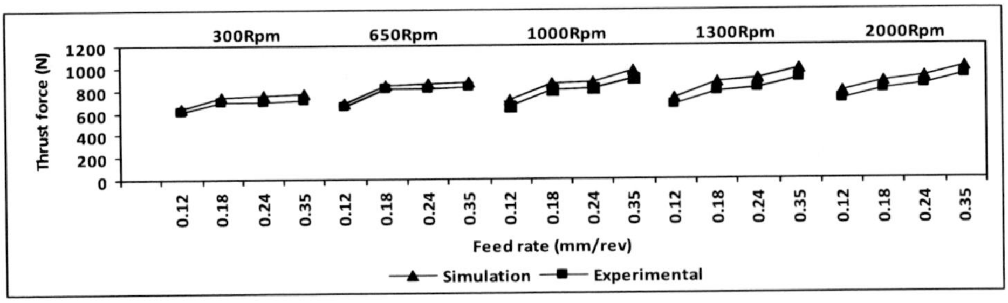

Figure 5. Comparison of FE simulation and measured thrust force in drilling process at various cutting condition.

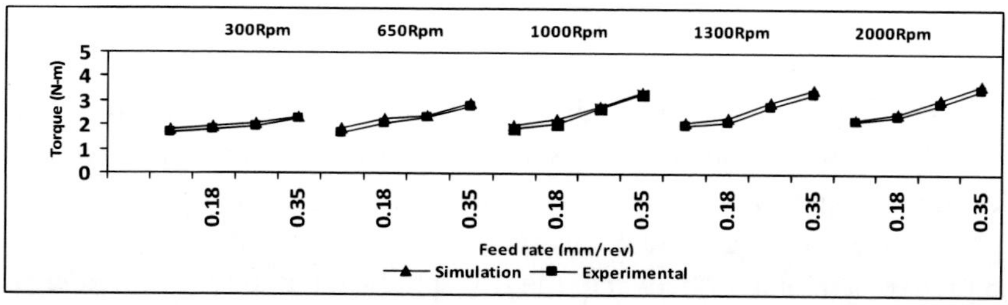

Figure 6. Comparison of FE simulation and measured torque in drilling process at various cutting condition.

Similarly the measured and FE predicted torque at various cutting condition are shown in Figure 5. In all the experiments, the cutting speed was kept constant and the feed rates were varying. A good agreement was found between the measured and FE predicted torque. Average errors of about 5%-6% was observed between the measured and FE Predicted torque.

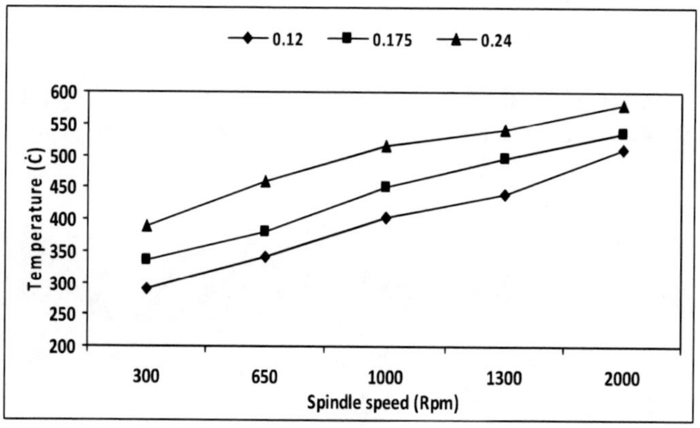

Figure 7. The effect of cutting speed (Rpm) on measured temperature distribution of the cutting region at constant feed rate (mm/rev).

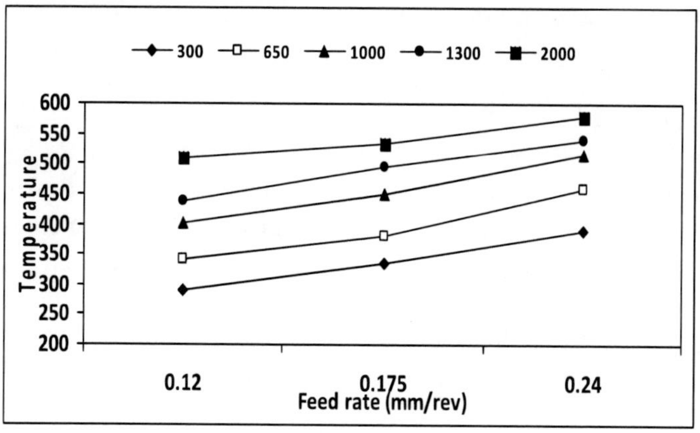

Figure 8. The effect of feed rate (mm/rev) on measured temperature distribution of the cutting region at constant cutting speed (Rpm).

The experimental and FEA study were performed to measure the temperatures changes in dry drilling for different spindle speeds and feed rates. The effect of cutting speeds and feed rates on measured temperature distribution in the cutting region is shown in Figure 7 and Figure 8. Similarly the effect of cutting speeds and feed rates on the FE predicted temperature are shown in Figure 9 and Figure 10. The simulation results were obtained for a very short duration of time (0.035 sec), therefore the temperature did not reached a steady state, however a good quantitative agreement between the FE predicted and experimentally measured temperature distribution was observed during the drilling process.

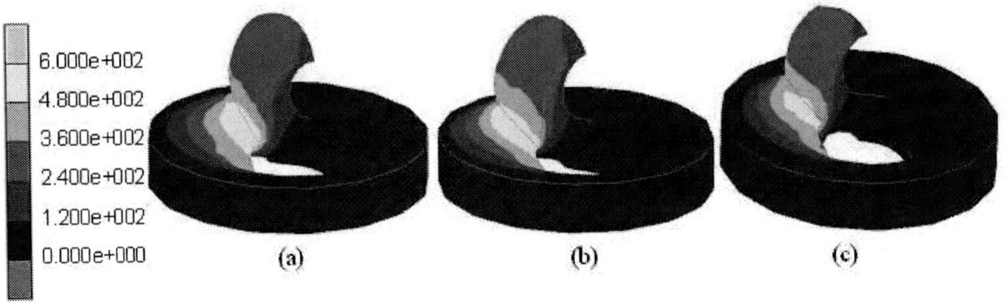

Figure 9. Effect of feed rate on temperature distribution of the cutting region (a) 1000Rpm, 0.24mm/rev (b) 1000 Rpm, 0.175mm/rev (c) 1000Rpm, 0.12mm/rev.

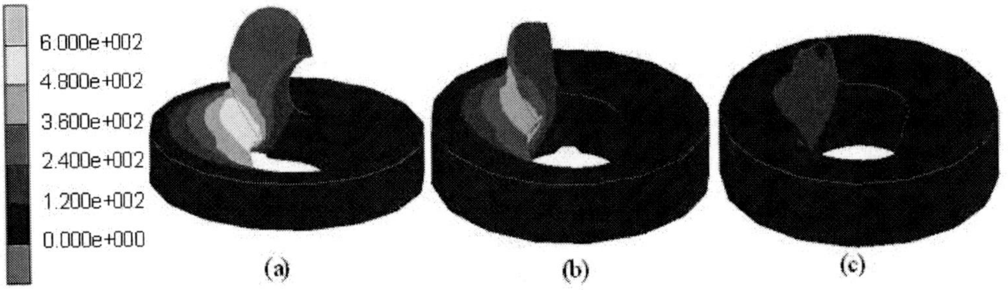

Figure 10. Effect of cutting speed on temperature distribution of the cutting region (a) 1000Rpm, 0.24mm/rev (b) 650 Rpm, 0.24mm/rev (c) 300Rpm, 0.24mm/rev.

4.1. Effect of Feed Rate

In Drilling, the variation of the feed rate has significant effects on thrust force, torque and temperature of the cutting region. The thrust force increases with the increase of the feed rate as shown in Figure 11, which is in good agreement with (Ozcelik, 2006) (Elhachimi, 1999) (Hamade, 2006). The simulations were carried out for both dry and fully lubricated drilling condition. When the feed rate increases from 0.12 to 0.175 mm/rev (approximately 31%), a 10% increase in thrust force is observed for fully lubrication case and 13% increase was observed in thrust force for dry drilling condition.

The torque during drilling process is also affected by the feed rate both in dry drilling and fully lubrication case. Figure 12 shows the effect of feed rate on torque produced during drilling process which is in good agreement to (Elhachimi, 1999) (Hamade, 2006). When the feed rate increases from 0.12 to 0.175 mm/rev (approximately 31%) an 8% increase in thrust force is observed for fully lubrication case and a 12% increase is observed in torque for dry drilling condition.

Similarly an increase in temperature was observed with the increase of feed rate (Ozcelik, 2006) (Shen, 1997). The temperature generated during drilling process increases with the increase of feed rates as shown in Figure 7 and Figure 9. Increasing the feed rate from 0.175 to 0.24 mm/rev (approximately 27%) caused a 4.5% increase in temperature when observed experimentally and a 5% increase in temperature during simulations.

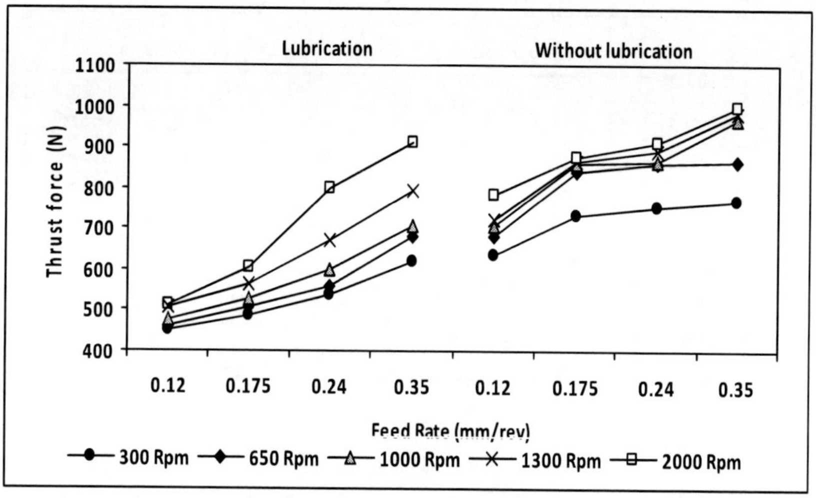

Figure 11. Variation of thrust force with feed rate (with lubrication, without lubrication).

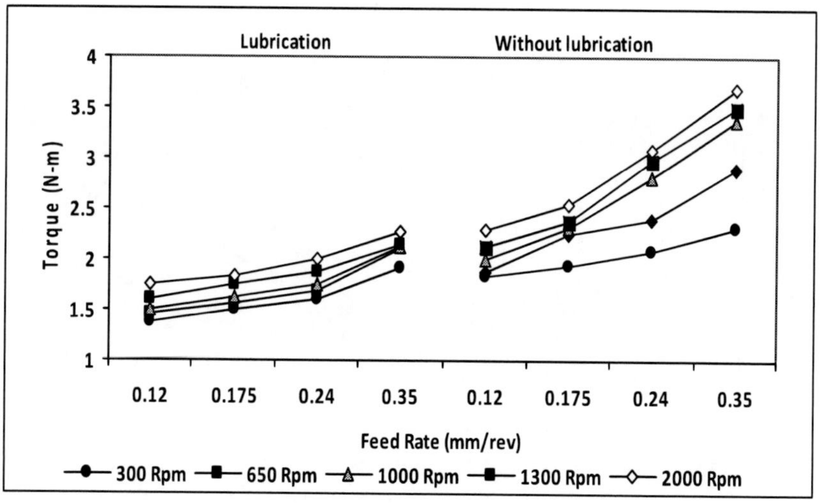

Figure 12. Variation of torque with feed rate (with lubrication, without lubrication).

4.2. Effect of Cutting Speed

Thrust force and torque are greatly affected by change in cutting speed. A rise in thrust force is observed with the increase of spindle speed as shown in Figure 13 both in dry drilling and fully lubrication case (Elhachimi, 1999).

It can be seen that the feed rate has a larger influence on thrust force than does the cutting speeds. For instant, with an increase in cutting speed from 300 to 650 Rpm (approximately 53%), an average increase of about 5% is observed in thrust force, while increasing feed rate from 0.12 to 0.175 mm/rev (approximately 31%) causes a 10% increase in thrust force.

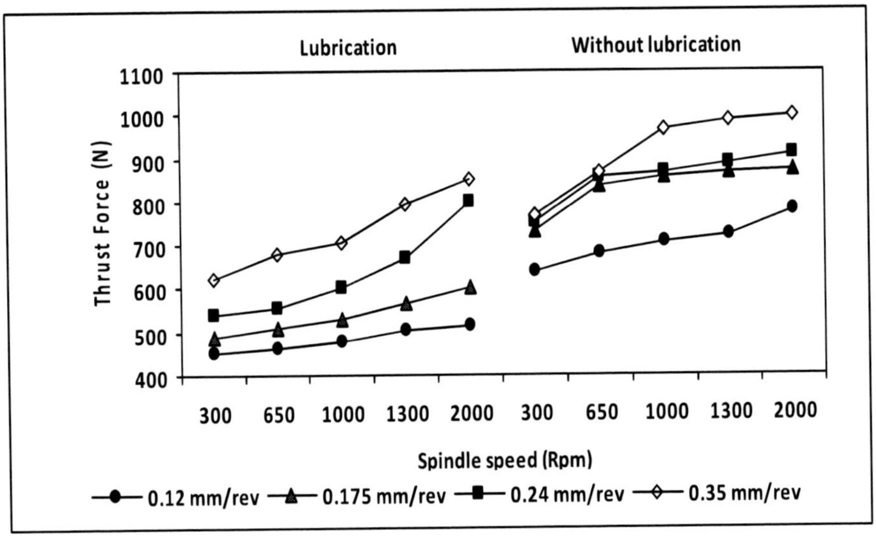

Figure 13. Variation of thrust force cutting speed (with lubrication, without lubrication).

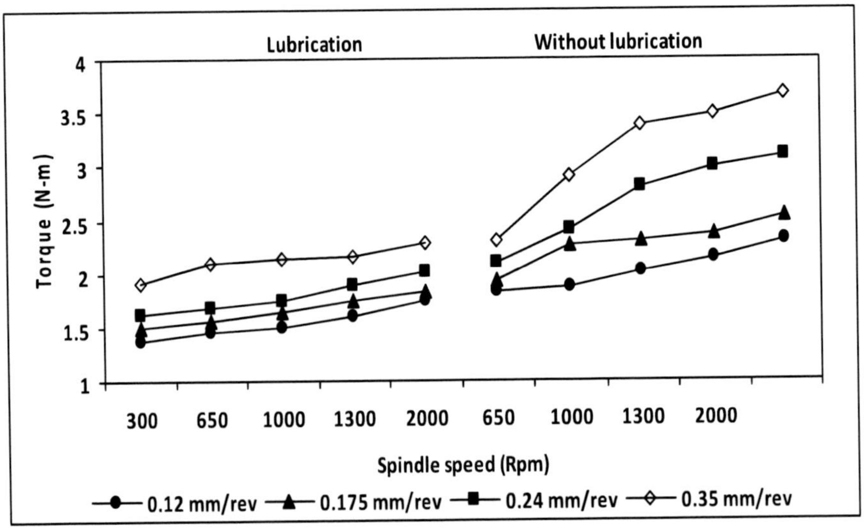

Figure 14. Variation of torque with cutting speed (with lubrication, without lubrication).

Similarly a change in torque is also observed during drilling process with the change in cutting speed in the fully lubrication case and dry drilling condition as shown in Figure 14. A larger influence of feed rate was observed on torque as compared to the cutting speed. For instant, with an increase in cutting speed from 300 to 650 Rpm (approximately 53%), an average increase in torque of about 5.5% is observed, while increasing feed rate from 0.12 to 0.175 mm/rev (approximately 31%) an 8% increase in torque is observed which is in good agreement with (Elhachimi, 1999).

In the same way the temperature generated during drilling process is affected by the cutting speed as shown in Figure 8 and Figure 10. The temperature in cutting region increases with the increase of spindle speed (Ozcelik, 2006). With an increase in cutting speed from 300 to 650 Rpm (approximately 53%), an average increase of about 15% is observed in temperature generated during drilling process. Therefore temperature generated during drilling process increases with the increase of cutting speeds as well as with feed rates but the effect of cutting speed is more dominant as compared to the cutting speeds.

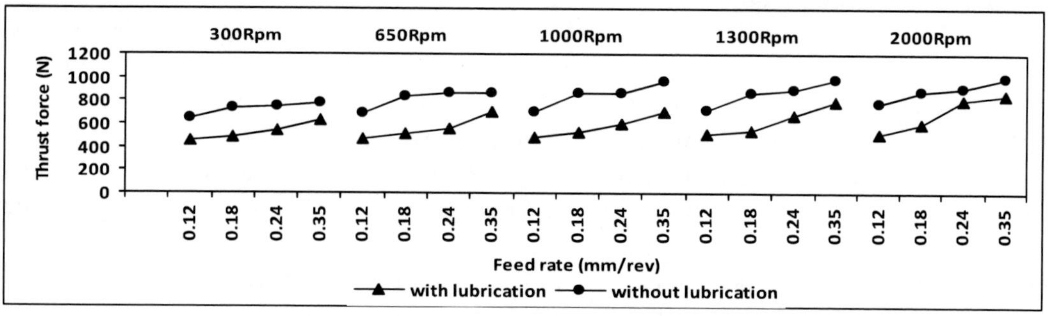

Figure 15. Comparison of thrust force at various cutting condition in fully lubricated and dry drilling condition.

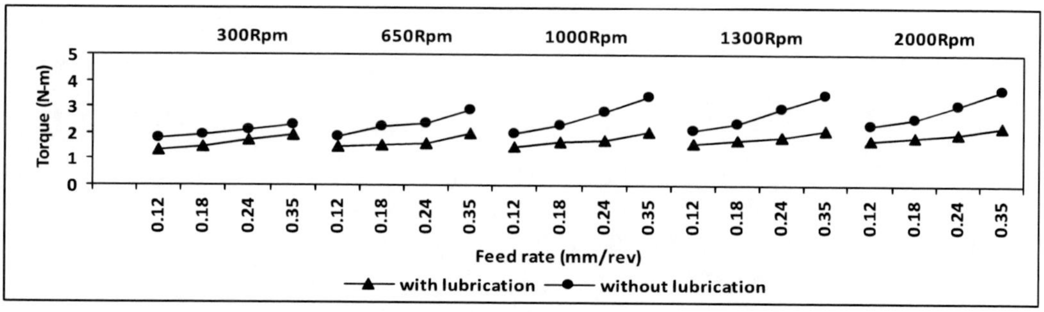

Figure 16. Comparison of torque at various cutting condition in fully lubricated and dry drilling condition

4.3. Effect of Friction

Lubrication in drilling process also greatly influences the thrust force and torque. Figure 15 shows the effect of lubrication on the thrust force. It is observed that an average of 20-30% reduction in thrust force is observed with the introduction of lubrication in drilling process.

Similarly Figure 16 shows the effect of lubrication on the torque during drilling process. An average of 25 to 35% reduction in torque is observed due to the introduction of lubrication in drilling process.

5. Conclusion

3D thermomechanically coupled finite element model of drilling process is used to study the effect of various cutting condition on thrust force, torque and temperature generated during drilling process. A comparison of feed rate shows that the thrust force increases with an increase of feed rate. Approximately 31% increase in feed rate causes a 10% increase in thrust force at fully lubricated case and 13% increase in thrust force for dry drilling case. Variation in torque is also observed with the variation in feed rate. Approximately 31% increase in feed rate causes an 8% increase in torque in fully lubricated case and 12% increase in torque in dry drilling condition.

The effect of cutting speed on thrust force and torque is also investigated. 53% increase in cutting speed causes an average of 5% increase in thrust force; similarly 53% increase of cutting speed causes an average increase of 6% in torque. However, the effect of feed rate on thrust force and torque is more dominant as compared to the cutting speed.

The temperature generated during drilling process increases with the increase of feed rate as well as the cutting speed. 27% increase in feed rate from 0.175 to 0.12 mm/rev causes an 8.5% increase in temperature generated during drilling process. Similarly an increase in cutting speed from 300 to 650 Rpm (53%) causes an average increase of about 15% in temperature. Therefore cutting speed has got more influential effect on temperature generated during drilling process as compared to the feed rate.

A significant reduction in thrust force and torque was also observed with the introduction of lubrication in drilling process. An average reduction of 20-30% and 25-35% in thrust forces and torque were observed respectively in simulation when lubrication is introduced during drilling process.

References

ASM-Hand-Book. (1990) ASM Hand book, *Properties and Selection: Irons, Steels and High-Performance Alloys.* 10th Edition.

Bagci, E., and Ozcelik, B. (2006) Investigation of the effect of drilling conditions on the twist drill temperature during step-by-step and continuous dry drilling. *Materials and Design,* Vol.27, pp.446-454.

Baker M., Rosler, J. and Siemers, C. (2002) Finite element simulation of segmented chip formation of Ti6Al4V. *J Manuf Sci Eng (Trans ASME),* Vol.124, pp.485-488.

Bono M., and Ni, J. (2006) The location of the maximum temperature on the cutting edges of a drill. *International Journal of Machine Tools & Manufacture,* Vol.46, pp.901-907.

Bono M., and Ni, J. (2002) A model for predicting heat flow into the work peice in dry drilling. *Journal of Manufacturing Science and Engineering,* Vol.124, pp.773-777.

Elhachimi, M., Torbaty, S, and Joyat, P. (1999) Mechanical modelling of high speed drilling. 1: predicting torque and thrust. *International Journal of Machine Tools & Manufacture,* Vol.39, pp.553-568.

Elhachimi, M., Torbaty, S. and Joyat, P. (1999) Mechanical modelling of high speed drilling. 2: predicted and experimental results. *International Journal of Machine Tools & Manufacture,* Vol.39, pp.569-581.

Fassi, H., Bousschine, I., Chaaba, A. and Elharif, A. (2003) Numerical simulation of orthogonal cutting by incremental elastoplastic analysis and finite element method. *J Mater Process Technol,* Vol.141, pp.181-188.

Hamade, R., Seif, C. F. and Ismail, F. (2006) Extracting cutting force coefficients from drilling experiments. *International Journal of Machine Tools & Manufacture,* Vol.46, pp.387-396.

Johnson, G., and Cook, W. (1985) Fracture characteristics of three metals subjected to various strains, strain rates, temperature and pressures. *Engineering Fracture Mechanics,* Vol.2, pp.31-48.

Li, R., Hegde, P. and Shih, A. J. (2007) High-throughput drilling of titanium alloys. *International Journal of Machine Tools & Manufacture,* Vol.47, pp.67-73.

Lin S. H., and Chen, I. K. (1996) Drilling carbon fiber-reinforced composite material at high speed. *Wear,* Vol.194, pp.156-162.

Liu, C., and Guo, Y. (2000) Finite element analysis of the effect of sequential cuts and tool - chip friction on residual stresses in a machined layer. *Int J Mech Sci,* Vol.42, pp.1069-1086.

MSC-Marc-Users-Guide (2005) MSC.MARC User,s Guide, Version 2005. MSC software corporation. Los Angeles 2005.

Ng, E., El Wardany, T., Dumitrescu M., and Elbestawi, M. (2002) Physics-based simulation of high speed machining. *Machin Sci Technol,* Vol.6, pp.301-329.

Oden, J. T. and Pires, E. B. (1983) Nonlocal and Nonlinear Friction Laws and Variational Principles for Contact Problems in Elasticity, *J. of Applied Mechanics,* V. 50.

Ozcelik, B., and Bagci, E. (2006) Experimental and numerical studies on the determination of twist drill temperature in dry drilling: A new approach. *Materials and Design,* Vol.27, pp.920-927.

Pirtini, M., and Lazoglu, L. (2005) Forces and hole quality in drilling. *International journal of Machine Tool & Manufacturing,* Vol.45, pp.1271-1281.

Shen, Q., Lee, T. C. and Lau, W. S. (1997) A finite element analysis of teperature distribution in spade drilling. *Journal of Materials Processing Technology,* Vol.66, pp.112-122.

Shi, G., Deng, X. and Shet, C. (2002) A finite element study of the effect of friction in orthogonal metal cutting. *Finite Elements in Analysis and Design,* Vol.38, pp.863-883.

Strenkowski. J. S., Hsieh. C. C. and Shih. A. J. (2004) An analytical finite element technique for predicting thrust force and torque in drilling. *International Journal of Machine Tools & Manufacture,* Vol.44, pp.1413-1421.

Ueda, T., Nozaki, R. and Hosokawa, A. (2007) Temperature Measurement of Cutting Edge in Drilling-Effect of Oil Mist. *Annals of the CIRP,* Vol.56, pp.93-96.

Umeda, T., Umeki, H. and Mimura, K. (2005) Numerical evaluation of measurment accuracy of non coaxial Hopkinson bar method. *JSME,* Vol.48, pp.215-221.

Yang, A. J., Jaganathan, V. and Du, R. (2002) A new dynamic model for drilling and reaming processes. *International Journal of Machine Tools & Manufacture,* Vol.42, pp.299-311.

Yen, Y., Jain, A. and Altan, A. (2004) A finite element analysis of orthogonal machining using different tool edge geometries. *J Mater Process Technol,* Vol.146, pp.72-81.

In: Machining and Forming Technologies, Volume 2
Editor: J. Paulo Davim
ISBN: 978-1-61122-820-5

Chapter 12

PCA Based NN Approach for Drill Wear Prediction in Drilling Mild Steel Specimen

S.S. Panda and S.S. Mahapatra*
Department of Mechanical Engineering. National Institute of Technology
Rourkela 769008, INDIA.

Abstract

In the present work, principal component has been fused with artificial neural network (ANN) for prediction of drill wear so as to reduce the redundancy of the network. Large numbers of experiments have been conducted and sensor signals have been acquired using data acquisition system. Cutting forces, torque, vibrations along with other process parameters such as spindle speed, feed rate, drill diameter, chip thickness and surface roughness have been used as input signals to train a standard back propagation neural network (BPNN) with flank wear as output. These input parameters with different combination of inputs have been tried and it has been observed that inclusion of chip thickness and surface roughness leads to a better trained network and hence a better prediction of wear. Principal component of these input parameters has been derived thereafter and has been used to predict the flank wear using BPNN.

Keywords: Neuron; sensor integration; signal analysis; flank wear; BPNN; PCA.

1. Introduction

The reason for acquiring the drill wear state information is to enhance the predictive capability to allow the machine operator to schedule tool change or regrind just in time to avoid under use or overuse of tools, avoid shutdown of machines due to damage, and to minimize scrap or rework. Drill wear also affects the ability of the hole cutting system to satisfy specified performance characteristics, such as hole roundness, centering, burr

* Corresponding author: Email: mahapatrass@yahoo.com

formation at drill exit, and surface finish. Wearing action in the tool is an inevitable phenomenon leading to the loss of dimensional accuracy and possible damage to the work piece. Hence, in order to alleviate such problems, online prediction of cutting tool wear becomes an important issue for today's manufacturing industries. Application of an appropriate monitoring technique can predict the failure of cutting tool; thus avoiding possible damage. There have been lots of efforts in the direction of prediction of wear using indirect methods and the following paragraph describes some of the relevant research in this direction.

Many early works on deterioration and failure of drill have been reported in literature. Altogether different types of drill wear can be recognized as outer corner wear, flank wear, margin wear, crater wear, chisel wear and chipping at the lip as reported by Kanai and Kanda (Kanai and Kanda 1978). They reported that because of the adhering nature of material, it is difficult to measure the extent of wear. Inasaki (Inasaki 1998) reported that detection of cutting tool state is more complex in multipoint cutting tool due to non-constant chip thickness and more than one active cutting edge. Some of the previous works (Chungchoo and Saini 2002; Haili et al. 2003) reported that average width of crater and maximum depth of flank wear could be used as to assess tool failure. Using roughness profile, they reported that the tool-chip contact consists of three zones: sticking, transition and sliding zones in turning operation. Elhachimi et al. (1999a) used analytical approach for determination of thrust force and torque in drilling in terms of the geometry of the tool, thermal and mechanical properties of the work material and the cutting conditions in order to detect the progress of drill wear. They even concluded that with quite large cutting speed values and a weak feed rate, good surface quality and a dimensional accuracy of the holes could be achieved. The concepts of dynamic chip thickness and dynamic chip area were introduced by Gong et al. (2005). They reported that chip thickness and chip area varied along the motion of the drill due to drill deflection. Yeo et al. (2000) estimated the amount of downtime due to cutter breakage. Gao et al. (2000) investigated the chip deformation, cutting forces and wear in the deep hole drilling using boring and trepanning association (BTA) drill which constituted of inner edge, middle edge and outer edge. They reported that the chip deformation at the inner edge is the largest among the other cutting edges. They studied the chip deformation process with feed rate, rake angle and cutting speed. They developed an empirical equation of the axial force based on the trial data.

Among the two approaches for tool condition monitoring, direct method has a limitation in continuous drilling operations. In indirect technique, factors affecting the wear are used to assess the tool condition. There are many work reported on multi sensor based tool condition monitoring (TCM). It is relatively common to measure only the thrust or axial component of force for TCM in drilling as suggested by Thangaraj and Wright (1998). The spindle and feed motor current are also used as inputs to the TCM system by Ramamurthi and Hough (1993). Liu and Chen (1998) used eight indices like average thrust force, average torque, peak thrust force, peak torque, root mean square (RMS) thrust force, RMS torque, area under thrust force in time domain and area under torque in time domain which are extracted from thrust and torque measurements in the drilling process. Stephenson and Agapiou (1992) formulated a model for calculation of the main cutting edge contributions to the torque, thrust force and radial forces generated when drilling a cast iron with carbide drills. Torque and thrust forces for sharp drills vary with feed in the same way as they do for classical materials as given in literature (Morin et al. 1995; Elhachimi et al. 1999b). Further, it is reported that feed rate affects drilling forces and so affect wear, but cutting speed has no significant effect over the range of speeds used. Stiffness and damping properties of the drilling system has no effect on wear as suggested by Abu-Mahfouz (2003). He predicted wear rate in drilling using Fast Fourier Transformation (FFT) of vibration signature as an input to ANN. Vibration monitoring techniques applied for detection of drill failure have been reported by several investigators (Wang et al 2003; Ko and Cho 1994; Li et a. 1994; Liu and Wu 1990). Scheffer and Heyns (2001) used eight different features extracted from vibration signature as input to the self-organizing map (SOM) for wear classification in turning operation. Bahr et al. (1997) combined an on-line, indirect technique (tool fracture detection based on vibration signals) with an offline, direct technique (wear area measurement by means of a Charge-Coupled Device (CCD) camera.

Many works on neural network based tool condition monitoring have been reported in the literature. Lin and Ting (1996) used six back propagation neural network (BPNN) architectures for the on-line detection of drill wear. They concluded that success rate of the on-line drill wear detection system depends on the BPNN architecture. However, increase of number of neurons does not necessarily increase the success rate. Lin and Ting (1996) used the neural network model to study the drill wear and observed that the training error in case of sample mode converges faster than that in case of batch mode. Liu and Anantharaman (1994) used nine features representing drill condition as input to multi layer perceptron and found 100% accuracy for correct classification of drill wear. Sanjay et al. (2005) used drill diameter, feed, cutting speed, time, force and torque as inputs and flank wear was estimated using different structures of artificial neural network (ANN). They suggested that wear estimation by neural network structures is better than that by statistical method (regression analysis). Wong and Hamouda (2003) used new type of randomized rounding (RR) scheme instead of steepest descent method to find out the global minimum instead of trapping at local minima in back propagation network. Brophy et al. (2002) used BPNN model to detect the condition of the drill state, either normal or abnormal for short term and long term experiment and reported that short run predicted better than long run. Wardany et al. (1996) used discernment features like kurtosis value in time domain and cepstrum in frequency domains which are insensitive to the location of sensor and cutting condition to monitor the tool wear in drilling. Singh et al. (2006a) used back propagation neural network for prediction of flank wear of high speed steel drill in a copper work piece using spindle speed, feed rate, drill diameter, thrust force and torque as input parameters and maximum flank wear as output parameter in a neural network. Influence of network paradigm due to inclusion of chip thickness as extra parameter has been reported by Singh et al. (2006b). Panda et al. (2006) have used fuzzy back propagation network with chip thickness as fuzzy input parameter for prediction of flank wear of HSS drill. Panda et al. (2008) had used different back propagation neural network (BPNN) and radial basis function network (RBFN) for predicting the flank wear of HSS drill in drilling a cast iron work-piece and reported that RBFN converges faster compare to BPNN but yielding a higher prediction error.

Since wear of drill related to number of direct and indirect factors and hence can be considered as dimension reduction problem in which large number of interrelated factor could be grouped into a smaller set of independent factors, the proposed approach applies factor analysis (Principal component analysis), a dimension reduction technique, to identify the wear on drill. Rummel (1988) used factor analysis using a mathematical model to study the human ability. Detailed description of factor analysis can be found in the relevant literature such as Kleinbaum et al. (1986) approach. Fisher (1936) had developed a linear classification algorithm to transform n dimensions pattern to 2 dimensions. Hong and Yang (1991) have used a method for constructing a classifier on the optimal discernment plane by using minimal distance criterion for multi classification problems. Kittler and Young (1973) have used Principal component mapping for n dimensional vector space R^n onto a 2 dimensional space. A large number of input variables can be conveniently transformed into smaller number of principal components retaining information expressed by original variables to obtain computational advantage when neural networks are used for online tool wear monitoring (Purushothaman and Srinivasa, 1998).

Literature review reveals that sensor signals give a definitive indication of wear. However, no single sensor signal could predict wear alone. In the present work, it was thought to look for proper combination of number of sensor signals and other process parameters in drilling which will lead to a better prediction of drill wear. Therefore, in the present work, different process parameters along with sensor signals have been used to train a BPNN to obtain the best network architecture for future prediction. Principal component analysis has been used to transform nine dimensional vectors to smaller number of vectors.

2. Back Propagation Neural Network

The basic structure of a BPNN having input, hidden and output layers has been considered in the present work. The input layer receives information from external sources, and passes this information to the network for processing. The hidden layer receives information from the input layer, and does the information processing. The output layer receives processed information from the network, and sends the results to an external receptor. The sum of modified signals (total activation) is then modified by a sigmoid transfer function. Batch mode type of supervised learning has been used in the present case, where, interconnection weights are adjusted using delta rule algorithm after sending the entire training sample to the network. During training, the predicted output has been compared with the desired output, and the mean square error has been calculated. If the mean square error is more than a prescribed limiting value, it is back propagated from output to input, and weights are further modified till the error or number of iteration are within a prescribed limit.

Each neuron is connected to all other neurons in the adjacent layer through the weighted connections. In the forward pass the synaptic weights remain unaltered through the network and function signals of the network are computed on a neuron-by-neuron basis. Specifically, the function signal appearing at the output of neuron, j, is computed as:

$$y_j(n) = \Phi\left(v_j(n)\right) \tag{1}$$

where $v_j(n)$ is the net internal activity level of neuron, j, defined by:

$$v_j(n) = \sum_{i=1}^{p} w_{ji}(n) y_i(n) \tag{2}$$

where p is the total number of inputs applied to neuron, i, and $w_{ji}(n)$ is the synaptic weight connecting neuron, i, to neuron, j, and $y_i(n)$ is the input signal of neuron, j or, equivalently, the function signal appearing at the output of neuron, i. If neuron, i, is in the first hidden layer of the network, then the index , i, refers to the ith input terminal of the network, for which we write

$$y_i(n) = x_i(n) \tag{3}$$

where $x_i(n)$ is the ith element of the input pattern. If, on the other hand, neuron, i is in the output layer of the network, the index, i, refers to the ith output terminal of the network, for which we write

$$y_i(n) = z_i(n) \tag{4}$$

where $z_i(n)$ is the ith element of the output pattern. This output is compared to the desired response, $d_i(n)$, for obtaining the error signal for the ith output neuron. Thus forward pass begins with presenting the input pattern at the input layer and ends with finding the error signal for each neuron at the output. The backward pass starts at the output layer by passing the error signals left-ward through the network, layer by layer, and recursively calculating the local gradients, δ, for each neuron. This recursive process permits the synaptic weight of the network to undergo changes in accordance with the delta rule as discussed in the next section. For a neuron located in the output, the 'δ' is simply equal to the error signal of that neuron multiplied by the first derivative of its nonlinearity. In this way calculate the changes to the weights of all the connections feeding into the output layer. Using the 'δ' for the neurons in the output layer, 'δ' for the neurons in the penultimate layer and thus the changes to the weights of all connections feeding into it can be calculated. The recursive computation is further computation is further continued, layer by layer, by propagating the changes to all synaptic weights.

In the back-propagation algorithm, we approximate the trajectory in the weight space computed by the steepest descent method. The value of the learning rate constant η is to be decided before starting the learning process. When a smaller value is used, the changes to the synaptic weights in the network will also be smaller and the trajectory in the weight space will be smoother. This, however, results in a slower rate of learning. And if a higher value of η is used, the resulting changes in the synaptic weights assume such a form that the network may time avoiding the instability is by introducing an extra term in the equation for finding the changes in weights as

$$\Delta w_{ji}(n) = \alpha \Delta w_{ji}(n-1) + \eta \delta_j(n) y_i(n) \tag{5}$$

where α is the momentum constant. The use of the moment constant is known to provide beneficial effects on the learning behaviour of the network. It may also prevent the learning process from terminating in shallow local minima on the error surface.

Entire experimental data set is divided into training set, testing set and validation set. The error on the testing set is monitored during the training process. The testing error normally decreases during the initial phase of training, as does the training set error. However, when the network begins to over fit the data, the error on the testing set will typically begin to rise.

When the testing error starts increasing for a specified number of iterations, the training is stopped; and the weights at the minimum value of the testing error are returned. With the trained network, unseen data set (validation set) are verified and percentage of variation of predicted output (flank wear) with respect to the actual output is thus evaluated.

3. Principal Component Analysis

Principal component analysis (PCA) is the most widely used and straightforward and quantitatively rigorous method, which transforms a given set of interrelated variables into a new set of variables called the principal components (corresponding to factors in factor analysis). The set of principal components generated are uncorrelated linear combinations of the original variables and account for the total variance of the original data. In this method, all the principal components are generated in such a way that they are orthogonal to each other; hence, correlation between them is zero. The principal components are generated in a sequentially ordered manner with decreasing contributions to the variance i.e. the first principal component explains most of the variations present in the original data, and successive principal components account for decreasing proportions of the variance. This property means that the data points can be rigorously separated into distinct clusters when projected into a space spanned by the first few principal components, which are called factors. This achieves the dimensionality reduction objective of factor analysis. In short, PCA is a statistical technique which generates uncorrelated linear combinations of original variables and account for the total variance

Principal component analysis is an analytical technique whereby a complex data set containing p variables is transformed to a smaller set of new variables, which maximize the variance of the original data set. All of the new variables are independent *i.e.* are not correlated (whereas the original, untransformed variables may have been correlated to a lesser or greater extent). Since the new principal component (PC) axes (Z_1, Z_2, ... , Z_p) are uncorrelated (*e.g.* Z_1 and Z_2 are perpendicular as shown in Figure 1).

In principle, each of the principal components is a linear combination of the original Y values for the p variables;

$$
\begin{aligned}
PC_1 &= c_{11}Y_1 + c_{12}Y_2 + c_{13}Y_3 + \dots\dots + c_{1p}Y_p \qquad (axis\ Z_1) \\
PC_2 &= c_{21}Y_1 + c_{22}Y_2 + c_{23}Y_3 + \dots\dots + c_{2p}Y_p \qquad (axis\ Z_2) \\
&\dots\dots\dots\dots\dots\dots \\
PC_p &= c_{p1}Y_1 + c_{p2}Y_2 + c_{p3}Y_3 + \dots\dots + c_{pp}Y_p \qquad (axis\ Z_p)
\end{aligned} \tag{6}
$$

where $c_{a,b}$ is the component score coefficient for variable b on PC axis Z_a, and Y_b is the Y score for variable b.

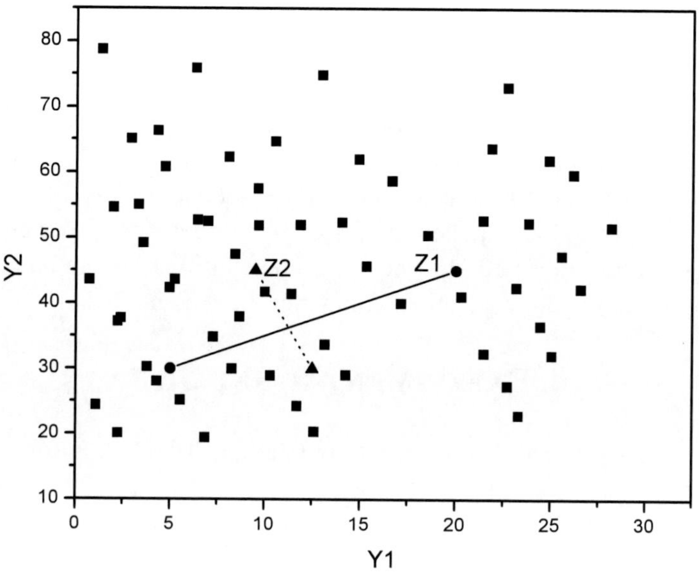

Figure 1. Diagram of uncorrelated principal component axis.

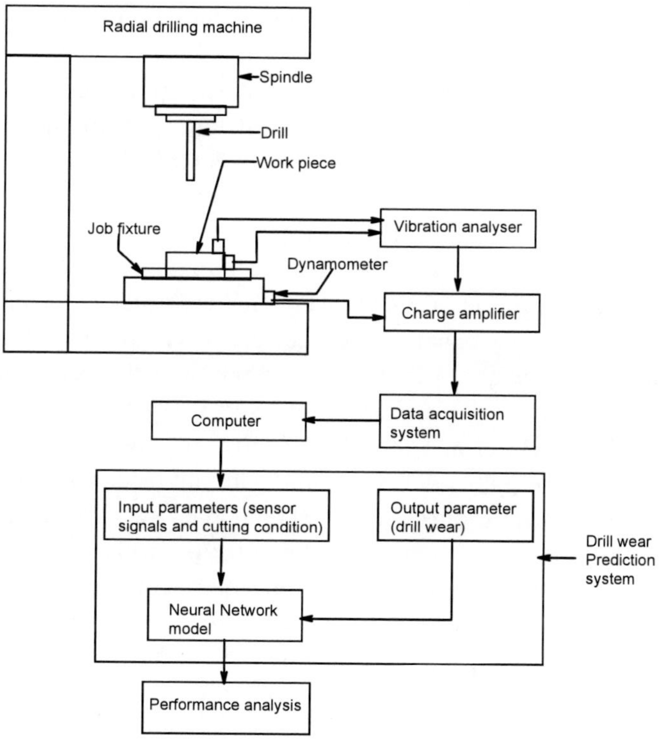

Figure 2. Schematic diagram of the experimental set-up.

Principal Component Analysis transforms a multivariate set of variables (Y_1, Y_2, ... , Y_p) to new variables (Z_1, Z_2, ... , Z_p) , which are uncorrelated. The first principal component consists of a principal component coefficient (α_i) for each variable (p) such that there is maximal variance in the calculated score for each case (n); the factor score for each case is calculated as $\alpha_1.Y_1 + \alpha_2.Y_2 + \ldots + \alpha_i.Y_i + \ldots\alpha_p.Y_p$ where Y_i is the centered value for the ith (Y_i - mean Y for the ith variable). The first principal component axis for the raw variables Y_1 is the new axis Z_1. The second principal component consists of the next set of principal component coefficients (α_i) such that there is a maximal remaining variance in the calculated score for each case (n), and there is no correlation of the second principal component with the first. Z_2 is the second principal component axis for Y_2. Further sets of principal components (third, fourth, etc.) can be calculated until no statistical significance can be attributed to that component (e.g. by χ^2 test) or all m principal components are computed.

4. Experimental Set Up

In the present work, a radial drilling machine (Batliboi Limited, BR618 model) is used for the drilling operation and Figure 2 shows the schematic diagram of the present experimental setup. Details of the equipment, sensors and the cutting conditions for the drilling operation performed has been presented in Table 1. Table 2 and 3 show the detail properties of the drill and work piece material used in the present study. In all the drilling operations performed in the present work, no coolant has been used. Chips are collected during each cutting condition and average thickness is measured. Similarly, after each cutting operation surface roughness of drill hole is measured with help of pocket surface roughness tester (make Mahr) with maximum roughness depth 0.2 μm to 25.3 μm and maximum stylus force 15.0 *mN*. In all the drilling operations performed in the present work, no coolant has been used. Root mean square (RMS*)* values of thrust force and torque signal are recorded through a piezo-electric dynamometer (Kistler, 9272). Signals from the dynamometer were passed through low pass filter, amplified through charge amplifier (Kistler, type 5015 model), and stored in the computer through a data acquisition system (Advantech, PCL 818 HG, 16 channel analog to digital (A/D) converter in 16 bit digital time-discrete, and 100 *kHz* sampling rate). Two piezo electric accelerometers (model Bruel & Kjaer, type 4396) have been used to capture vibration signals. One accelerometer has been attached on the top surface of the mild steel specimen to extract feed vibration and other on the side surface of the mild steel specimen to extract radial vibration. Signals from accelerometer were passed through vibration analyzer (Bruel & Kjaer, type 3560 D) in the frequency range 7 *Hz*-25.6 *kHz*. RMS of maximum amplitude of vibration both in feed and radial direction were collected through Bruel & Kjaer pulse software (version 7) and is stored in the computer through data acquisition system through data recorder (type Bruel & Kjaer, Type 7701). The optical microscopes along with Carl-Zeiss software interfacing have been used to measure flank wear. The maximum depth of flank wear has been used as the criterion to characterize the drill condition, and is obtained by measuring the wear at four different points randomly on either of the cutting edges.

Table 1. Details of the cutting conditions and experimental set up

Ranges of spindle speed (rpm)	Ranges of feed rate (mm/rev)	Ranges of drill diameter (mm)	Experimental Set-up	Features collected/ Measured	Specification of Experimental Set-up
315,500, 630, 1000	9,10,1 1,12		Radial drilling machine (Batliboi Limited, BR618 model)		50-1600 rpm in 16 steps, 0.13-1.4 mm/rev in 8 steps
			Dynamometer (Kistler type 9272)	Thrust force, torque	Sensitivity 3.8 Pc/N and -1.6 Pc/N-cm
			Charge amplifier (Kistler, 5015)		25Khz sampling rate, 1000N/v and 10Nm/v
			Data acquisition system for cutting force signals (Advantech, PCL 818 HG, 16 channel A/D)		10 Khz sampling rate
			Accelerometer probes (Bruel & Kjaer 4396)	Feed and radial vibrations	25.6 Khz frequency range
			Vibration analyzer (Bruel & Kjaer, 3560 D)		7 Hz-25.6 kHZ
			Data acquisition system for vibration signals (Bruel & Kjaer, 7701)		65.5 kHz
			Optical microscope (model Carl Zeiss Axiotech)	Flank wear	25×25 mm travel range, height of 440 mm with a 30° viewing angle
			CCD color camera (WAT-201A model)		280 K pixel and 20 magnification
			Surface roughness tester	Drilled hole roughness	Average surface roughness (0.2 μm -25.2 μm)
			Digital micrometer (model IP54)	Chip thickness	Accuracy $\pm 0.001''$, resolution $\pm 0.00005''$
			Scanning electron microscope	Composition study	Resolution 4 nm, Accelerating Voltage 0.5 to 30 kV, Magnification x15 to 200,000
			Brinell hardness tester	Hardness of HSS drill and work piece	Hydraulic, 1500-3000 kgf load range, 450 mm throat depth
			Too maker's microscope	Geometry of HSS drill	30 times magnification

Table 2. HSS drill geometry and chemical composition
(a) Geometry of HSS drill bit (long series)

Tool diameter (mm)	Flute length (mm)	Total length (mm)	Point angle (degree)	Helix angle (degree)
9.5	115	175	118	30
10.5	121	184	118	30
11.5	128	195	118	30
Flute 2 flutes Flute type parabolic Shank type straight cylindrical Coating any no				

(b) Chemical Composition of HSS drill materials (wt%)

Tungston	Cromium	Vanadium	Cobalt	Molybdenum	Carbon	Hardness
18	4.3	1.1	5	0.65	0.75	290 BHN

Table 3. Mild steel chemical composition and mechanical properties
(a) Chemical composition (wt%)

Carbon	Manganese	Silicon	Sulfur	Phosphorous	Other	Rest
0.07	1.1	0.5	0.035	0.025	0.08 Ti 0.07 Zr	Iron (Fe)

(b) Mechanical properties

Ultimate tensile stress (MN/m^2)	Yield stress (MN/m^2)	Density (Kg/m^3)	Elongation (%)	Vicker's Hardness
300	170	7850	42	140

4.1. Static Calibration of the Dynamometer

In order to determine induced voltage under static load, the dynamometer was calibrated. The calibration was made for feed force and torque. An explanation to calibration of feed force is described in the context. The loads up to 2100 N with 25 N intervals were applied and the induced voltage was recorded for each load intervals. The induced voltage is then transformed to force (N) through amplifier. Thus a calibrated curve for feed force obtained is as shown in Figure 3. In order to verify the consistency, the measurements were repeated three times and very close values within an error of $\pm 2\%$ were obtained. A nonlinear regression model of 2nd order with $R^2=1$ has been fitted with the calibrated graph to check the responses.

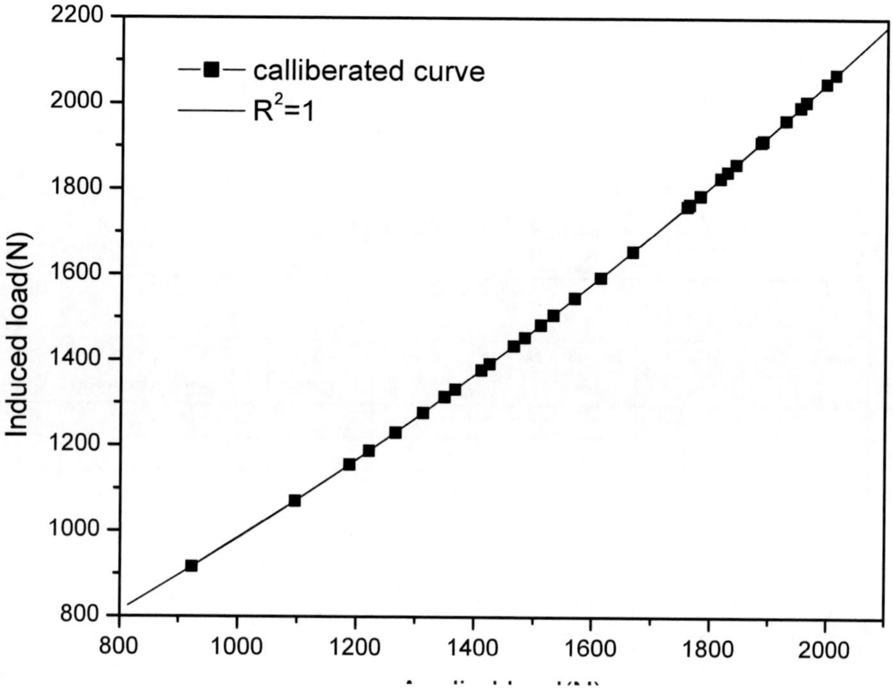

Figure 3. Static calibration of dynamometer.

4.2. Dynamic Calibration of the Dynamometer

Frequencies of the dynamometer determine its dynamic nature. A test experiment has been carried out while the machine was running idle and during cutting operation so as to collect the frequency of the vibration. It has been observed that the frequency during machine idle time is more than that during cutting time and hence the recorded force is not influenced by the dynamic nature of dynamometer.

5. Results and Discussion

Drilling operations have been conducted over a wide a range of cutting condition. Spindle speed has been varied in the range 315 *rpm* to 1000 *rpm* in four steps. Feed rate has been varied from 0.13 to 0.36 *mm/rev* in four steps. High-speed steel (HSS) drills of four different diameters of (9*mm*, 10*mm*, *11mm* and 12*mm*) have been used for drilling through holes of 15*mm* depth. A full factorial design has been considered of three different factors with four different level thus conducting 64 different drilling operation in mild steel plates.

5.1. Wear Prediction in BPNN

In order to understand the relevance of signal features and process parameters in describing the tool condition, different tool condition monitoring strategies have been studied to extract the best combination. Table 4 shows the eleven different strategies studied in the present work.

Table 4. Decision strategy

	Input to ANN									Output of ANN
Strategy	Speed	Feed	Diameter	Thrust force	Torque	Feed vibration	Radial Vibration	Chip thickness	Surface roughness	Flank wear
I	+	+	+	+	+					+
II	+	+	+			+	+			+
III	+	+	+	+	+	+				+
IV	+	+	+		+	+	+			+
V	+	+	+	+	+	+	+			+
VI	+	+	+	+	+			+		+
VII	+	+	+			+	+	+		+
VIII	+	+	+	+	+	+		+		+
IX	+	+	+		+	+	+	+		+
X	+	+	+	+	+	+	+	+		+
XI	+	+	+	+	+	+	+	+	+	+

+ indicate inclusion of parameters

In the present work 64 sets of data have been obtained at different cutting conditions in drilling a mild steel specimen. The different sensor signals obtained are thrust force, torque, feed vibration, radial vibration and in addition chip thickness and surface roughness are also measured. The neural network architecture has been modeled with any possible combination of thrust force, torque, feed vibration, radial vibration, drill diameter, feed rate, spindle speed, chip thickness and surface roughness as input and flank wear as the output of the network. 64 data sets have been partitioned into 40 training set, 18 testing set (used to evaluate the testing error to be used as stopping criteria for the network) and rest 6 validation set (unseen data set) which are used for validation to all different strategy in the present work. Back propagation neural network architectures have been prepared using various combination of input parameter such as spindle speed, feed rate, drill diameter, thrust force, torque, feed vibration, radial vibration, chip thickness and surface roughness and in all the case the output of the network has been flank wear. In the present case eleven different strategies has been tried. The normalized data sets in the range of 0.1 to 0.9 are used for training the network.

Large number of architectures has been tried corresponding to each of the strategies to arrive at the best drill wear prediction system using BPNN. Table 5 shows only the best and the second best network in terms of the validation error and the number of iteration corresponding to each strategy.

Table 5. Optimum network architectures for BPNN

Architecture L-M-N	η	α	Iteration	MSE training	MSE testing	Maximum validation error (%)	Strategy
5-3-1	0.9	0.7	895	0.000332	0.000599	5.9357	I
5-5-1	0.1	0.1	19814	0.000299	0.000534	5.0772	
5-5-1	0.1	0.5	19849	0.000293	0.000452	5.2618	II
5-8-1	0.9	0.9	2177	0.000155	0.000390	5.9784	
6-3-1	0.1	0.5	19830	0.000319	0.000557	5.0577	III
6-8-1	0.5	0.3	6811	0.000241	0.000420	6.8109	
6-3-1	0.3	0.3	19460	0.000254	0.000367	4.6904	IV
6-5-1	0.5	0.9	870	0.000233	0.000478	5.3619	
7-12-1	0.5	0.7	5636	0.00017	0.000319	4.8929	V
7-12-1	0.9	0.3	565	0.00034	0.000575	6.2404	
6-3-1	0.3	0.7	4101	0.000273	0.000506	5.1334	VI
6-5-1	0.1	0.7	19707	0.000231	0.000516	4.8900	
6-3-1	0.9	0.5	9612	0.000294	0.000353	4.6780	VII
6-8-1	0.7	0.7	2837	0.000224	0.000393	5.3707	
7-3-1	0.1	0.3	19826	0.000288	0.000485	4.5703	VIII
7-8-1	0.7	0.7	8994	0.000102	0.000239	6.4241	
7-5-1	0.1	0.1	19952	0.000289	0.000472	4.4593	IX
7-5-1	0.7	0.9	3387	0.000197	0.000394	6.3360	
8-5-1	0.9	0.3	4060	0.000251	0.000414	5.3678	X
8-8-1	0.5	0.3	14363	0.000157	0.000262	4.2557	
9-8-1	0.5	0.7	8703	0.000131	0.000158	3.2197	XI
9-8-1	0.9	0.3	1736	0.00023	0.000333	6.1009	

L= Number of neuron at input layer
M= Number of neuron at Hidden layer
N= Number of neuron at output layer

In comparing the performances of strategy I and strategy II, it could be observed that strategy II gives better prediction of wear. Therefore, it could be inferred that use of vibration signal along with other process parameters give better prediction of wear compared to that using thrust force and torque. It could be observed from Table 5 that inclusion of chip thickness as one of the input parameters in the neural network reduces the testing as well as validation error of the network compared to case where chip thickness is not used as an input parameter. Table 5 also shows that inclusion of surface roughness as an additional input to the network further reduces the testing and validation error. So in general increase in number of sensor signals in the network leads to better training of the network and hence a better tool wear prediction.

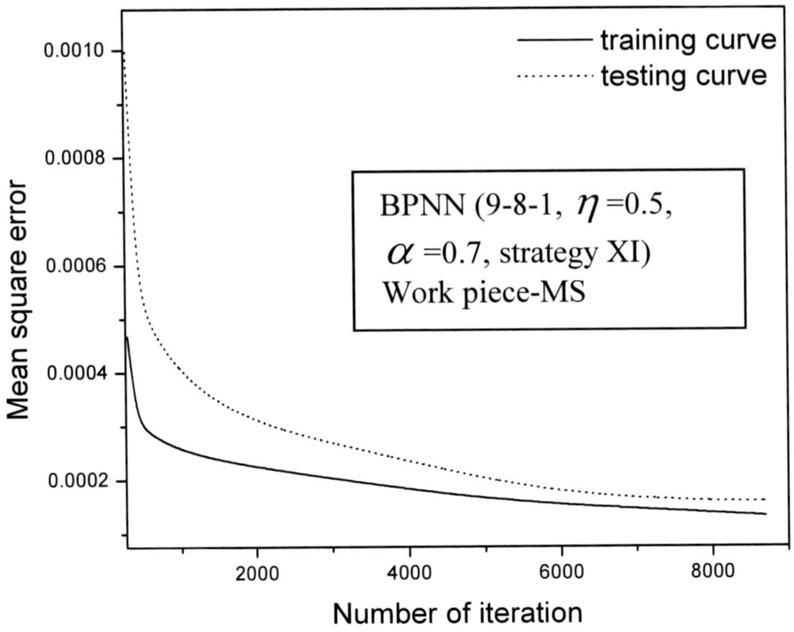

Figure 4. Variation of mean square error with number of iteration of 9-8-1 with $\eta = 0.5$ and $\alpha = 0.7$.

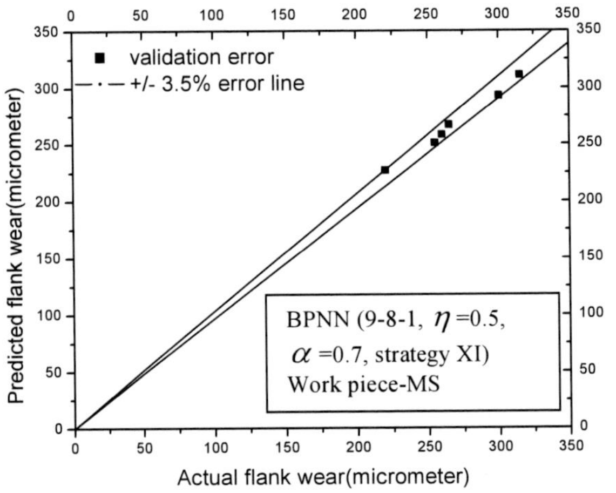

Figure 5. Comparison of Predicted value with actual value of 9-8-1 with $\eta = 0.5$ and $\alpha = 0.7$.

It could be observed that the MSE training in the case of strategy XI is much less compared to all other strategies and also the same has been achieved at a much less number of iteration. Therefore it is clear that inclusion of surface roughness value in the neural network input increases the efficacy of the network. Therefore, after running a large number of runs, the optimum network architecture in the present case has been observed to be 9-8-1 with $\eta = 0.5$ and $\alpha = 0.7$. Variation of mean square error in training and testing with number of iteration for the optimum network is shown in Figure 4. After the network has been trained, it has been validated with unknown data sample. Figure 5 shows percentage of error between maximum depth of actual value and the predicted value of flank wear and it can be observed that present neural network of 9-8-1 with $\eta = 0.5$ and $\alpha = 0.7$ predicts the results within $\pm 3.21\%$ error.

Table 6. Variance of Eigen values.

Value	PC 1	PC 2	PC 3	PC 4	PC 5	PC 6	PC 7	PC 8	PC 9
Eigen value	0.302	0.149	0.096	0.006	0.003	0.002	0.000	0.000	0.000
Variance (%)	54.127	26.799	17.189	1.056	0.523	0.271	0.024	0.009	0.003
Cumulative variance (%)	54.127	80.926	98.115	99.170	99.694	99.964	99.988	99.997	100.00

Table 7. Correlation coefficient

	Diameter	Spindle speed	Feed rate	Thrust Force	Torque	Feed Vibration	radial vibration	Chip thickness	Surface Rough-ness
PC 1	0.181	-0.213	0.958	0.972	0.974	-0.521	-0.517	0.831	0.973
PC 2	0.819	0.561	-0.02	0.172	0.165	0.835	0.837	0.348	0.057
PC 3	-0.536	0.795	0.280	0.000	-0.006	0.087	0.082	-0.317	0.201

5.2. Wear Prediction by PCA Based NN

In order to extract the principal component, a test statistics of different variable of sample size of 64 have been studied. XLSTAT version 6 has been used for finding of principal components. Table 6 provides the Eigen values for all nine principal components, with the percentage of the variance which they explain and their cumulative percentage of variation explained. Eigen value for the first principal component is 0.302, which is 54.12% of the sum of the Eigen values, hence 54.12% of the total variance. The second principal component has an Eigen value of 0.149, which is 26.79% of the total, and cumulatively Eigen values one and two explain 80.92% of the total variance. The third principal component has an Eigen value of 0.096%, which is 17.18% of the total, and cumulatively Eigen values one, two and three explain 98.11% of the total variance. Rest of the component explains only about 1.89% of cumulative variance. The component loadings are listed for the original variables and each principal component in Table 7. These loadings are the correlations between the original variable and the principal component and indicate roughly the relative contributions/ importance of each variable in each principal component *e.g.* feed rate has highest loading for PC1 whereas drill diameter has largest loading for PC2 and spindle speed has the highest loading for PC3. Total three principal component axes have been considered from the present work is gauged from the Figure 6 as these three components explain about 98.11% of the total variance.

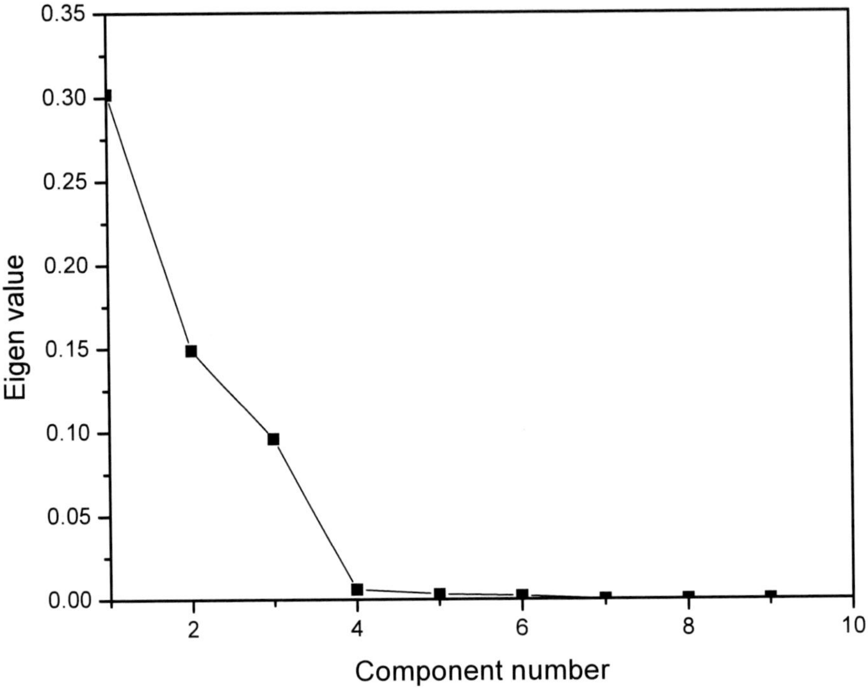

Figure 6. Scree plot.

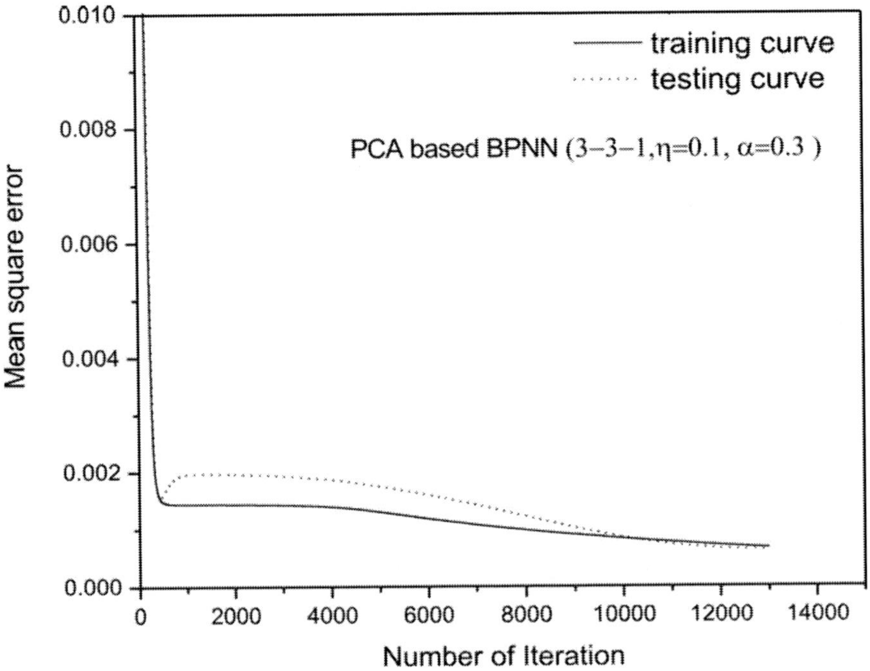

Figure 7. Variation of mean square error with number of iteration of 3-3-1 with $\eta = 0.1$ and $\alpha = 0.3$.

Table 8. Optimum network architectures for PCA based BPNN

Architecture L-M-N	η	α	Iteration	MSE training	MSE testing	Maximum validation error (%)
3-3-1	0.1	0.1	16728	0.000664	0.000631	6.42
	0.1	0.3	13010	0.000664	0.000631	6.42
	0.1	0.5	9293	0.000665	0.000631	6.43
	0.1	0.7	5578	0.000666	0.000631	6.44
	0.3	0.3	4336	0.000665	0.000631	6.43
	0.3	0.5	3099	0.000666	0.000631	6.44
	0.3	0.7	1868	0.000668	0.000631	6.46
	0.5	0.1	2426	0.000871	0.000944	8.51
	0.5	0.3	2511	0.000686	0.000637	6.67
	0.5	0.5	1861	0.000667	0.000631	6.45
	0.7	0.5	946	0.000891	0.00099	8.69
	0.9	0.1	4139	0.001182	0.001041	12.13
	0.9	0.7	401	0.001028	0.00132	9.79
3-5-1	0.1	0.1	13665	0.000469	0.000479	9.45
	0.1	0.3	10617	0.000472	0.000479	9.46
	0.1	0.5	7685	0.000465	0.000477	9.43
	0.1	0.7	4670	0.000471	0.000474	9.46
	0.1	0.9	1486	0.000506	0.000529	7.06
	0.3	0.1	4614	0.000464	0.000477	9.42
	0.3	0.3	3596	0.000469	0.000476	9.44
	0.3	0.5	2610	0.000472	0.000473	9.47
	0.5	0.3	1926	0.000638	0.000617	9.96
	0.5	0.5	1686	0.000478	0.000464	9.52
	0.5	0.7	938	0.000486	0.000498	8.03
	0.5	0.9	327	0.000516	0.000676	7.80
	0.7	0.1	1667	0.000523	0.000561	6.96
	0.7	0.3	1587	0.00049	0.000489	7.66
	0.7	0.5	1114	0.000493	0.000497	7.59
	0.7	0.7	564	0.000545	0.000591	6.78
	0.9	0.1	1299	0.000512	0.000546	7.62
	0.9	0.3	1001	0.000516	0.000534	7.39
	0.9	0.5	728	0.000534	0.000587	6.87
	0.9	0.7	419	0.00051	0.000556	7.36
3-8-1	0.1	0.1	11773	0.000406	0.000512	7.39
	0.1	0.3	9173	0.000405	0.000512	7.38
	0.1	0.5	6591	0.000401	0.000512	7.35
	0.1	0.7	3921	0.000407	0.000514	7.39
	0.1	0.9	1343	0.000399	0.000535	7.22
	0.3	0.1	3944	0.000402	0.000512	7.36
	0.3	0.3	3045	0.000408	0.000513	7.40
	0.3	0.5	2178	0.000407	0.000514	7.39

Table 8. (Continued)

Architecture L-M-N	η	α	Iteration	MSE training	MSE testing	Maximum validation error (%)
	0.3	0.7	1312	0.000406	0.000518	7.36
	0.3	0.9	449	0.000409	0.000596	7.25
	0.5	0.1	2124	0.000525	0.000518	7.90
	0.5	0.3	1756	0.000518	0.00055	7.45
	0.5	0.5	988	0.000674	0.000784	9.20
	0.5	0.7	782	0.000411	0.000524	7.37
	0.5	0.9	275	0.00043	0.000594	7.46
	0.7	0.1	1529	0.000409	0.000503	8.68
	0.7	0.3	1192	0.000452	0.000483	8.63
	0.7	0.5	838	0.000531	0.000523	7.84
	0.7	0.7	509	0.000529	0.000522	7.83
	0.9	0.1	1223	0.000437	0.000564	9.17
	0.9	0.3	926	0.00041	0.000512	8.79
	0.9	0.5	659	0.000424	0.000494	8.83
	0.9	0.7	392	0.000381	0.00053	8.83
3-12-1	0.1	0.1	11084	0.00045	0.000523	7.26
	0.1	0.3	8722	0.000441	0.000522	7.17
	0.1	0.5	6151	0.000451	0.000522	7.26
	0.1	0.7	3681	0.000454	0.000519	7.26
	0.1	0.9	1201	0.000553	0.000523	7.32
	0.3	0.1	3613	0.000491	0.000499	7.23
	0.3	0.3	2868	0.000452	0.000517	7.23
	0.3	0.5	2061	0.000449	0.000511	7.15
	0.3	0.7	1216	0.000486	0.000498	7.19
	0.5	0.1	2053	0.000462	0.000445	8.19
	0.5	0.3	1613	0.000497	0.000454	7.95
	0.5	0.5	1158	0.000534	0.00048	7.74
	0.7	0.1	1509	0.000514	0.000456	9.07
	0.7	0.3	1158	0.000449	0.000442	8.53
	0.7	0.5	829	0.000515	0.000452	8.90
	0.9	0.5	650	0.000484	0.000443	8.81

Taking consideration of these three principal component (PCA1, PCA2 and PCA3) which is having all the variables at different score coefficient. So these three principal components is now acting as input to neural network and output is drill flank wear. Now in the similar procedure network has been trained at different condition of learning rate, momentum coefficient and number of neuron in hidden layer. Large numbers of runs were given for selecting the best architecture and Table 8 shows only the best and the second best network in terms of the validation error and the number of iteration. The optimum network architecture in the present case has been observed to be 3-3-1 with $\eta = 0.1$ and $\alpha = 0.3$. Variation of mean square error in training and testing with number of iteration for the optimum network is shown in Figure 7. After the network has been trained, it has been validated with unknown data sample. Figure 8 shows percentage of error between maximum depth of actual value and the predicted value of flank wear and it can be observed that present neural network of 3-3-1 with $\eta = 0.1$ and $\alpha = 0.3$ predicts the results within ± 6.42 % error. This optimum network is later on used for classifying the new set of pattern of low wear and high wear. Hence in the same experimental environment another 64 test data (32 data in low range of wear and 32 in high range of wear) were selected. The classification accuracy of the network was calculated by taking the number of correctly classified sample by the network and divided by the total number of sample into the test data is represented as confusion matrix in Table 9. Table 9 shows that classification accuracy of the network for low wear rate is 81%, high wear rate 96% and overall accuracy is approximately 89%.

Table 9. Confusion matrix for wear classification

	No of sample	Low wear	High wear	Accuracy
Low wear	32	26	6	81.25%
High wear	32	1	31	96.87%
	64			89.06%

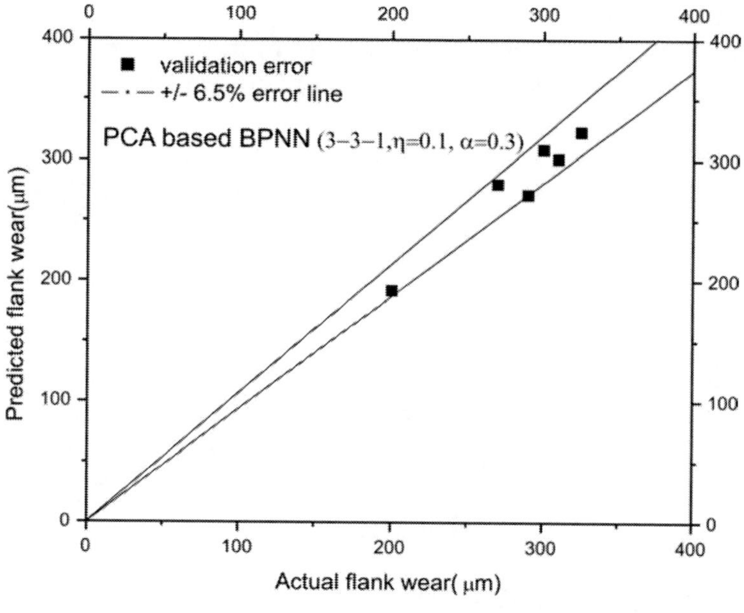

Figure 8. Comparison of Predicted value with actual value of 3-3-1 with $\eta = 0.1$ and $\alpha = 0.3$.

6. Conclusion

Combination of different process parameters in conjunction with different ANN architectures has been studied in search of best prediction of drill wear. From the present study following conclusion have been made.

Generalized Conclusion

(i) Direct process parameters such as spindle speed, feed rate, and drill diameter has definitive effect on progressive flank wear of the drill.
(ii) Wear on the drill (condition of the drill) affects the sensor signals such as thrust force, torque, vibration and hence these could be used in the drill wear prediction,
(iii) Drill wear (condition of the drill) also affects the chip thickness as well as the surface roughness and hence thickness of the chip and the surface roughness could also be used as indicative parameters for prediction of drill condition.

Conclusion Derived from Observation

(i) thrust force and torque signals shows better correlation with drill wear compared to vibration signals
(ii) Inclusion of more sensor signal as input to network results in a better-trained network, which can predict the wear more accurately. Hence, multi sensor integration is a better stand compared to single sensor approach in wear prediction capability.
(iii) Inclusion of surface roughness and chip thickness to BPNN, which leads to more precise and better prediction of wear.
(iv)Nine input variable could be better represented by three principal component
(v)Fusing these three principal components as input to neural network drastically reduces the number of iterations required for predicting the drill flank wear
(vi) There is no more significant improvement in predicting the drill flank wear by both BPNN and PCA based BPNN architectures but number of iteration required in case of PCA based BPNN drastically reduces. The best network configuration predicts the tool wear within $\pm 3.21\%$ error when BPNN is used. However, error is marginally increased to $\pm 6.42\%$ in case of using PCA based BPNN but computational efforts have been drastically reduced.
(vii) Network could able to classify low wear and high wear within an accuracy of 89%. However, high wear is accurately predicted compared to low wear rate.

REFERENCES

Abu-Mahfouz, I., 2003, Drilling wear detection and classification using vibration signals and artificial neural network, *Int. J. Mach. tools & Manuf.,* 43, 707–720.

Bahr, L B., Motavalli, S. and Arfi, T., 1997, Sensor fusion for monitoring machine tool conditions, *Int. J. Comput. Integr. Manuf.,* 10, 314–323.

Brophy, B., Kelly, K., and Byrne, G., 2002, AI-based condition monitoring of the drilling process, *J. of Mater. Process. Technolo*., 124, 305-310.

Chungchoo, C. and Saini, D., 2002, A computer algorithm for flank and crater wear estimation in CNC turning operations, *Int. J. of Mach. Tools & Manuf*., 42, 1465–1477.

Elhachimi, M., Torbaty, S. and Joyot, P., 1999b, Mechanical modeling of high speed drilling 2: predicted and experimental results, *Int. J. of Mach. tools & Manuf.,* 39, 569-581.

Elhachimi, M., Torbaty, S.and Joyot, P., 1999a, Mechanical modeling of high speeds drilling 1: Predicting torque and thrust, *Int. J. of Mach. Tools & Manuf.*, 39, 553–568.

Fisher, R. A., 1936, The use of multiple measurement in taxonomic problems, *Annals of Eugenis,* 7, 178-188.

Gao, C. H., Cheng, K., and Kirkwood, D., 2000, The investigation on the machining process of BTA deep hole drilling, *J. of Mater. Process. Technolo.,* 107, 222-227.

Gong, Y., Lin, C. and Ehmann, K. F., 2005, Dynamics of Initial Penetration in Drilling: Part 1-Mechanistic Model for Dynamic Forces, ASME, *J. of Manuf. Sci. and Eng.*, 127,280-288.

Haili, W., Hua, S., Ming, C. and Dejin, H., 2003, On-line tool breakage monitoring in turning, *J. of Mater. Process. Technolo.,* 139, 237–242.

Hong, Z. Q. and Yang, J. Y., 1991, Optimal discernment plane for a small number of samples and design method of classifier on the plane, *Pattern Recognition,* 24 (4), 317-324.

Inasaki, I., 1998, Application of acoustic emission sensor machining processes, *Ultrasonic,* 36, 273-281.

Kanai, M. and Kanda, Y., 1978, Statistical characteristic of drill wear and drill life for standardized performance tests, *Annals of CRIP*, 27(1), 61-66.

Kittler, J. and Young, P. C., 1973, Approach to feature selection based on Karhunen-Loeve expansion, *Pattern Recognition,* 5(5), 335-352.

Kleinbaum, D., Kupper, L. and Vannelli, A., 1986, Applied regression analysis and other multi variable methods, California: *International J. of Production Research,* 24, 387-397.

Ko, T. J. and Cho. D. W., 1994, Cutting state monitoring in milling by a neural network, *Int. J. of Mach. & Tools Manuf.,* 34(5), 659.

Li, X., Dong, S. and Venuvinod, P. K., 2000, Hybrid Learning for Tool Wear Monitoring, *Int. J. of Adv. Manuf. Technolo.*, 16, 303-307.

Lin, S. C., and Ting, C. J., 1996, Drill wear monitoring using neural network, *Int. J. of Mach. tools & Manuf.,* (1996), 36, 465-475.

Lin, S.C. and Ting, C.J., 1996, Drill wear monitoring using neural network, *Int. J. Mach. Tools Manuf.* 36 (4), 465–475.

Liu, T. I. and Ananthataraman, K. S., 1994, Intelligent classification and measurement of drill wear, ASME, *J. Eng. Ind.,* 116, 392.

Liu, T. I. and Chen, W. Y. and Anatharaman, K.S., 1998, Intelligent detection of drill wear, *Mech. Syst. & Signal Process.*, 12(6), 863-873.

Liu, T. I. and Wu, S. M., 1990, On-line detection of drill wear. *J Eng. Ind.,* 112, 299–302.

Morin, E., Masounave, J. and Laufer, E. E., 1995, Effect of drill wear on cutting forces in the drilling of metal-matrix composites, *J. Wear,* 184, 11-16.

Panda, S. S., Pal, S. K. and Chakraborty, D., 2006, Monitoring of drill flank wear using fuzzy back propagation neural network, *Int. J. of Adv. Manuf. Technolo.,* DOI 10.1007/s00170-006-0589-0.

Panda, S. S., Pal, S. K. and Chakraborty, D., 2008, Flank wear prediction in drilling cast iron using Back Propagation Neural Network and Radial Basis Function Network, *J. of Appl. Soft Comput.,* 8, 2, 858-871.

Purushothaman, S. and Srinivasa, Y. G., 1998, A procedure for training an artificial neural network with application to tool wear monitoring, *International J. of Production Research,* 36(3), 635-651.

Ramamurthi, K. and Hough, C. L., 1993, Intelligent real-time predictive diagnostics for cutting tools and supervisory control of machining operations, ASME, *J. of Eng. for Ind.*, 115, 268–277.

Rummel, R. J., 1988, *Applied factor analysis*, Evanston, Northwestern University Press.

Sanjay, C., Neema, M. L. and Chin, C. W., 2005, Modeling of tool wear in drilling by statistical analysis and artificial neural network, *J. of Mater. Process. Technolo.*, 170, 494-500.

Scheffer, C. and Heyns, P. S., 2001, Wear monitoring in turning operations using vibration and strain measurements, *J. Mech. Syst. and Signal Process.*, 15(6), 1185-1202.

Singh, A. K., Panda, S. S., Pal, S. K. and Chakraborty, D., 2006a, Predicting drill wear using an artificial neural network, *Inter. J. of Adv. Manuf. Technolo.*, 28, 456-462.

Singh, A. K., Panda, S. S., Pal, S. K. and Chakraborty, D., 2006b, Drill Wear Monitoring using Back Propagation Neural Network, *J. of Mater. Process. Technolo.*, 172, 283-290.

Stephenson, D. A. and Agapiou, J. S., 1992, Calculation of main cutting edge forces and torque for drills with arbitrary point geometries, *Int. J. of Mach. Tools & Manuf.*, 32(4), 521-538.

Thangaraj, A and Wright, P. K., 1998, Computer assisted prediction of drill failure using in-process measurements of thrust force, *J. of Eng. Ind.*, 110,192–200.

Wang, L., Kannatey-Asibu, E. and Mehrabi, M. G., 2003, A Method for Sensor Selection in Reconfigurable Process Monitoring, ASME, *J. of Manuf. Sci. and Eng.*, 125, 95-99.

Wardany, E. L., Gao, T. I D., and Elbestawi, M. A., 1996, Tool condition monitoring in drilling using vibration signature analysis, *Int. J. of Mach. tools & Manuf.*, 36(6), 687-711.

Wong, S. V. and Hamouda, A. M. S., 2003, Machinability data representation with artificial neural network, *J. of Mater. Process. Technolo.*, 138, 538-544.

Yeo, S. H., Khoo, L. P. and Neo, S. S., 2000, Tool condition monitoring using reflectance of chip surface and neural network, *J. of Intell. Manuf.*, 11, 507–514.

In: Machining and Forming Technologies, Volume 2
Editor: J. Paulo Davim
ISBN: 978-1-61122-820-5

Chapter 13

2D FINITE ELEMENT ANALYSIS OF SLOT MILLING

***Muhammad Imran**[*1]**, Naseer Ahmed**[2] **and V.V. Silberschmidt**[3]*
[1]Ghulam Ishaq Khan Institute of Engineering Sciences and Technology Topi, Swabi, Pakistan
[2] Faculty of Mechanical Engg. Ghulam Ishaq Khan Institute of Engineering Sciences and Technology, Topi, Swabi, Pakistan
[3] Mechanics of Advanced Materials Research Group, Wolfson School of Mechanical and Manufacturing Engineering Loughborough, University Loughborough Leicestershire LE11 3TU, U.K.

ABSTRACT

Milling is used to manufacture a large number of components. Studying the thermomechanics of the milling process by means of purely experimental approach is difficult and time consuming, especially when accounting for various parameters, e.g. tool geometry, material's properties and cutting conditions etc. In the present study fully thermomechanically 2D coupled finite element analysis of slot milling operation is performed. The cutting forces, temperature distribution in the workpiece and in the cutting tool are calculated for selected cutting conditions. For the measurement of cutting forces a strain gauges-based three-component dynamometer was designed and constructed. The cutting forces measured for the case of dry slot milling were compared with results of the finite element analysis. The effect of friction on cutting forces was also studied with two contact conditions.

Keywords: Slot milling, cutting forces, finite element analysis, temperature, friction conditions

[*] Corresponding author: E-mail Address: mehrarh@yahoo.com

1. Introduction

Material removal processes such as metal cutting are used to achieve a desired shape, tolerances, and surface finishes that are difficult to obtain by other means. The economical importance of, and the competitive environment in, the machining industry necessitates optimization of the manufacturing processes aimed at improving the dimensional accuracy, process efficiency, and product quality.

The area of research in machining is broad because there are many factors affecting the machining process. Theses factors include the cutting parameters (cutting speed, feed rate, depth and width of cut), material's properties, and geometry of a workpiece and a tool.

Many researchers have used finite element analysis to study the machining processes; however, there are only few studies on implementation of the finite element analysis of the milling process. The bibliography of research on the finite element formulation of milling and other machining processes is given in (Mackerle, 1999 and Mackerle, 2003).

Still, the authors found no literature for the type of milling investigated in this study. The techniques and methodologies used in some works on finite element analysis of milling are discussed below.

Özel and Altan (2000) performed simulations of the process using the finite element method (FEM) to predict cutting forces, tool stresses and temperatures in high speed flat-end milling. The FEM-based commercially available software, DEFORM-2D, was used for the process simulations. A plane strain deformation model was used to simulate chip flow at a primary cutting edge. Only the tip of the tool and an arch-shaped part (with a thickness of 5 mm) of the workpiece were meshed in order to have a suitable number of elements for calculations. Cutting edge preparation was also modelled with a tool tip radius of 0.012 mm in the simulations. Only rotational motion of the tool was considered in this model.

A two-dimensional model of the end milling using the commercial software Advant Edge was presented in (Casalino *et al.*, 2007). In that paper an analysis of the machining errors due to the cutting forces in peripheral milling was performed with the use of a numerical model based on the finite element method. The distribution of dynamic cutting forces was evaluated for several cutting conditions with varying levels of the effective rake angle, the axial depth of cut, the radial depth of cut, and the feed per tooth per revolution. The results obtained with the model were compared with the cutting forces measured by Yucesan and Altintas (1994) during in-field trials on a titanium alloy.

A two-dimensional finite element analysis of the mechanical action of the residual stresses in milling using FEM- software ANSYS was developed in (Litao *et al.*, 2006). The following assumptions were made:

The effects of temperature and the strain rate were ignored for the formation of the residual stresses;

The tool was assumed to be rigid and sharp all the time;

The viscous and visco-plastic deformations of the workpiece materials were neglected;

Plastic deformations were described with von Mises formulation;

Plane-strain formulation was used since the milling layer was much thinner than workpiece length;

The removal of material was modelled using the element elimination technique.

The milling force that was measured in experiments was set as input to the finite element analysis of milling. Moreover the modelling process was divided into two steps - the transition state and the steady state - because it was shown that there was a sudden change in the milling force when the tool was cutting into the work piece.

A two-dimensional finite element analysis of the ball end-milling process using the explicit Lagrangian software Third Wave Advant Edge was presented in (Ginting and Nouari, 2006). The purpose of that study was to evaluate the performance of alloyed carbide tool in dry milling of aerospace material. In the analysis the friction was based on the Columb's friction law and the thermal properties were described by a polynomial function. The material behaviour including the strain-rate effect was introduced in terms of the power strain hardening law. During the simulation the mesh, which became significantly distorted, was updated refining both small and large elements. The material deformed in the primary shear zone, which was assumed to be a thin layer inclined at a shear angle to the free surface of the workpiece material. The dissipation of energy at tool-chip interface occurred as the material moved from the primary shear zone to the secondary shear zone.

Another two-dimensional finite element model similar to that in (Ozel and Altan, 2000), was presented in (Ozel *et al.*, 1998). In milling the tool tip travels along a trochoidal path but in that model it was considered circular due to small value of maximum chip thickness. The flow stress behaviour was modelled with the Maekawa equation. The continuous chip formation was simulated, and the free formed chip flowed for cutter rotation of 65°. The chip touched the uncut workpiece surface at 75° of cutter rotation. Tool temperatures and stresses were calculated for the latter point and presented for each set of cutting conditions.

Dhanorker and Ozel (2006) developed, a mechanistic and finite element model of micro/meso-end milling process predicting distribution of forces, temperature and stresses in the presence of tool edge ploughing. FEM-based commercially available software Deform 2D was used for the process simulation. The Johnson-Cook material model was used to model the elasto-plastic deformation of the workpiece material AL2024-T6. A tool edge radius of 30 μm was used as suggested by the manufacturer. The predicted values of forces and temperatures were in good agreement with the experimental results.

2. Finite Element Modelling of Slot Milling

The milling process modelled here is slot milling. The cutter used in this study is side and face milling cutter with straight teeth. It has teeth on both sides as well as on the periphery. For making slots in the workpiece only the teeth on the periphery are used, while the teeth on the sides are used in face milling operation.

The actual view of slot milling operation is shown in Figure 1a. The workpiece is fed linearly against the rotating cutter (Figure 1b) Main parameters of the process are given in the magnified side view (Figure 1c) for one tooth of the cutter.

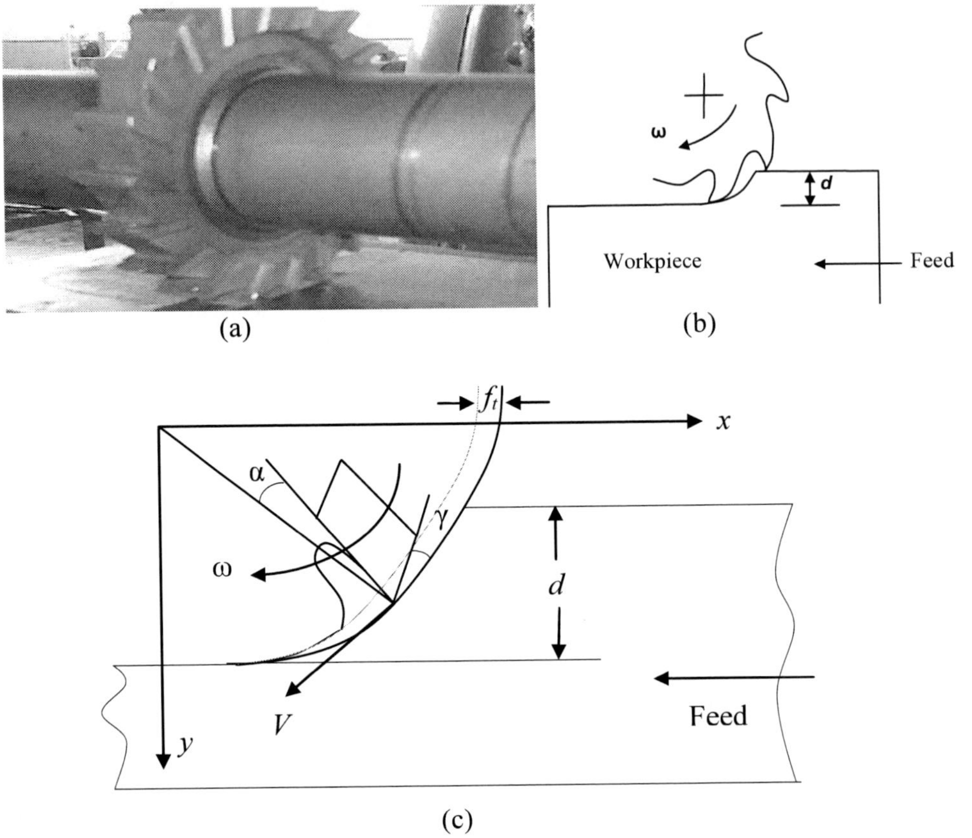

Figure 1. Slot milling: (a) actual process; (b) side view; (c) magnified view.

2.1. General Features

The finite element code MSC MARC/MENTAT (MSC.MARC user guide, 2005) was used to perform two-dimensional finite element analysis of slot milling. This code uses the updated Lagrangian analysis procedure to provide a transient analysis for an elasto-plastic material and also accounts for the frictional contact interaction between the cutter and workpiece as well as the material separation in front of the cutting edge.

During slot milling (or any other machining operation) the excessive plastic deformation and friction at the tool-chip interface lead to generation of high temperatures in the cutting region. This results in thermal stresses and expansion and also affects material properties of the workpiece, such as thermal conductivity and specific heat. Therefore, a fully thermomechanically coupled finite element model was used in the simulations. Such model can properly reflect interconnections between the thermal and mechanical processes in the cutting zone (Ahmed, 2007).

Since this is a thermo-mechanically coupled model, thermal boundary conditions are also defined. The workpiece is loosing heat to the environment, which is assumed to be at 20°C, at a rate of 0.4 N/mm °C (Bil *et al*, 2004). The workpiece loses heat to the environment due to convection according to the following formula:

$$q=h(T_w-T_o) \qquad (1)$$

Where h is the convection heat transfer coefficient of the workpiece (0.4 N/mm °C), T_w is the workpiece surface temperature, and T_o is the ambient temperature. On the other hand, the heat is generated due to two processes:

Plastic work done at the shear zone. It is assumed that 90% of all plastic work is converted into heat (Bil *et al*, 2004, *Ozel, 2006,* Shih, 1996):

Friction at the interface of tool and chip. Heat generated due to friction is given by the following formula (Bil *et al*, 2004):

$$q_f=F_{fr}V_rM \qquad (2)$$

Where F_{fr} is the friction force, V_r is the relative sliding velocity between tool and chip, and M is the mechanical equivalent of heat (M = 1.0).

The rate of specific volumetric flux due to plastic (R) work is given by the following relation (Bil *et al*, 2004):

$$R=\frac{M.f.W^p}{\rho}, \qquad (3)$$

Where W^p is the rate of plastic work, f is the fraction of plastic work converted into heat (assumed to be 0.9) (Bil *et al*, 2004, Ozel, 2006) and ρ is the density of workpiece material (7.870 g/cm^3).

The mechanical behaviour of the workpiece material AISI 1010 at high strains, strain rates and elevated temperature can be adequately described by the Johnson-Cook material model (Johnson and Cook, 1985). That model was used in the present study:

$$\sigma_y=(A+B\varepsilon_p^{\mathrm{n}})(1+C\ln(\frac{\overset{\circ}{\varepsilon}_p}{\overset{\circ}{\varepsilon}_o}))(1-\frac{T-T_{room}}{T_{melt}-T_{room}}), \qquad (4)$$

where σ_y is flow stress, A = 250 MPa is the initial yield stress at the reference strain rate, B = 275 MPa is the hardening modulus, C=0.078 is strain rate sensitivity coefficient, n = 0.36 is the hardening coefficient, m is the thermal softening coefficient, T_{melt} is the melting temperature of the AISI 1010 steel, T_{room} is the room temperature , T is the absolute temperature, ε_p, $\overset{\circ}{\varepsilon}_p$ and $\overset{\circ}{\varepsilon}_o$ =1 s^{-1} are the equivalent plastic strain, equivalent plastic strain rate and reference strain rate, respectively. All those parameters were determined experimentally using the Split Hopkinson Pressure Bar method and were adopted from (Umeda *et al.*, 2005).

Quite often in contact problems, neutral lines develop. This means that along a contact surface, the material flows in one direction on one part of the surface and in the opposite direction on another part of the surface. Therefore, the friction stress has a step-function behaviour based upon the value of the relative sliding velocity V_r. This discontinuity in the value of σ_{fr} can result in numerical difficulties. Besides this (Oden et al. 1983) pointed out that for metals; there is an elasto-plastic deformation of asperities at the microscopic level (termed as 'cold weld') which leads to a nonlocal and nonlinear frictional contact behavior. The arctan representation of the friction model is a mathematical idealization of this nonlinear friction behavior and more over cutting forces are accurately predicted as compared to other friction models (Bil *et al*, 2004), so shear friction between the tool and workpiece was modelled with a modified constant shear friction model. This model states that the frictional stress is a fraction of the equivalent stress in the material (MSC.MARC users' guide, 2005; Ahmed, 2007). Thus the friction stress is introduced in the following form:

$$\sigma_{\text{fr}} \leq -\mu \frac{\bar{\sigma}}{\sqrt{3}} \frac{2}{\pi} \text{sgn}(V_r) \arctan\left(\frac{V_r}{V_{\text{cr}}}\right), \tag{5}$$

Where $\bar{\sigma}$ is the equivalent stress, V_{cr} is the critical sliding velocity below which sticking is simulated, μ is the friction coefficient and $\text{sgn}(x)$ is the signum function. This model can be implemented using two approximations of the theoretical step function. First, the arctangent function to smooth out the step function is applied. Second, the arctangent model is used based on the assumption that the shear (friction) stress in a node is proportional to the applied shear (friction) force.

Physically, V_{cr} has the value of the relative velocity when sliding occurs. The value of V_{cr} is important in determining how closely the mathematical model represents the step function. A large value of V_{cr} results in a reduced level of the effective friction. On the other hand, a small value results in poor convergence. It is recommended (MSC.MARC users' guide, 2005) that the value of V_{cr} be 1% or 10% of a typical relative sliding velocity V_r. Because of this smoothing procedure, a node in contact always has some slipping.

There are a number of methods to separate material from the workpiece to form the chip as the tool penetrates into it. In this model, the method of remeshing was applied. Therefore, the workpiece is re-meshed whenever the tool penetrates into the workpiece by a predefined threshold value.

By default MSC.MARC takes penetration value as 0.05 of the minimum element edge length, however it is possible to change it. In addition, the moment of penetration control can also be altered, being either at the end of an increment, or at every iteration. Although the penetration value does not introduce a significant difference, the type of penetration control leads to major changes in the results.

2.2. Finite Element Model

A two dimensional finite element model of slot milling was developed. In earlier studies orthogonal cutting in milling process and turning process was simulated by utilizing plain strain deformation models (Özel and Alatn, 2000).

In this study a plain strain deformation model was used to simulate the chip flow in slot milling operation.

The element type 11 was used to model the workpiece and was not used in simulation due to its poor representation of shear behaviors and lack of support for remeshing. The element type 11 is automatically converted by remeshing to triangular element type 155 of MSC.MARC (MSC.MARC users' guide, 2005). This triangular element is isoparametric plain strain, 3+1-node element with an additional degree of freedom at each of the corner node. This is a Herrmann type element, which typically uses pressure as well as displacement in the FEM analysis. This element can be used to model incompressible material undergoing large shear deformations. The standard quadrilateral element can not be used in this situation, because it locks and lacks flexibility.

The cutting tool was modelled with element type 39, a four-node, isoparametric, arbitrary quadrilateral used for two dimensional heat transfer application and is recommended for modelling coupled thermomechanical analysis (MSC.MARC users' guide, 2005). Only a small portion of the workpiece and one tooth from the cutting tool were meshed in order to have practical number of elements for calculations. The dimensions of the workpiece used in simulations were 2.15 mm in length and 0.125 mm in height.

The properties of workpiece material were those of AISI 1010 steel (Table 1). The cutting tool was considered as rigid body with a heat transfer capability. The thermal and mechanical properties of the cutting tool used in the simulations were those of high speed steel AISI M2 (Table 2).

There were a total of 40 elements representing the workpiece, which increased to 600 immediately in the first iteration, and 110 elements in the cutting tool. The rake angle (α) and clearance (γ) angle of the tool were 13° and 10°, respectively (Figure 2).

Table 1. Properties of the AISI 1010 steel (ASM metal hand book, 1990)

Property	Magnitude
Elastic modulus (Gpa)	205
Poisson ratio	0.29
Tensile strength (Mpa), Ultimate	365
Tensile strength (Mpa), Yield	305
Thermal conductivity (W/mK) at 100°C, 200 °C , 300°C, 400°C, 500°C	60.2, 55.5, 50.7, 46, 41.5
Specific heat capacity (J/kgK) at 100°C, 300°C, 450°C	448, 536, 649
Density (Kg/ m^3)	7870
Coefficient of thermal expansion(10^{-6}/K) at 20°C, 250°C, 500°C	12.2, 13.5, 14.2

Table 2. Properties of the AISI M2 steel (ASM metal hand book, 1990)

Property	Magnitude
Elastic modulus (Gpa)	210
Poisson ratio	0.3
Compressive strength (Mpa), Yield	3250
Thermal conductivity (W/mK) at 95°C, 200°C, 400°C, 540°C	21.3, 23.5, 25.6, 27
Specific heat capacity (J/kgK)	55
Density (kg/ m^3)	8150
Coefficient of thermal expansion (10^{-6}/k) at 20°C-540°C	11.9

In milling the cutter travels on a trochoidal path as a result of superposition of a linear movement with a feed rate and rotation of the cutter. But for small value of maximum undeformed chip thickness it can be considered circular (Özel and Altan, 2000). In the present FE model the path of cutting tool was considered circular.

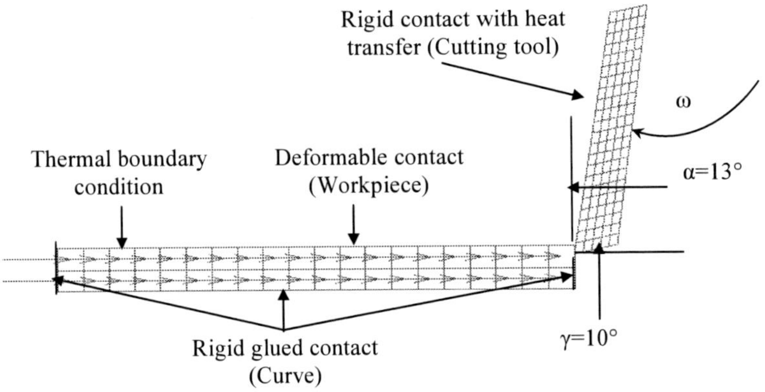

Figure 2. Initial FE model with contacts and boundary condition.

The workpiece was modelled stationary by defining three position-controlled rigid contacts, one at the base and two at the opposite sides of the workpiece (Figure 2). Those rigid contacts were 'glued' to the workpiece that was also defined as deformable contact. The rotational speed (radian per second) was applied to the cutting tool, defined as rigid contact with heat transfer. The rotation of the cutting tool is clear from the difference in height of the two sides of workpiece as shown in the Figure 3.

3. Results of Finite Element Analysis and Discussion

In all simulations the analysis time was kept at 1.5 ms. That analysis time corresponded to a fully formed chip. That time interval was divided into 1200 increments each of 1.250×10^{-6} second. Successful completion of the analysis required about one hour on Pentium IV computer.

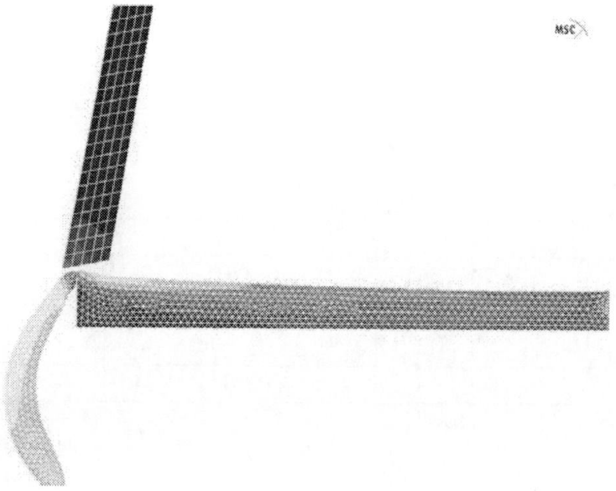

Figure 3. Fully formed chip.

In order to get good understanding of the effects due to different process parameters of slot milling process - forces, temperature and the effect of different variables on theses process parameters - a series of different simulations were performed.

Table 3. Variables used in simulation of slot milling process

Parameter	Magnitude used in FEA
Friction coefficient	0, 0.5
Feed rate (mm/tooth/rev)	0.0201, 0.032, 0.05, 0.064
Cutting speed (m/min)	30, 47, 74, 94

Table 4. Matrix of analyzed cases to study effect of cutting speed

Simulation no	Cutting speed (m/min)	Feed rate (mm/tooth/rev)	Coefficient of friction
1	30	0.05	0.5
2	47	0.05	0.5
3	74	0.05	0.5
4	94	0.05	0.5
5	30	0.05	0
6	47	0.05	0
7	74	0.05	0
8	94	0.05	0

Two cases of friction were selected for simulations, with $\mu = 0$ representing a well lubricated case and $\mu = 0.5$ for a case with out lubrication. Those cases were selected since the coefficient of friction ranging from 0 to 0.5 had already been used in simulations (Ahmed, 2007).

Table 5. Matrix of analyzed cases to study effect of feed rate

Simulation no	Cutting speed (m/min)	Feed rate (mm/tooth/rev)	Coefficient of friction
9	74	0.0201	0.5
10	74	0.0320	0.5
3	74	0.0500	0.5
11	74	0.0640	0.5
12	74	0.0201	0
13	74	0.032	0
7	74	0.05	0
14	74	0.064	0

3.1. Comparison of Calculated and Measured Forces

Finite element simulations of the slot milling process can predict forces in the x and y direction with out conducting physical experiments. In the current study the effect of tool wear upon the cutting forces was neglected. This section presents the comparison of calculated and measured forces for different cutting conditions given in Tables 3 and 4.

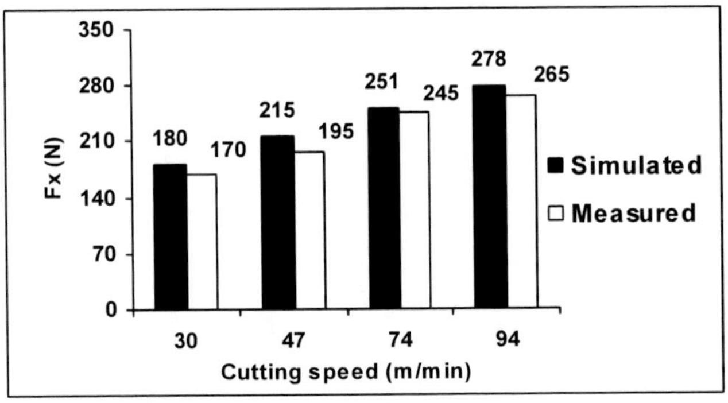

(a)

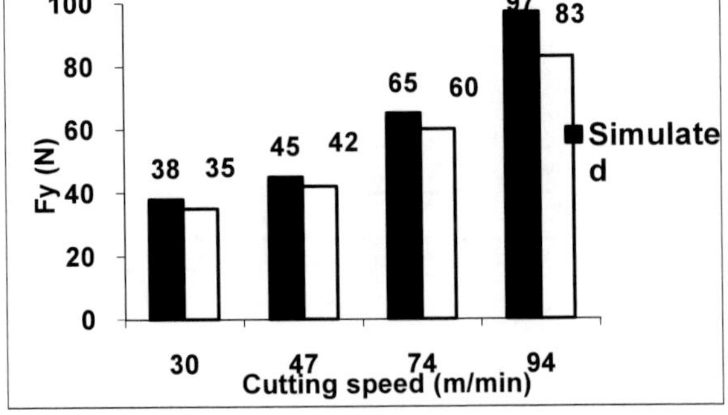

(b)

Figure 4. Comparison of forces in the cutting tool along x-direction (a) and y-direction (b) (f_t = 0.05 mm/tooth/rev, μ = 0.5).

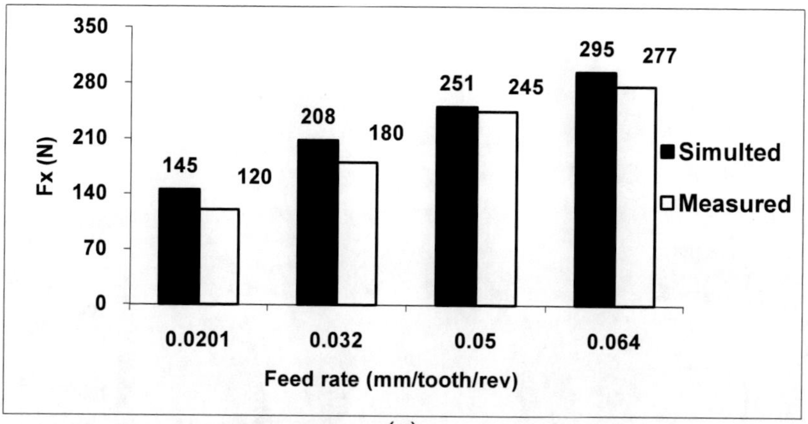

(a)

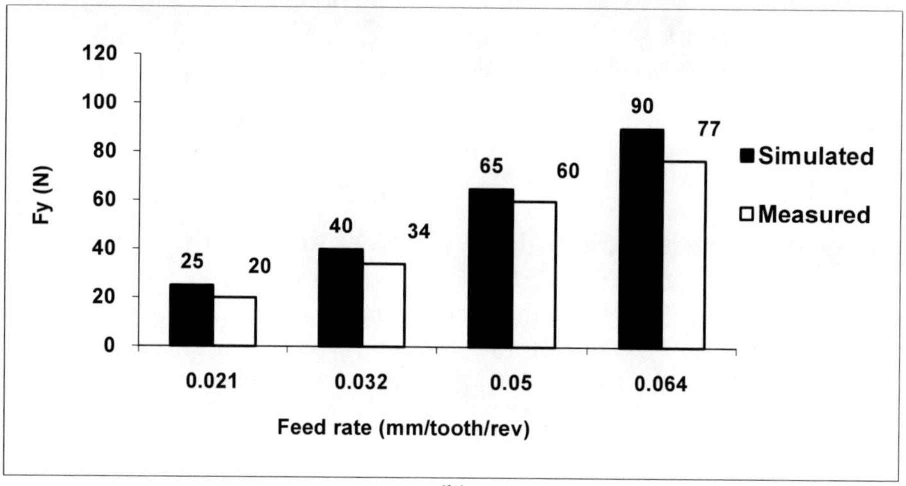

(b)

Figure 5. Comparison of forces in cutting tool along x-direction (a) and y-direction (b) (V=74 m/min, μ = 0.5).

As can be seen from Figure 4 the cutting forces both in the x and y-direction increase with the increasing cutting speed. Let us note here that it is true only for intermediate and low cutting speeds used in our simulations. At higher cutting speeds the cutting forces decrease as in the case of high speed machining (Özel and Altan, 2000). The cutting forces obtained in finite element simulations are in good agreement with the measured forces, slightly over predicting them.

The cutting forces also increase with the increasing feed rate which also corresponds to the maximum undeformed chip thickness, both in the x and y-direction as can be seen from Figure 5. The increase in the feed rate increases the chip's cross sectional area of the resisting material resulting hence in higher cutting forces.

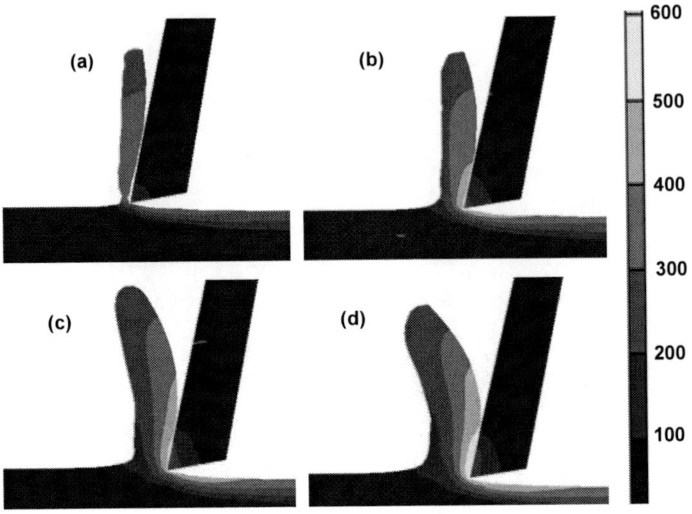

Figure 6. Temperature distribution (in °C) in the workpiece and cutting tool for various cutting speeds (in m/min): (a) 30; (b) 47; (c) 74; (d) 94 (feed rate is 0.05 mm/tooth/rev, μ=0.5 and t=1.5 ms)

3.2. Temperature Distribution in the Workpiece and Cutting Tool

The FE model was fully thermomechanically coupled in order to obtain the temperature distribution in the cutting region and in the cutting tool. The calculated temperature distributions for the first four cases of Tables 3 and 4 are shown in Figures 6 and 7, respectively. The temperature predicted here are not the one achieved in the workpiece and tool at the advanced stage of milling since the simulation time is 1.5 ms, and in real situations a larger amount of time is required for the temperature to saturate. However, the relative difference in the predicted temperatures is useful for improving the tool life, tool wear prevention and quality of the machined surface.

As can be seen from Figure 6 the temperature in the workpiece and cutting tool increases with increasing the cutting speed (see Table 6).

Table 6. Maximum temperature in the workpiece and cutting tool for various cutting speeds (feed rate is 0.05 mm/tooth/rev, μ=0.5 and t=1.5 ms)

Cutting speed (m/min)	Max. temperature in the workpiece (°C)	Max. temperature in the cutting tool (°C)
30	374	187
47	456	228
74	528	264
94	548	274

Similarly, from the Figure 7 it is clear that the temperature increases as the feed rate is increased (see Table 7).

Table 7. Maximum temperature in the workpiece and cutting tool for various feed rates (V=74 m/min, μ=0.5 and t=1.5 ms).

Feed rate(mm/tooth/rev)	Max. temperature in the workpiece (°C)	Max. temperature in the cutting tool (°C)
0.0201	434	217
0.032	515	257
0.05	528	264
0.064	555	280

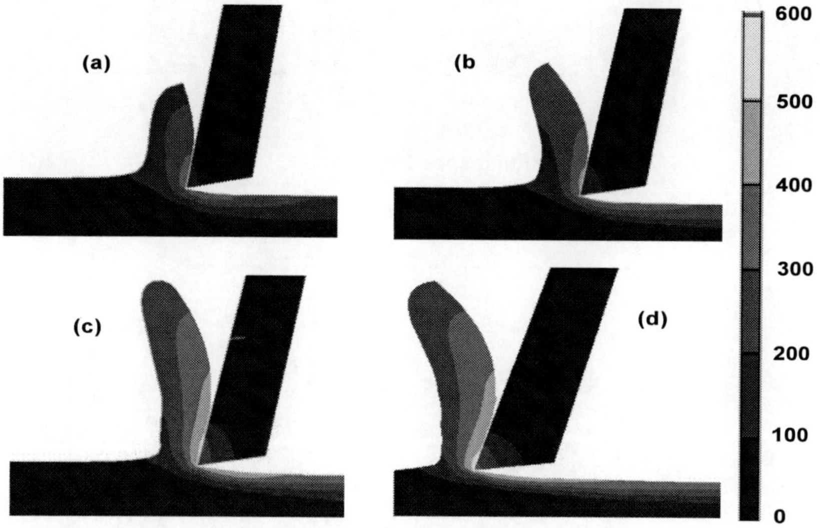

Figure 7. Temperature distribution (in °C) in the workpiece and cutting tool for various cases of feed rate (in mm/tooth/rev): (a) 0.0201; (b) 0.032; (c) 0.05; (d) 0.064 (V=74 m/min, μ=0.5 and t=1.5 ms).

3.3. Effect of Friction on Cutting Forces

In all the FE simulations conducted for slot milling process in the present study, the friction condition defined between the tool and workpiece greatly affected the numerical predictions. The correct definition of contact and friction at the tool workpiece interface helps to simulate the lubrication conditions (Ahmed, 2007). A significant difference was observed when the friction coefficient was changed from 0.5 to 0 (Figures 8 and 9). The friction coefficients of 0 and 0.5 were selected to see the difference in the cutting forces at the two extremes. The same bounds for friction coefficients were used in (Grzesik *et al.*, 2005 and Lin et al, 1992). The cutting forces in the x and y directions in case of cutting with friction (μ = 0.5) are 8-20% and 20-45%, respectively, greater than those for cutting with out friction.

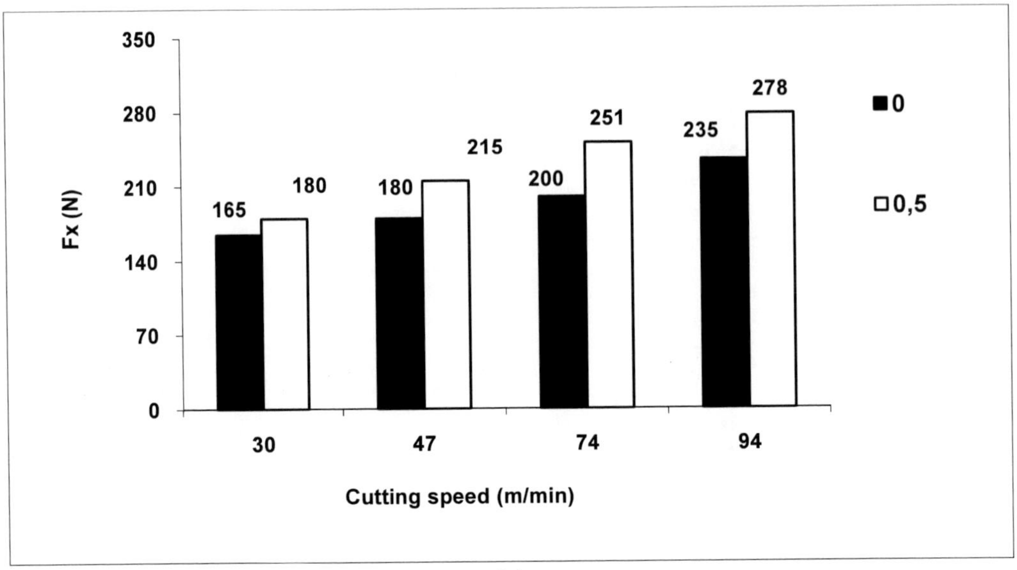

(a)

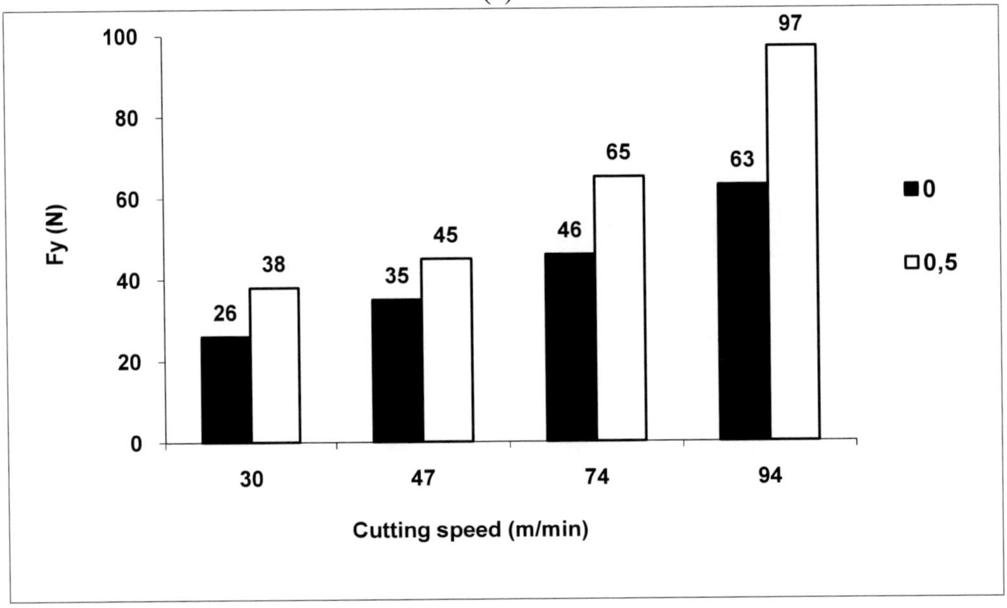

(b)

Figure 8 Effect of friction on cutting forces along x-direction (a) and y-direction (b) (f_t=0.05 mm/tooth/rev)

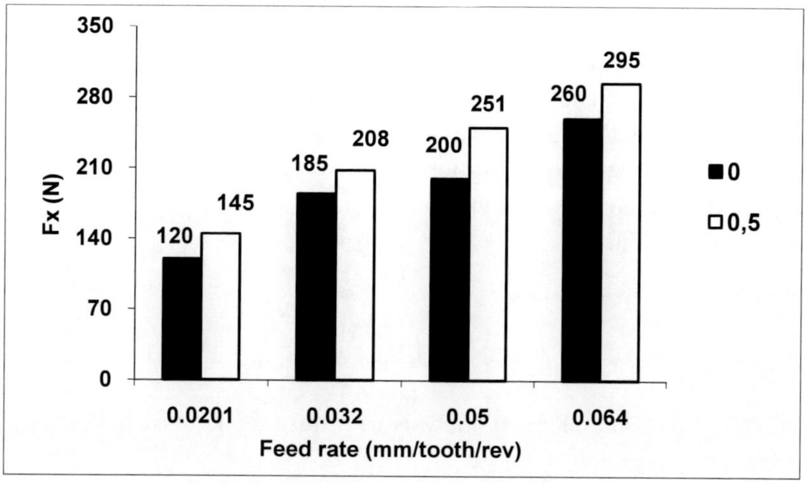

(a)

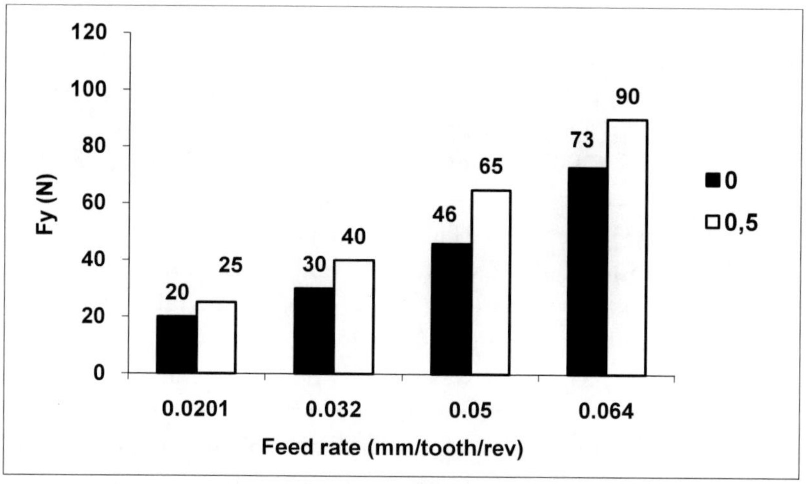

(b)

Figure 9 Effect of friction on cutting forces along x-direction (a) and y-direction (b) (V=74 m/min)

4. Conclusion

The aim of this study was to develop a 2D thermomechanically coupled finite element model of slot milling to investigate the effect of feed rate and cutting speed on the cutting force and temperature distribution in the workpiece and in the cutting tool. For measurement of cutting forces a strain gauges-based three-component dynamometer was designed and built. Various experiments were carried out to measure the cutting forces for selected cutting conditions similar to those used in the finite element simulations of the slot milling process. The cutting forces in the *x* and *y* directions and temperature in the workpiece and cutting tool increases with the increasing cutting speed or the increasing feed rate. A comparison of the results of the FE simulations for frictionless milling with experimentally measured magnitudes (Figures 4 and 5) demonstrates that simulations either slightly under predicted the cutting forces or provide the same level (Simulation 12). It means that the chosen bonds properly characterise the real-life conditions. The cutting forces in the *x* and y directions in case of cutting with friction ($\mu = 0.5$) are 8-20% and 20-45%, respectively, greater than those for cutting with out friction.

REFERENCES

Ahmed, N. (2007) '3D finite element analysis of ultrasonically assisted turning', *PhD thesis, Loughborough University.*

ASM metal hand book (1990) ' *ASM international'*, 10th edition, Vol.1

Bil, h., Kilic, S. E, and Tekkaya, A. E (2004) 'A comparison of orthogonal cutting data from experiments with three different finite element models' *International journal of machine tools and manufacture*, Vol. 44, pp.933-944.

Casalino, G, Campanelli, S. L., Ludovico, A. D. and Launi, I. (2007) 'Finite element analysis of the dynamic cutting force distribution in peripheral milling' *CP907 10th ESAFORM Conference on material forming, American Institute of Physics*, pp.733-738.

Dhanorker, A. and Ozel, T. (2006) 'An experimental and modelling Study on meso/micro end milling process', *Proceeding of ASME international conference on manufacturing science and engineering*, paper no.21127, Ypsilanti, MI, USA

Ginting, A. and Nouari, M. (2006) 'Experimental and numerical studies on the performance of alloyed carbide tool in dry milling of aerospace material', *International journal of machine tools and manufacture*, Vol. 46, pp.758-768.

Grzesik, W., Bartoszuk, M. and Nieslony, P. (2005) 'Finite element modelling of temperature distribution in the cutting zone in turning processes with differently coated tools', *Journal of materials processing technology,* Vol.164-165, pp.1204-1211.

Johnson, G., and Cook, W. (1985) 'Fracture characteristics of three metals subjected to various strains, strain rates, temperatures and pressures', *Engineering Fracture Mechanics*, Vol.2, pp.31-48.

Lin, Z. C. and Lin, S. Y. (1992) 'A Coupled Finite Element Model of Thermo Elastic Plastic Large Deformation for Orthogonal Cutting', *International journal of engineering materials and technology,* Vol.114, p.218

Litao, W., Hongfeng, W. and Huan, Y. (2006) 'Analysis and Computation of Mechanical Action of Residual Stress in Milling', *Proceedings of the 5th WSEAS international conference on telecommunications and informatics; Istanbul*, pp.452-456.

Mackerle, J. (1999) 'Finite element simulation of machining: a bibliography (1976-1996)', *Journal of material processing technology*, Vol. 86, pp.17-44.

Mackerle, J. (2003) 'Finite element simulation of machining: a bibliography (1996-2002)', *Journal of material Processing Technology*, Vol.43, pp.103-114.

MSC.MARC User Guide. Los Angeles (2005), *MSC Software Corporation*

Oden, J. T. and Pires, E. B (1983) 'Nonlocal and nonlinear friction laws and variational principles for contact problems in elasticity' *Journal. of applied mechanics*, Vol. 50

Ozel, T. (2006) 'The influence of friction models on finite element simulations of machining' *International journal of machine tools and manufacture*; Vol.46, pp.518-530.

Ozel, T., and Altan, T. (2000) 'Process simulation using finite element method - prediction of cutting forces, tool stresses and temperatures in high-speed flat end milling', *International journal of machine tools and manufacture*, Vol. 40, pp.713-738.

Ozel, T., Lucchi, M., Rodríguez, C. A. and Altan, T. (1998) 'Prediction of chip formation and cutting forces in flat end milling, Comparison of process simulations with experiments', *Transaction of NAMI/SAE* , Vol.26, pp.231-237

Shih, A. J, (1996) 'Finite element analysis of orthogonal metal cutting mechanics' International journal of machine tools and manufacture, Vol. 36, pp.255-273.

Umeda, T., Umiki, H. and Mimura, K. (2005) 'Numerical evaluation of measurement accuracy of non-coaxial Hopkinson bar method', *JSME international journal 2005*, Vol.48, pp.215-221

Yucesan, G. and Altintas, Y. (1994) 'Improved modelling of cutting force coefficients in peripheral milling' *International journal of machine tools and manufacture*, Vol.34, pp.463-487.

In: Machining and Forming Technologies, Volume 2
Editor: J. Paulo Davim
ISBN: 978-1-61122-820-5

Chapter 14

MODELING FRACTAL DIMENSION IN CNC MILLING OF BRASS USING ARTIFICIAL NEURAL NETWORK

T K Barman and P Sahoo*

Department of Mechanical Engineering, Jadavpur University,
Kolkata 700032, India

ABSTRACT

The objective of the present study is to develop a multi-layer feed forward neural network model to predict fractal dimension in CNC milling of brass material. Experiments are carried out using full factorial design and spindle speed, feed rate and depth of cut are considered as cutting parameters. These cutting parameters are used as the input neurons and corresponding fractal dimension is considered as the output of the network. Neural networks are trained with Levenberg-Marquardt algorithm. Minimum mean squared error is considered as the criterion for selecting the best network. It is observed that the network with 3-16-1 architecture gives the best performance. The network predicted and experimental fractal dimension are very close to each other. Mean absolute percentage errors are 0.54% and 1.26% for training and testing pattern respectively.

Keywords: ANN, Fractal Dimension, Milling, Brass

1. INTRODUCTION

CNC milling is a popular machining process in the modern industry because of its ability to remove materials with a multi-point cutting tool at a faster rate with a reasonably good surface quality. Any machining surface has some irregularities which is termed as the surface roughness. The surface roughness is a very important parameter because it affects the

* Corresponding author: E-mail:psahoo@vsnl.net

performance of the mechanical parts as well as production cost. It also has impact on the mechanical properties of the part like friction, wear, light reflection, heat transmission, lubrication, electrical conductivity etc. In order to get specified surface roughness, selection of controlling parameters is necessary. There has been a great many research developments in modeling surface roughness and optimization of the controlling parameters to obtain a surface finish of desired level since only proper selection of cutting parameters can produce a better surface finish. But such studies are far from complete since it is very difficult to consider all the parameters that control the surface roughness for a particular manufacturing process.

It is now well established that to express roughness the machining surfaces can be characterized by the concept of fractal. The concept is based on the self-affinity and self-similarity of surfaces at different scales. Roughness measurements on a variety of surfaces show that the power spectra of the surface profiles follow power laws. This suggests that when a surface is magnified appropriately, the magnified image looks very similar to the original surface. Fractals may retain all the structural information and are characterized by single descriptor, the fractal dimension, *D*. This fractal dimension, which forms the essence of fractal geometry, is both scale-invariant and is closely linked to the concepts of self-similarity and self-affinity (Mandelbrot, 1982). On the other hand, the conventional roughness parameters are strongly depend on the resolution and filter processing of the instrument and therefore these roughness parameters are scale depended. As the surface topography is a non-stationary random process, the slope and curvature are used for characterization and the variances of slope and curvature depend strongly on the resolution of the roughness-measuring instrument or any other form of filter. It is therefore essential to use fractal dimension to characterize rough surfaces and provide the geometric structure at all length scales. In a material removal process, mechanical intervention happens over length scales, which extend from atomic dimensions to centimeters. The machine vibration, clearances and tolerances affect the outcome of the process at the largest of length scales (above 10^{-3} m). The tool form, feed rate, tool radius in case of single point cutting and grit size in multiple point cutting affect the process outcome at the intermediate length scales (10^{-6} to 10^{-3} m). The roughness of the tool or details of the grit surfaces influence the final topography of the generated surface at the lowest length scales (10^{-9} to 10^{-6} m). It has been shown that surfaces formed by electric discharge machining, milling, cutting or grinding (Zhang et al, 2001; Jiang et al, 2001; Jahn and Truckenbrodt, 2004; Kang et al, 2005; Han et al, 2005, Sahoo et al, 2008), and worn surfaces have fractal structures, and fractal parameters can reflect the intrinsic properties of surfaces to overcome the disadvantages of conventional roughness parameters.

There are many researchers who have developed models in milling operations using several tools like response surface methodology, Taguchi analysis, soft computing etc. But now-a-days, application of artificial neural network to model surface roughness is more popular in machining as this tool can be applied efficiently and satisfactorily. The available literatures in milling operations using neural network as a tool are presented here. Tsai et al. (1999) have developed an artificial neural network (ANN) model using spindle speed, feed rate, depth of cut, and the vibration average per revolution as four input neurons, to predict the output neuron-surface roughness R_a values for end milling operation. Balic and Korosec (2002) have developed a model with the help of ANN, for the prediction of milling tool-path strategy to achieve the best surface quality. To define the surface quality, they have used the center line average roughness, R_a and they have shown the NN predicted and experimental R_a

values are in good agreement. Benardos and Vosniakos (2002) have developed a neural network model to predict the surface roughness (R_a) in CNC face milling operation. They have also showed that the developed model accurately predicts the surface roughness. Jesuthanam et al. (2007) developed a hybrid neural network trained with GA and Particle Swarm Optimization (PSO) for the prediction of surface roughness in CNC end milling operation. Oktem (2008) has done an integrated study of surface roughness (R_a) for modeling and optimization of cutting parameters during end milling of AISI 1040 material. El-Sonabaty et al. (2008) have used fractal geometry approach for predicting surface roughness using neural network. Munoz-Escalona and Maropoulos (2009) have recently addressed the application of neural network model for prediction of surface roughness (R_a) in face milling operation. Samanta (2009) has also applied different soft computing techniques including ANN, GA, Fuzzy logic etc. to model surface roughness (R_a). Recently, Zain et al. (2010) have reported the usefulness of artificial neural network for predicting R_a in end milling operations. From the literature survey, it is seen that there are few literatures available in milling which deal with surface roughness modeling using artificial neural network. Moreover, almost all of the available literatures include the conventional roughness parameters to describe the surface roughness and there is a scarcity of literatures which deal with fractal dimension.

In the present study, fractal dimension is used to describe the surface roughness, and for predicting fractal dimension a multi-layer back-propagation neural network model trained with Levenberg-Marquardt algorithm is developed in end milling of brass material. Spindle speed, feed rate and depth of cut are considered as the process parameters and these parameters are used as the input neurons to the network and the corresponding fractal dimension is used as the output neuron of the network. The developed network is tested with a new testing set to check the generalization capability of the network.

2. Basics of Fractal Dimension

Most rough surfaces including machining ones and corresponding profiles are multi-scale in nature. This multi-scale property is better expressed as self-similarity or self-affinity in fractal geometry implying that when the surface or the profile is magnified more and more details emerge and the magnified image is statistically similar to the original topography. Statistical self-similarity means that the probability distribution of a small part of a profile will be congruent with the probability distribution of the whole profile if the small part is magnified equally in all directions. On the other hand, self-affinity implies unequal scaling in different directions.

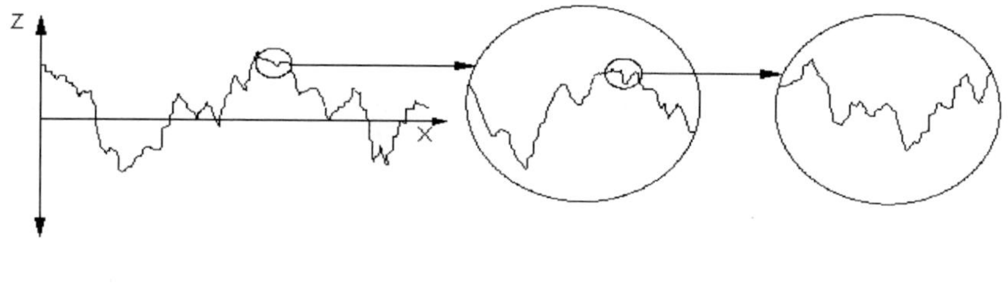

Figure 1. Qualitative description of self-affinity of a surface profile.

The qualitative description of statistical self-affinity for a surface profile is shown in Figure 1. The property of self-affinity can be characterized by the profile fractal dimension D ($1< D <2$). In the present paper, isotropic and homogeneous rough engineering surface of dimension D_s ($2 < D_s < 3$) is considered. The property of isotropy means that the probability distribution of heights is invariant when the coordinate axes are rotated and the surface is reflected on any plane. The property of homogeneity of a surface indicates that the probability distribution of the height is independent of the location on the surface. Therefore, the profile, $z(x)$, of such a surface along a straight line and in any arbitrary direction is of dimension $D = D_s - 1$ and is a statistically valid representation of the surface. Thus the profile fractal dimension D is adopted to characterize the fractal nature of the surface in this paper. The profile $z(x)$ in Figure 1 has the mathematical properties of being continuous everywhere but non-differentiable and is self-affine in roughness structure. These properties are satisfied by the Weierstrass–Mandelbrot (W-M) fractal function, which can be used to characterize the roughness of surface profile and is given as

$$z(x) = G^{(D-1)} \sum_{n=n_1}^{\infty} \frac{\cos 2\pi\gamma^n x}{\gamma^{(2-D)n}} \quad ; 1 < D < 2; \gamma > 1 \tag{1}$$

where G is a characteristic length scale, γ where L is the sampling length. $\gamma^n = \omega$, where frequency ω is the reciprocal of wave length and n is called wave number. To provide both the phase randomization and high spectral density, γ is selected to be 1.5. The parameters G and D form the set to characterize profile $z(x)$. The methods for calculating profile fractal dimension mainly include the yard-stick, the box counting, the variation, the structure function and the power spectrum methods. Out of these, the power spectrum and structure function methods are most popular. The power spectrum of W–M function is given as

$$S(\omega) = \frac{G^{2(D-1)}}{2\ln\gamma} \frac{1}{\omega^{(5-2D)}} \tag{2}$$

It can be seen that both G and D are independent of ω, therefore they are scale-invariant. Equation (2) shows that the power spectrum follows a power law behavior. If $S(\omega)$ is plotted versus ω on a log–log plot, then the power law behavior would result in a straight line. Using Fast Fourier Transform (FFT), the power spectrum of profile can be calculated. Comparing the power spectrum of W–M function with that of a real surface, D relates to the slope of the spectrum on the log–log plot, G correlates with the intercept of the log–log plot. The structure function $S(\tau)$ of sampling data on the profile curve $z(x)$ can be described as

$$S(\tau) = \langle [z(x+\tau) - z(x)]^2 \rangle = C\tau^{(4-2D)} \tag{3}$$

where τ is any displacement along the x direction, $\langle\ \rangle$ is temporal average, C is a constant. Different τ and the corresponding $S(\tau)$ can be plotted verses the τ on a log–log scale. Then the profile fractal dimension D can be related to the slope of the fitting line, m, on log-log plot as $D = (4 - m)/2$. In comparison with power spectrum method, the structure function is more exact and easy to operate. For this reason the structure function method is adopted to characterize the fractal character of the surface profiles in this paper. The continuous analogue electric signals of profiles from a measuring basis are put into a computer. After analogue to digital (A/D) translation, the profiles are given with digitalized height values during a traverse over a sampling length. From the dispersed digital signals, the mean line $z(x) = 0$ is fitted by use of least squares method and the deviations of profile $z(x)$ from its mean line are determined. Let the sampling spacing of the computer system is Δt with each sampling length L having $N+1$ evenly spaced data points with $z(x_i) = z_i\,(i = 0,1,2,\ldots\ldots, N)$ and $\tau = n\Delta t$ (n=1,2,3..). Then the structure function takes the form:

$$S(\tau) = S(n\Delta t) = \left\langle [z(x_i + n\Delta t) - z(x_i)]^2 \right\rangle = \frac{1}{N-n}\sum_{i=0}^{N-n}(z_{i+n} - z_i)^2 \tag{4}$$

By virtue of the above equation, D can be determined from the structure function plot on log–log coordinates.

3. Artificial Neural Network

Artificial neural networks are simplified models of biological nervous system which mimic the computational architecture of a human brain. A neural network is a massively parallel distributed processor that has a natural propensity for storing experimental knowledge and making it available for use (Haykin, 2003). In general, a neural network comprises of processing elements called neuron. These neurons are connected via connection links that carry signals between processors and generally grouped into input layer, a variable number of hidden layers and an output layer. The input layer receives the information from an external source. Hidden layer will receive information from input layer and create internal representation. Finally, output will be available via output layer. The network will acquire knowledge through a learning process. The learning is done by examples. The connection links posses some weights and these synaptic weights are used to store the knowledge in the network. The weights are adjusted during the learning stage to get the desired result for a given set of inputs. The greater is the weight between neurons, the stronger is the influence of the connection. There are several learning methods developed for ANN. Back propagation learning method is frequently used by many researchers successfully in manufacturing sciences. This learning method is simple, most representative and commonly used. The basic structure of a back-propagation neural network consists of an input layer, hidden layer (s) and output layer. There may be one or more hidden layers consisting variable neurons or nodes. The basic neural architecture is presented in Figure 2.

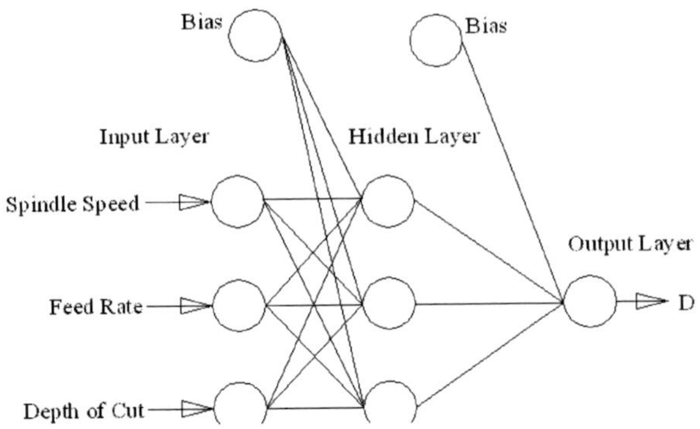

Figure 2. Architecture of the neural network used in the study.

The input layer is used to feed the data in the network. The inputs are subsequently modified based on the interconnection weights between the layers. The net input to each neuron from the preceding layer will be as

$$net_j = \sum_{i=1}^{N} w_{ij} x_i + b_j \tag{5}$$

where, net_j is the net input, N is the number of neurons of the inputs to the jth neuron in the hidden layer. w_{ij} is the interconnected weight from i^{th} neuron in the forward layer to j^{th} neuron in the hidden layer, x_i is the input from the ith neuron in the hidden layer and b_j is the bias value to the corresponding neuron. The output signal is obtained by applying activations to the net input.

The performance of a particular problem in neural network depends on various factors like number of training data sets, performance goal, complexity of the data set, network architecture which includes number of neurons in the hidden layers and number of layers etc. It is very difficult to predict which training algorithm will provide the best performance among several training algorithms which can be applied to train the network. Standard back-propagation is a gradient descent algorithm in which the network weights and biases updates along the negative of the gradient of the performance function. The problem with standard gradient descent algorithm is that it converges very slowly. With standard steepest descent, the learning rate is held constant throughout training. But the performance of the algorithm is very sensitive to the proper setting of the learning rate. The performance of the algorithm can be improved using adaptive learning rate. The learning rate is made responsive to the complexity of the local error surface. Newton's method is an alternative to the conjugate gradient methods for fast optimization. The basic step of Newton's method is to compute the Hessian matrix (second derivatives) of the performance index at the current values of the weights and biases and the Hessian matrix of second derivatives of the function to be minimized. Newton's method often converges faster than conjugate gradient methods. But it is complex and expensive to compute the Hessian matrix for feedforward neural networks. Levenberg-Marquardt can approach second-order training speed without having to compute the Hessian matrix (second derivatives) (Demuth and Beale, 1998).

4. EXPERIMENTAL SCHEME

4.1. Design of Experiment

The design of experiments technique is a very powerful tool, which permits us to carry out the modeling and analysis of the influence of process variables on the response variables. The response variable is an unknown function of the process variables, which are known as design factors. There are a large number of factors that can be considered for machining of a particular material in end milling. Surface roughness is generally dependent on many parameters such as the tool geometry (tool nose radius and flank width, run-out error), tool material, work material, machine-tool rigidity and various cutting conditions including feed rate, depth of cut and cutting speed. The presence of chatter or vibration of the machine tool, defects in the surface of work material, wear in the tool, or irregularities of chip formation contribute to the surface damage in practice during actual machining operations. It is difficult to consider all the factors that affect the surface finish in any experimental study. Thus, it is necessary to relate the surface roughness with primary machining variables. In the present study, variation of three primary machining parameters, viz. depth of cut (d, mm), spindle speed (N, rpm) and feed rate (f, mm/min) are considered (Alauddin et al. 1995, 1997;Lou et al, 1998;Wang et al., 2005; Vivancos et al., 2004; Oktem, 2008; El-Sonbaty et al., 2008). In the present study these are selected as design factors while other parameters have been assumed to be constant over the experimental domain. A full factorial design is used with five levels of each of the three design factors. Thus the design chosen was five level-three factor (5^3) full factorial design consisting of 125 sets of coded combinations. The upper and lower limits of a factor were coded as +1 and -1 respectively, the coded value being calculated from the following relationships:

$$x_i = \frac{[2x - (x_{max} + x_{min})]}{(x_{max} - x_{min})} \tag{6}$$

where x_i is the required coded value of a variable x. The process variables / design factors with their values on different levels are listed in Table 1. The selection of the values of the variables is limited by the capacity of the machine used in the experimentation as well as the recommended specifications for different work-tool material combinations (Oberg et al, 2000).

Table 1. Variable levels used in the experimentation

Levels	*Depth of cut (d) (mm)*	*Spindle speed (N) (rpm)*	*Feed rate (f) (mm/min)*
-1	0.10	1500	550
-0.5	0.15	1800	600
0	0.20	2100	650
0.5	0.25	2400	700
1	0.30	2700	750

4.2. Equipment Used

The machine used for the milling tests is a 'DYNA V4.5' CNC end milling machine (Figure 3) having the control system SINUMERIK 802 D with a vertical milling head and equipped with maximum spindle speed of 9000 rpm, feed rate 10 m/min and 10 kW driver motor. For generating the milled surfaces, CNC part programs for tool paths are created with specific commands.

Figure 3. Photograph of CNC End milling machine used for the experiment.

4.3. Cutting Tool Used

Coated carbide tools are known to perform better than uncoated carbide tools. Thus commercially available CVD coated carbide tools are used in this investigation. The tools used are flat end mill cutters (8 mm diameter, 30^0 helix angle, TiAlN coated solid carbide, parallel shank, 4 flutes) produced by WIDIA (EM-TiAlN). The tools are coated with TiAlN coating having hardness, density and transverse rupture strength as 1570 HV, 14.5 g/cc and 3800 N/mm^2. The compressed coolant servo-cut is used as cutting environment.

4.4. Work Piece Material

The present study is carried out with medium leaded Brass C34000. All the specimens are in the form of 100 mm x 75 mm x 25 mm blocks.

4.5. Response Variable Selected

The response variable used to accomplish the present study on surface topography characterization is the profile fractal dimension D.

4.6. Fractal Dimension Measurement

Roughness profile measurement is done using a portable stylus-type profilometer, *Talysurf* (Figure 4). The profilometer is set to a cut-off length of 0.8 mm, filter 2CR, traverse speed 1 mm/sec and 4 mm traverse length. Roughness measurements, in the transverse direction, on the work pieces are repeated four times and average of four measurements of surface roughness parameter values is recorded. The measured profile is digitized and processed through the dedicated advanced surface finish analysis software *Talyprofile*. Then fractal dimension is evaluated following the structure function method as detailed in section 2.

5. Neural Network Modeling

According to the full factorial design of experiments, total 125 experiments are carried out. These datasets are randomized and 60% of data are chosen for training, 20% for testing and rest 20% for validation purpose. Three cutting parameters viz., spindle speed, feed rate and depth of cut are used as the input neurons and corresponding fractal dimension as the output neuron. The data points are normalized so that the datasets range in -1 to +1.

Figure 4. Photograph of stylus-type profilometer, Talysurf (Taylor Hobson, Surtronic 3+).

5.1. Training of the Network

To construct the models, the hyperbolic tangent sigmoid function in the hidden layer and linear activation function in the output layer is considered. To train the network, Levenberg-Marquardt (L-M) training algorithm is applied. This Levenberg-Marquardt algorithm is a faster and powerful training algorithm than the gradient descent algorithm. Initial learning rate for the network is 0.001 with learning rate decrease factor as 0.10 and learning rate increase factor as 10. The maximum learning rate is 10^{10}. MATLAB 6.5 is used to develop the neural network, train and test the network.

5.2. Initialization of Weights

Initialization of weights in the network has a great impact on the performance of the network. Nguyen-Widrow weight initialization algorithm has been applied which generates initial weight and bias values for a layer so that the active regions of the layer's neurons will be distributed roughly evenly over the input space (Demuth and Beale, 1998).

5.3. Generalization of the Models

The main target of a properly trained neural network is to give reasonable answers when presented with inputs that they have never seen. That means the network should be generalized well to take new inputs. Here "early stopping" technique is implemented. There are different techniques to improve generalization. For this, complete set of data is divided into 3 subsets viz. training set, testing set and validation set. Training set is used to train the network. The validation set determines when the training should stop by monitoring the error. Testing set does not participate in the training of the network but is used to test the generalization of the trained network.

5.4. Selection Criterion of the Best Network

The determination of hidden layer(s) and numbers of neurons in the hidden layer(s) is a considerable task. But, generally, one hidden layer network is selected first. If it does not perform well then two hidden layers are used. For selecting number of neurons in the hidden layer, many researchers propose different algorithms/ criterions. Here, number of neurons in the hidden layer is decided by trail with increasing number of neurons. The neurons in the hidden layers are varied from 5 to find out the best neural network model. There are several criteria for selecting the best network. But in this study minimum mean squared error (MSE) is used as the selection parameter. Mean squared error is measured as

$$MSE = \frac{1}{N}\sum_{p=1}^{N}\sum_{k=1}^{n}(d_{k,p} - o_{k,p})^2 \tag{7}$$

where, N is the number of pattern, n is the number of node in the output layer, dk,p is the desired/experimental output for k[th] node of the pth pattern and $o_{k,p}$ is the calculated/predicted output of kth node of pth pattern.

Besides this parameter, for the network, correlation coefficient (R) and maximum absolute percentage error for the training data set and testing data set are calculated. Correlation coefficient (R) is an indication of agreement between experimental and predicted values. It ranges from -1.00 to +1.00. Correlation coefficient of -1.00 represents a perfect negative correlation whereas correlation coefficient of +1.00 represents a perfect positive correlation. Absolute percentage error for the training and testing data set is measured as

$$Absolute\_Percentage\_error = \left| \frac{y_{i,experimental} - y_{i,predicted}}{y_{i,experimental}} \right| \times 100 \quad (8)$$

where $y_{i,experimental}$ is the experimental value and $y_{i,predicted}$ is the network predicted value of the response for i[th] observation.

Table 2. Experimental results

Sl No	Depth of cut(*d*)	Spindle speed(*N*)	Feed rate(*f*)	Fractal Dimension, D
1	-1	-1	-1	1.28
2	-1	-1	-0.5	1.31
3	-1	-1	0	1.22
4	-1	-1	0.5	1.28
5	-1	-1	1	1.27
6	-1	-0.5	-1	1.30
7	-1	-0.5	-0.5	1.29
8	-1	-0.5	0	1.30
9	-1	-0.5	0.5	1.32
10	-1	-0.5	1	1.27
11	-1	0	-1	1.38
12	-1	0	-0.5	1.33
13	-1	0	0	1.31
14	-1	0	0.5	1.30
15	-1	0	1	1.31
16	-1	0.5	-1	1.36
17	-1	0.5	-0.5	1.33
18	-1	0.5	0	1.30
19	-1	0.5	0.5	1.31
20	-1	0.5	1	1.32
21	-1	1	-1	1.37
22	-1	1	-0.5	1.35
23	-1	1	0	1.34
24	-1	1	0.5	1.35
25	-1	1	1	1.30
26	-0.5	-1	-1	1.30
27	-0.5	-1	-0.5	1.27
28	-0.5	-1	0	1.26

Table 2. – (Continued)

29	-0.5	-1	0.5	1.25
30	-0.5	-1	1	1.28
31	-0.5	-0.5	-1	1.31
32	-0.5	-0.5	-0.5	1.29
33	-0.5	-0.5	0	1.31
34	-0.5	-0.5	0.5	1.29
35	-0.5	-0.5	1	1.29
36	-0.5	0	-1	1.38
37	-0.5	0	-0.5	1.34
38	-0.5	0	0	1.31
39	-0.5	0	0.5	1.28
40	-0.5	0	1	1.29
41	-0.5	0.5	-1	1.35
42	-0.5	0.5	-0.5	1.33
43	-0.5	0.5	0	1.32
44	-0.5	0.5	0.5	1.32
45	-0.5	0.5	1	1.29
46	-0.5	1	-1	1.37
47	-0.5	1	-0.5	1.37
48	-0.5	1	0	1.36
49	-0.5	1	0.5	1.32
50	-0.5	1	1	1.31
51	0	-1	-1	1.26
52	0	-1	-0.5	1.25
53	0	-1	0	1.26
54	0	-1	0.5	1.27
55	0	-1	1	1.29
56	0	-0.5	-1	1.36
57	0	-0.5	-0.5	1.27
58	0	-0.5	0	1.32
59	0	-0.5	0.5	1.27
60	0	-0.5	1	1.29
61	0	0	-1	1.35
62	0	0	-0.5	1.35
63	0	0	0	1.32
69	0	0.5	0.5	1.32
70	0	0.5	1	1.36
71	0	1	-1	1.37
72	0	1	-0.5	1.33
73	0	1	0	1.35
74	0	1	0.5	1.34
75	0	1	1	1.36
76	0.5	-1	-1	1.24
77	0.5	-1	-0.5	1.27
78	0.5	-1	0	1.23

79	0.5	-1	0.5	1.26
80	0.5	-1	1	1.28
81	0.5	-0.5	-1	1.27
82	0.5	-0.5	-0.5	1.27
83	0.5	-0.5	0	1.33
84	0.5	-0.5	0.5	1.25
85	0.5	-0.5	1	1.28
86	0.5	0	-1	1.38
87	0.5	0	-0.5	1.33
88	0.5	0	0	1.32
89	0.5	0	0.5	1.29
90	0.5	0	1	1.31
91	0.5	0.5	-1	1.39
92	0.5	0.5	-0.5	1.33
93	0.5	0.5	0	1.32
94	0.5	0.5	0.5	1.31
95	0.5	0.5	1	1.36
96	0.5	1	-1	1.36
97	0.5	1	-0.5	1.37
98	0.5	1	0	1.37
99	0.5	1	0.5	1.34
100	0.5	1	1	1.34
101	1	-1	-1	1.29
102	1	-1	-0.5	1.28
103	1	-1	0	1.26
104	1	-1	0.5	1.24
105	1	-1	1	1.26
106	1	-0.5	-1	1.31
107	1	-0.5	-0.5	1.28
108	1	-0.5	0	1.31
109	1	-0.5	0.5	1.27
110	1	-0.5	1	1.30
111	1	0	-1	1.37
112	1	0	-0.5	1.33
116	1	0.5	-1	1.39
117	1	0.5	-0.5	1.37
118	1	0.5	0	1.31
119	1	0.5	0.5	1.29
120	1	0.5	1	1.35
121	1	1	-1	1.37
122	1	1	-0.5	1.37
123	1	1	0	1.36
124	1	1	0.5	1.33
125	1	1	1	1.31

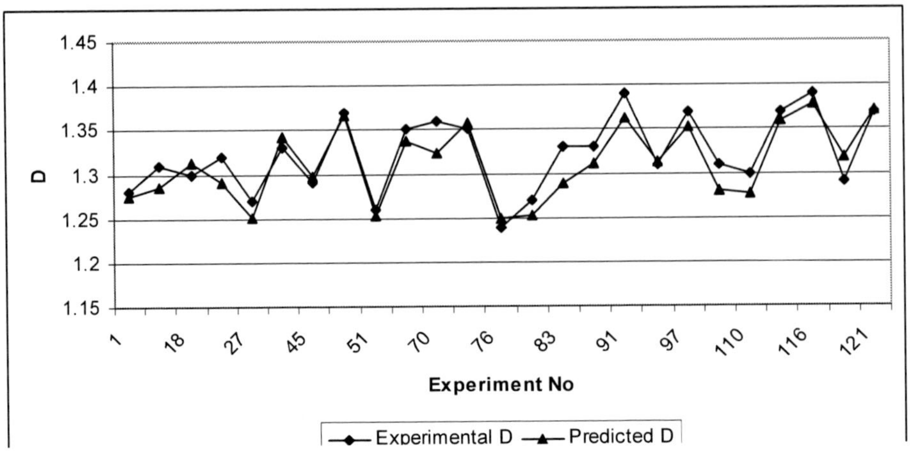

Figure 5. Comparative study of Experimental and ANN predicted D for testing set.

6. Results and Discussion

Experimental results of end milling operation are presented in Table 2. Using Matlab, several models are developed varying the number of neurons in the hidden layer. The performances of the different models are presented in Table 3. From the results, it is seen that with increasing the number of neurons in the hidden layer does not always ensure that the mean square error will decrease but it is clear that with increasing the number of neurons, the architecture gets complicated. Considering the minimum mean square error, architecture 3-16-1 is the best network with MSE value of 0.00008218. This network also provides the highest correlation coefficient and minimum mean absolute percentage error.

From Table 3, it is seen that the maximum absolute percentage error is 2.05% for training data set. The network is also tested with 25 new testing sets which the network has not seen before. For the testing pattern, Table 4 and Figure 5 present the comparative study of the experimental fractal dimension (D) and ANN predicted fractal dimension (D). Maximum absolute error and mean absolute error are 3.13 and 1.26 respectively for testing pattern (Table 4). It implies that the experimental value and ANN predicted values lie very close to each other. Thus, it can be said that the network has the requisite robustness and adaptability to the new pattern. Regression analysis is conducted to check the correlation between the experimental and predicted fractal dimension. It is observed from the regression analysis that correlation coefficient (R) is 0.967 and 0.913 for training and testing respectively (Figure 6 and Figure 7) and it is a good indication of the correlation between predicted and experimental outputs. It also proves that the generalization capability of the network. From the results, it is obvious that the network gives a good prediction of fractal dimension in milling of brass material.

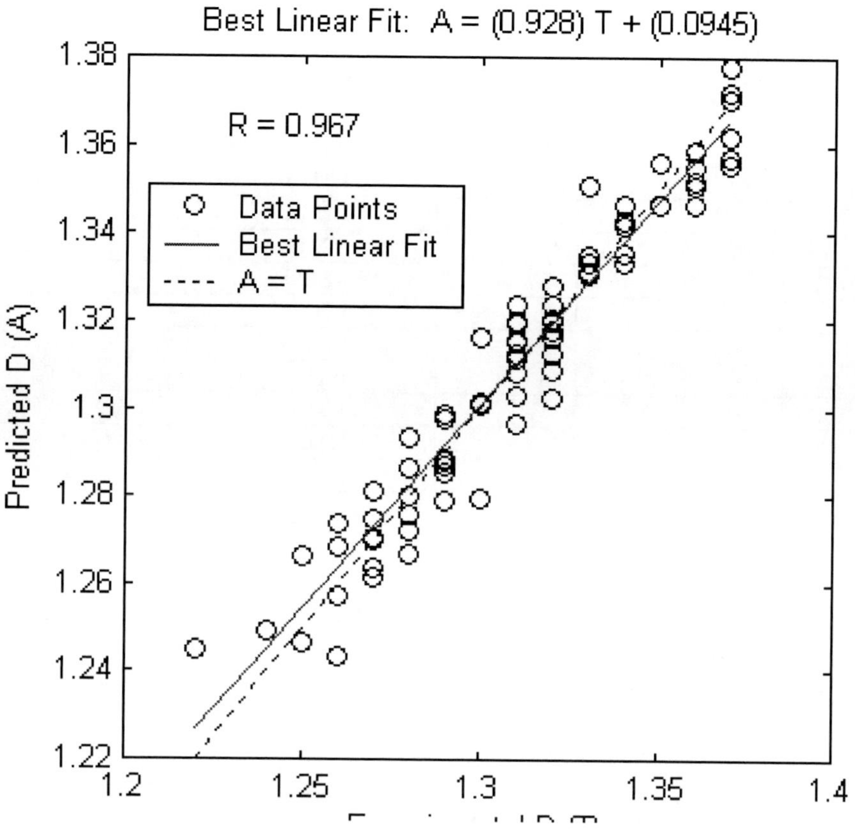

Figure 6. Performance of ANN model using regression analysis for training pattern.

Table 3. Performance comparison of different models

Model	MSE	Correlation Coefficient (R)	Max. Training Error (%)	Max. Testing Error (%)
3-5-1	0.00034129	0.874	3.49	3.47
3-6-1	0.00022272	0.932	3.37	3.19
3-7-1	0.00023735	0.924	3.34	4.38
3-8-1	0.00021034	0.922	3.12	3.17
3-9-1	0.00020804	0.929	2.57	3.47
3-10-1	0.00016868	0.946	3.20	3.61
3-11-1	0.00014909	0.941	2.69	3.48
3-12-1	0.00014721	0.949	3.62	2.95
3-13-1	0.00012751	0.953	2.39	3.14
3-14-1	0.00013875	0.949	2.49	3.02
3-15-1	0.00015600	0.946	2.78	4.07
3-16-1	0.00008218	0.967	2.05	3.13
3-17-1	0.00019180	0.945	2.63	2.58
3-18-1	0.00014573	0.945	2.12	3.49
3-19-1	0.00021403	0.933	3.48	3.35
3-20-1	0.00017428	0.939	3.41	3.35

Table 4. Comparison between Experimental and ANN predicted *D* for testing set

Cutting Parameters			Experimental D	Neural Network Predicted D	Absolute Percentage Error for NN
d	N	f			
-1	-1	-1	1.28	1.2759	0.32
-1	0	1	1.31	1.2848	1.92
-1	0.5	0	1.30	1.3135	1.04
-1	0.5	1	1.32	1.2906	2.23
-0.5	-1	-0.5	1.27	1.2513	1.47
-0.5	0.5	-0.5	1.33	1.3420	0.90
-0.5	0.5	1	1.29	1.2968	0.53
-0.5	1	-1	1.37	1.3663	0.27
0	-1	-1	1.26	1.2536	0.51
0	0	-1	1.35	1.3372	0.95
0	0.5	1	1.36	1.3225	2.76
0	1	0	1.35	1.3575	0.56
0.5	-1	-1	1.24	1.2493	0.75
0.5	-1	-0.5	1.27	1.2522	1.40
0.5	-0.5	0	1.33	1.2884	3.13
0.5	0	-0.5	1.33	1.3107	1.45
0.5	0.5	-1	1.39	1.3630	1.94
0.5	0.5	0.5	1.31	1.3137	0.28
0.5	1	-0.5	1.37	1.3527	1.26
1	-0.5	0	1.31	1.2800	2.29
1	-0.5	1	1.30	1.2773	1.75
1	0	-1	1.37	1.3595	0.77
1	0.5	-1	1.39	1.3785	0.83
1	0.5	0.5	1.29	1.3174	2.12
1	1	-1	1.37	1.3720	0.15

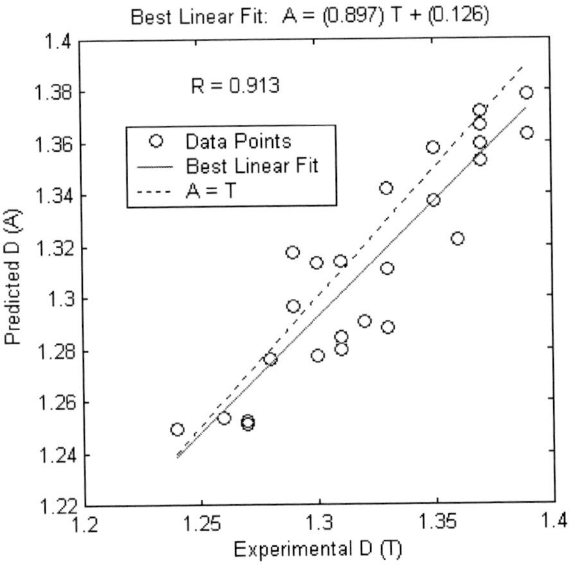

Figure 7. Performance of ANN model using regression analysis for testing pattern.

To see the variations of fractal dimension with process parameters, three dimensional surfaces as well as contour plots for fractal dimension (D) as functions of two independent machining parameters while the third machining parameter is held constant at central value are generated (Figures. 8-10). All these figures clearly depict the variation of fractal dimension with controlling variables within the experimental regime. Fractal dimension is found in general to increase with spindle speed and decrease with feed rate.

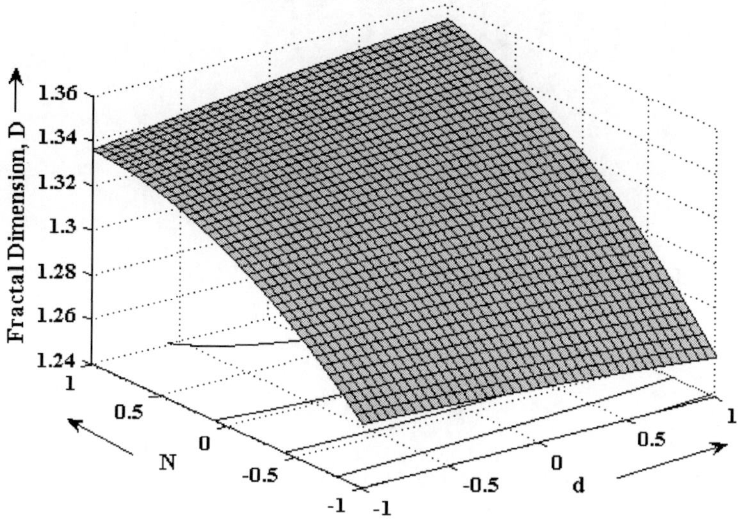

Figure 8. Surface and contour plots of fractal dimension with depth of cut and spindle speed at central level of feed rate.

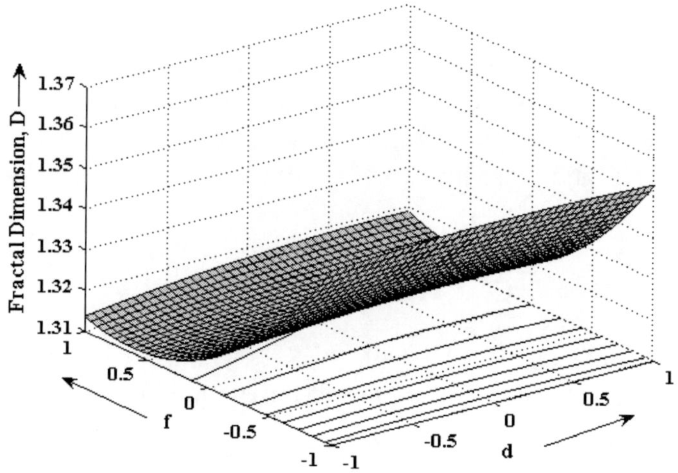

Figure 9. Surface and contour plots of fractal dimension with depth of cut and feed rate at central level of spindle speed.

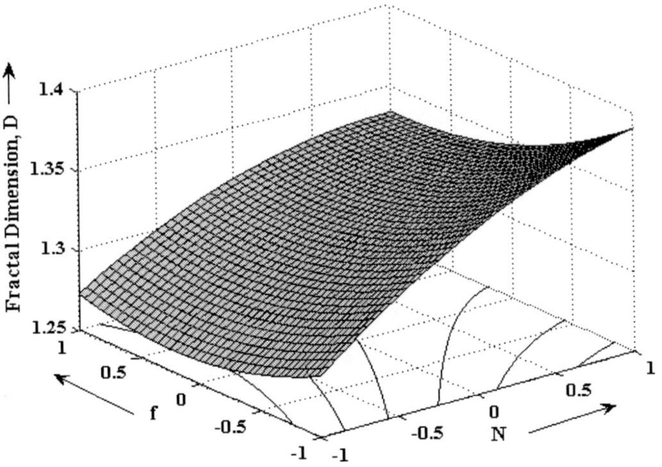

Figure 10. Surface and contour plots of fractal dimension with spindle speed and feed rate at central level of depth of cut.

7. CONCLUSIONS

In this study, back-propagation algorithm of artificial neural network technique is applied to predict fractal dimension in milling of brass material. Models are developed with one hidden layer but to get the best network based on the minimum mean square error the number of neurons in the hidden layer are varied. Levenberg-Marquardt training algorithm is used to train the network. The network 3-16-1 architecture gives the best performance. Results show that the developed network accurately predict fractal dimension in milling of brass material. The mean absolute percentage error for training pattern is only 0.54% and for testing pattern is 1.26%. From the regression analysis, it is also clear that the experimental and predicted fractal dimensions are in good agreement. From these results, it can be concluded that the developed neural networks can successfully predict the fractal dimension and the network also has a good generalization capability.

REFERENCES

Alauddin, M., Baradie, M. A., and Hashmi, M. S. J., (1995), "Computer aided Computer-aided analysis of a surface-roughness model for end milling", *J Mater Process Technol* 55,123-127.

Alauddin, M., El Baradie, M. A. and Hashmi, M. S. J., (1997), "Prediction of tool life in end milling by response surface methodology", *J. Mater Process Technol,* 71, 456-465.

Balic, J. and Korosec, M. (2002), "Intelligent tool path generation for milling of free surfaces using neural networks", *Int J. Mach Tools Manufact*, 42, 1171-1179.

Benardos, P. G. and Vosniakos, G. C. (2002), "Prediction of surface roughness in CNC face milling using neural networks and Taguchi's design of experiments", *Robot Comput Integr Manuf,* 18, 343–354.

Demuth, H. and Beale, M. (1998), *Neural network toolbox user's guide*, version 3.0. Natick, MA: Mathworks Inc.

El-Sonabaty, I. A., Khashaba, U. A., Selmy, A. I. and Ali, A. I. (2008), "Prediction of surface roughness profiles for milled surfaces using an artificial neural network and fractal geometry approach", *J Mater Process Technol,* 200, 271-278.

Han, J. H., Ping, S. and Shengsun, H. (2005), "Fractal characterization and simulation of surface profiles of copper electrodes and aluminum sheets," *Materials Science and Engineering* A, 403, 174-181.

Haykin, S. (2003), *Neural networks: a comprehensive foundation*, 2nd ed. Pearson Education.

Jahn, R. and Truckenbrodt, H. (2004), "A simple fractal analysis method of the surface roughness." *J Mater Process Technol*, 145, 40-45.

Jesuthanam, C. P., Kumanan, S. and Asokan, P., (2007), "Surface roughness prediction using hybrid neural networks", *Mach. Sc. Technol*, 11, 271-286.

Jiang, Z., Wang, H. and Fei, B. (2001), "Research into the application of fractal geometry in characterizing machined surfaces," *Int J Mach Tools Manufact*, 41, 2179-2185.

Kang, M. C., Kim J. S. and Kim, K. H. (2005), "Fractal dimension analysis of machined surface depending on coated tool wear," *Surf Coat Technol,* 193(1-3), 259-265

Lou, M. S., Chen, J. C. and Li, C. M., (1998), "Surface roughness prediction technique for CNC end milling". *J Ind Technol*,15(1),1-6.

Mandelbrot, B. B. (1982), *The fractal geometry of nature*. Freeman, New York.

Munoz-Escalona, P. and Maropoulos, P. G., (2009), "Artificial neural networks for surface roughness prediction when face milling Al 7075-T7351", *J Mater Eng Perf,* DOI: 10.1007/s11665-009-9452-4.

Oberg, E., Jones, F. D., Horton, H. L. and Ryffel, H. H. (2000) *Machinery's Hand Book*, Industrial Press Inc, New York

Oktem, H. (2008), "An integrated study of surface roughness for modeling and optimization of cutting parameters during end milling operation", *Int J Adv Manuf Technol,* DOI 10.1007/s00170-008-1763-3.

Sahoo, P, Barman, T. K. and Routara, B. C., (2008), "Fractal dimension modeling of surface profile and optimization in CNC end milling using response surface method", *Int J Manuf Res,* 3 (3), 360-377.

Samanta, B. (2009), "Surface roughness prediction in machining using soft computing", *Int J Compt Integr Manuf,* 22, 3, 257-266.

Tsai, Y. H., Chen, J. C. and Lou, S. J., (1999), "An in-process surface recognition system based on neural networks in end milling cutting operations", *Int J Mach Tools Manufact*, 39, 583-605.

Vivancos, J., Luis, C. J., Costa, L. and Ortiz, J. A., (2004), "Optimal machining parameters selection in high speed milling of hardened steels for injection moulds", *J Mater Process technol*, 155-156, 1505-1512

Wang, W., Kweon, S. H. and Yang, S. H., (2005), "A study on roughness of the micro-end-milled surface produced by a miniatured machine tool", *J Mater Process Technol* 162-163,702-708

Zain, A. M, Haron, H. and Sharif, S., (2010), "Prediction of surface roughness in the end milling machining using artificial neural network", *Expert Systems with Applications*, 37, 1755-1768.

Zhang, Y., Luo, Y., Wang J. F. and Li, Z. (2001), "Research on the fractal of surface topography of grinding", *Int J Mach Tools Manufact*, 41, 2045-2049.

In: Machining and Forming Technologies, Volume 2
Editor: J. Paulo Davim
ISBN: 978-1-61122-820-5

Chapter 15

Wire Electrode Wear in WEDM: An Experimental Investigation during Production of Key Chain Die Cavity

Alakesh Manna**[*] **and Nitin K. Lautre
Department of Mechanical Engineering, Punjab Engineering College, Deemed University; Chandigarh-160012, India

Abstract

In this study, the effect of cutting parameters on the wire electrode wear in wire electrical discharge machining (WEDM) was investigated experimentally. The experiments were carried out utilizing brass wire of 250 μm diameter on E0300 alloy steel. Experimentally it is found that wire feed rate (WF) is the most significant parameter for wire crater depth (d, μm). Wire tension (WT) is the most significant parameter for both the performance criteria e.g. electrode wear rate (EWR, mm3/min) and surface finish (Ra, μm). Mathematical models for WEDM parameters are developed utilizing the results of various machining performance characteristics. The confirmation test results show 37.52% improvement of surface finish when the machining test was conducted with optimum parametric setting as compared to the initial parametric setting..

Keywords: WEDM parameters, image analysis, wire crater, EWR, surface finish.

1. Introduction

Wire Electrical Discharge Machining (WEDM) is a thermoelectric spark erosion process which erodes material from the work-surface by a series of discrete sparks between wire electrode and work-piece immersed in a dielectric fluid. This process is used to produce any complex two and three dimensional shapes on electrically conductive material. However, wire wear as well as frequent breakage of wire electrode during machining may decline the

[*] Corresponding author: E-mail: kgpmanna@rediffmail.com

machined surface quality. The quality of machined surface may become poor due to the generation of crater wear on wire electrode. Hence, investigations on wire wear are crucial to understand the behavior of wire rupture and generation of surface roughness during machining.

The machining characteristic of WEDM was studied during cutting of polished single-crystal silicon (Takino, H. et. al., 2004). They investigated the surface profiles of single-crystal silicon with smooth surface cut by WEDM. The use of composite wires encompassing with high tensile core and several coating are introduced and experimentally validated (Kruth, J.P. et. al., 2004). An investigation was carried out to study the effect of cutting parameters on the size of crater formed during experimentation on WEDM (Tosun, N., Cogun, C. and Pihtili, H., 2003). The parametric effect of WEDM on material removal rate and surface roughness was investigated (Tosun, N. and Cogun, C., 2003). Experimental investigation during WED machining of T160G12 steel was carried out to determine the wire crater diameter, crater depth and surface finish separately (Lautre, N.K. and Manna, A.,2005). Optimize the WEDM parameters using Taguchi's design of experiments for machining of aluminium silicon carbide metal matrix composites (Manna, A. and Bhattacharyya, B., 2006).WEDM process can be utilized to machine Inconel 601 material under acceptable volumetric material removal rate and surface finish (Hewidy, M.S. et. al., 2005). A second order multi-variable regression model and a feed-forward back-propagation neural network model was developed to correlate the WEDM parameters with performance characteristic e.g. surface finish during machining of tungsten carbide-cobalt composite material (Saha, Probir. et.al., 2008). However, several works have been reported on the WED machining characteristics (Dauw, D.F. and Beltrami, ETHI.,1994; Taha Ali El-Taweel, 2006; Mahapatra, Swarup S. and Patnaik, A., 2006), but a little research investigation has been carried out on wire crater and electrode wear rate and their effect on the machined surface.

Keeping in view, the main objective of the study is to investigate the effect of cutting parameters on generation of electrode i.e. wire crater wear during machining. An image analysis technique is utilized to simulate the machining parameters for predicting the wire crater during machining of E0300 die steel. A program is generated and used in MATLAB to convert the electrode wear image into numeric value. The difference of unused electrode diameter and estimated numeric values obtained from the generated program using MATLAB represents the relative effect on wear of electrode at different parametric setting. Taguchi robust design, L_{18} (2^1 x 3^7) mixed orthogonal array, ANOVA and F-test are utilized to analyze the effect of different machining parameters on wire crater depth, electrode wear rate and machined surface finish R_a during machining of E0300 alloy steel.

2. Experimental Planning

Electronica Sprincut-734 Wire electrical discharge machine is used for experimental investigation. E0300 alloy steel and brass wire of 250 μm diameter, are used as work-piece (anode) and electrode (cathode) respectively. Table-1 shows the chemical composition of E0300 material used for experimental investigation. Initially, 28 trials are conducted to choose the level of the seven parameters considered for experimental investigation. According to research paper (Manna, A. and Bhattacharyya, B.; 2006), seven parameters are selected for experimental investigation. Table-2 shows the control factors and their levels with coded values which are considered for experimentation. Other parameters which are kept constant during machining are as follows: (i) Work-piece height = 40 mm, (ii) cutting length = 60 mm, (iii) die-electric fluid flow rate 10 litre per minute, (iv) average voltage gap 38 to 44 volts maintained during machining by adjusting servo feed setting, (v) die electric temperature maintain between 24 ± 2 ^{O}C.

Table 1. Composition of E0300 alloy steel used for experiment

Types of Material	% C	%Cr	% W	% Mn	% Mo	% Si	% V	% Ni	% Tin	%P	%S
E0300 alloy steel	0.87	1.34	0.15	0.76	0.03	0.25	0.01	0.03	0.01	0.05	0.08

Table 2. Coded values of various control factors and their levels.

Symbols: Control Factor	Units	Level		
		1	2	3
A: Open circuit voltage (VP);	Volt	1	2	-
B: Pulse ON Time (TON);	μ Sec	115	118	121
C: Pulse OFF Time (TOFF);	μ Sec	46	48	50
D: Pulse Peak Current (IP);	A	190	210	230
E: Wire Feed rate (WF);	m/min	4	5	6
F: Wire Tension (WT);	g	6	7	8
G: Spark gap set voltage (SV);	volt	16	18	20

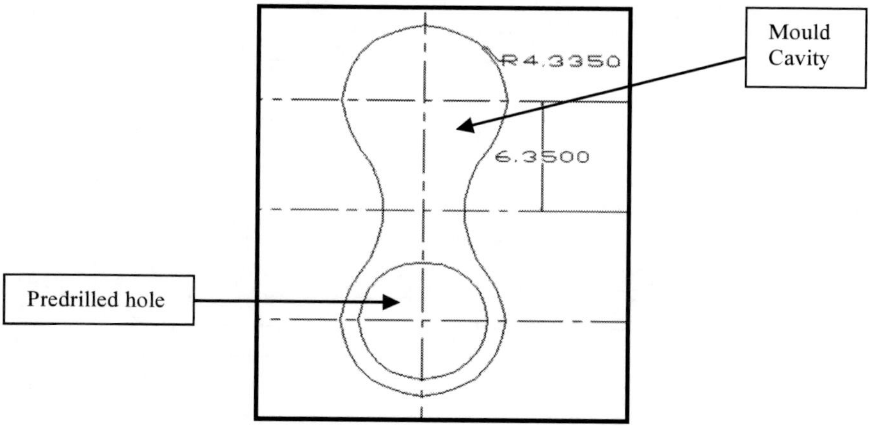

Figure 1. Wire Path Profile.

Surfcom 130A surface roughness measuring instrument is used to measure the machined surface roughness. Samsung SDC-310 digital camera of reflective microscope M20 x 0.4 lens is used to generate the images of electrode crater wear. The crater wear image is used to estimate the actual depth through developed program in the MATLAB platform. The used brass wire electrode is placed in a circular tube of thermoplastic fixture for microscopic analysis. DPI-RR cold cure acrylic paste is used to hold the wear electrode properly inside the thermoplastic fixture. Taguchi experimental design based methodology, L_{18} (2^1 x 3^7) mixed orthogonal arrays (Ross PJ, 1989). , ANOVA, and F test values are used for experimental investigation to analyze the effect of various machining parameters on wire crater depth, electrode wear rate and machined surface roughness height (Ra, μm) during machining of E0300 alloy steel. Total 54 experiments are conducted for L_{18} experimental setup with repeating each experiment thrice and taking average results from three for further analysis. Figure 1 shows the actual path of wire tool movement during cutting of key chain die cavity.

3. EXPERIMENTAL RESULTS AND ANALYSIS FOR PARAMETRIC OPTIMIZATION

Table-3 shows the parametric level as per the Taguchi design of experiment, L_{18} (2^1x3^7) mixed orthogonal array and S/N ratios (dB) for wire crater depth (d, μm), electrode wear rate (EWR, mm^3/min) and surface roughness height (Ra, μm). Table-4 represents the ANOVA and 'F' test values for wire crater depth, electrode wear rate and surface roughness height (R_a, μm). Table-4 also represents the percentage of contribution i.e. effectiveness of individual WEDM parameters on wire crater depth, electrode wear rate and surface roughness during machining of E0300 alloy steel respectively. The table also shows the most significant (*) WEDM parameters during machining of E0300 alloy steel.

Table-3. Design of experiment of Taguchi $L_{18}(2^1x3^7)$ and S/N ratios(dB)

Exp. No	Control factors and their levels							S/N Ratios		
	A:VP	B:TON	C:TOFF	D: IP	E:WF	F:WT	G: SV	'd'	'EWR'	'R_a'
1	1	1	1	1	1	1	1	-11.54	-12.49	-8.11
2	1	1	2	2	2	2	2	-10.49	-12.18	-8.08
3	1	1	3	3	3	3	3	-9.9	-11.19	-7.93
4	1	2	1	1	2	2	3	-7.39	-10.14	-8.15
5	1	2	2	2	3	3	1	-9.6	-12.6	-7.81
6	1	2	3	3	1	1	2	-11.07	-11.73	-8.17
7	1	3	1	2	1	3	2	-11.59	-10.26	-7.29
8	1	3	2	3	2	1	3	-11.88	-14.78	-8.52
9	1	3	3	1	3	2	1	-8.76	-13.02	-7.28
10	2	1	1	3	3	2	2	-11.22	-12.86	-8.76
11	2	1	2	1	1	3	3	-11.21	-10.48	-7.68
12	2	1	3	2	2	1	1	-11.57	-15.66	-7.91
13	2	2	1	2	3	1	3	-8.26	-13.96	-8.83
14	2	2	2	3	1	2	1	-11.98	-12.53	-7.83
15	2	2	3	1	2	3	2	-11.54	-10.27	-8.15
16	2	3	1	3	2	3	1	-11.95	-11.45	-7.69
17	2	3	2	1	3	1	2	-8.49	-12.79	-8.59
18	2	3	3	2	1	2	3	-11.26	-10.46	-6.69

'd' Wire crater depth; 'EWR' electrode wear rate; and 'R_a' surface roughness

Figure 2, 3 and 4 show the S/N ratio (dB) for wire crater depth, electrode wear rate (EWR) and machined surface roughness height (Ra) respectively. From figure 2 and experimentally obtained results, it is concluded that for minimum wire crater depth the optimal parametric combination is $A_1B_2C_1D_1E_3F_2G_3$. Similarly, from figure 3, for minimum electrode wear rate the parametric combination is $A_1B_2C_1D_1E_1F_3G_2$ and from figure 4, for minimum surface roughness height, R_a the parametric combination is $A_1B_3C_3D_2E_1F_3G_1$.

Table 4. ANOVA for crater depth (d), EWR and surface roughness (R_a, µm)

Control factors	d.f.	Crater depth (d)				Electrode wear rate (EWR)				Surface roughness (Ra, µm)			
		SS	MS	F_0 test	% of contribution	SS	MS	F_0 test	% of contribution	SS	MS	F_0 test	% of contribution
VP	1	1.54	1.54	2.03	4.63	0.237	0.237	0.22	0.55	0.037	0.037	2.31	0.74
TON	2	3.216	1.608	2.12	9.67	1.107	0.553	0.52	2.58	0.794	0.397	24.81	15.92
TOFF	2	0.425	0.212	0.28	1.27	1.569	0.785	0.74	3.65	0.727	0.363	22.69	14.57
IP	2	6.910	3.455	4.56	20.74	3.557	1.778	1.68	8.28	0.444	0.222	13.87	8.91
WF	2	13.464	6.732	8.88*	40.42	6.512	3.256	3.08	15.16	1.100	0.550	34.37	22.06
WT	2	1.873	0.936	1.23	5.62	19.922	9.996	9.42*	46.36	1.344	0.672	42.00*	26.95
SV	2	2.851	1.426	1.88	8.56	5.835	2.917	2.76	13.58	0.476	0.238	14.87	9.54
Error	4	3.031	0.758	-	9.09	4.23	1.057	-	9.84	0.065	0.016	----	1.31
Total	17	33.31	-	-	100.00	42.97	-	-	100.00	4.987	----	----	100.00

Where, * Most significant, SS = Sum of square; MS = Mean square; d.f.= Degree of freedom;

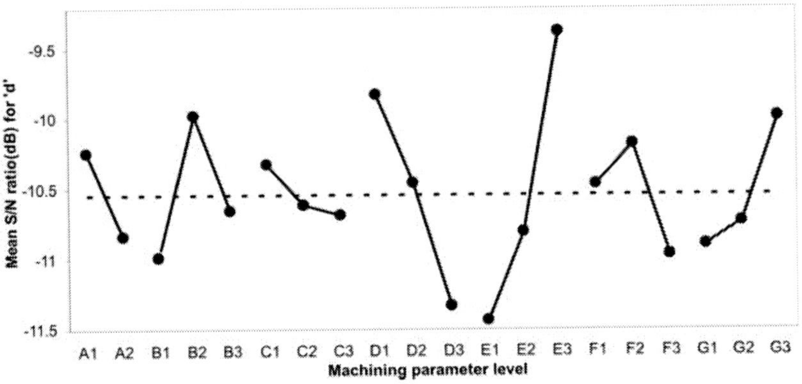

Figure 2. S/N ratio (dB) for electrode crater depth.

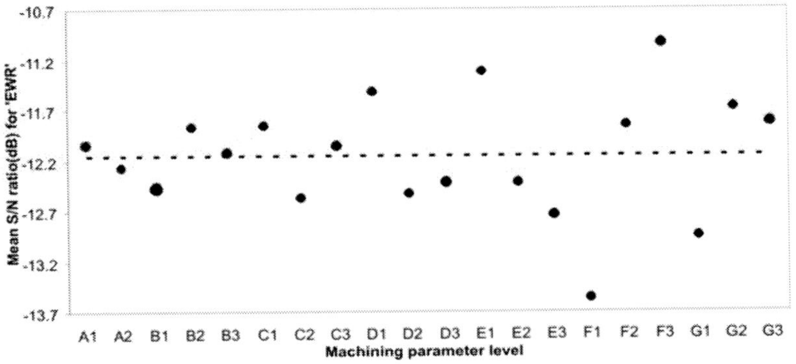

Figure 3. S/N ratio (dB) for electrode wear rate (EWR).

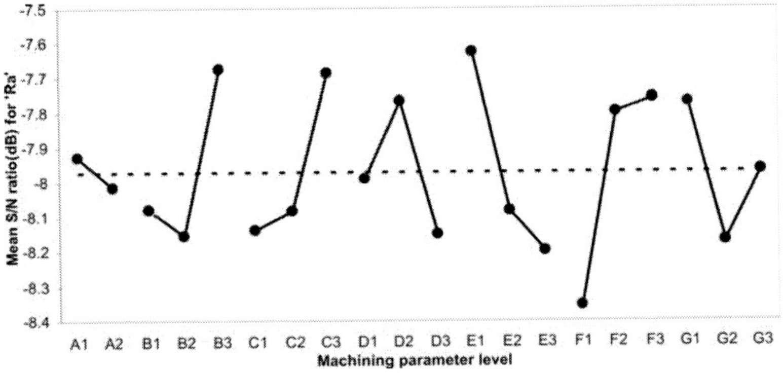

Figure 4. S/N ratio (dB) for machined surface roughness value, R_a

The actual wire craters are analyzed through various images of the used wire. The conditions of wire electrode surfaces are also analyzed through different cross-sectional images. Figure 5 shows a cross sectional view of the used wire which represents the actual condition of wire electrode and depth of crater formed on it. The crater recorded in cross section also depicts the material removed from the used wire during WEDM operation. The electrode wear rate is calculated from the difference of weight of the machined wire and fresh brass wire (i.e. electrode).

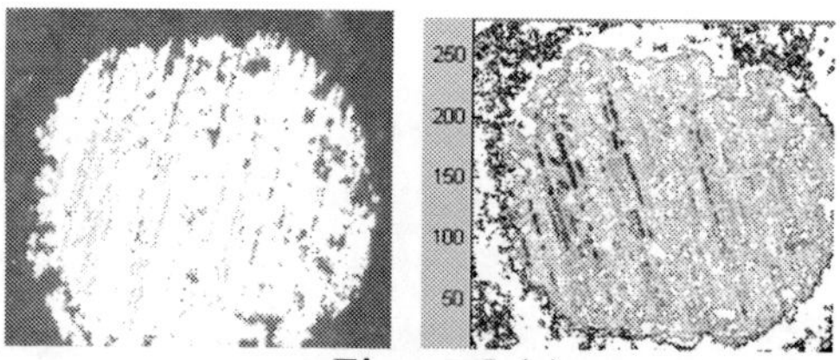

Figure 5 (a)

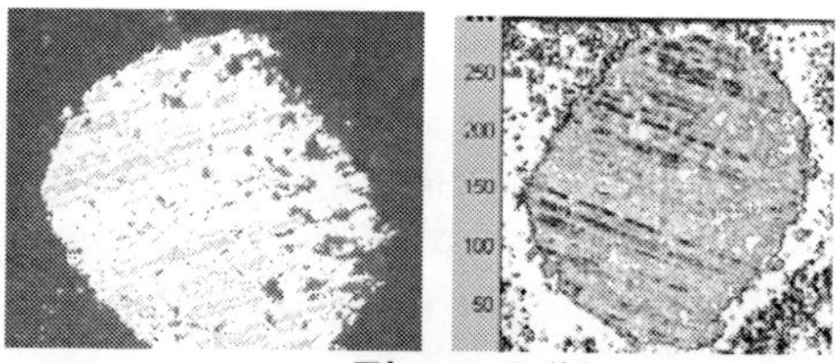

Figure 5 (b)

Figure 5. Crater formed after machining on the wire electrode at parametric setting level (a) $A_1B_3C_2D_3E_2F_1G_3$ and (b) $A_1B_3C_1D_2E_1F_3G_2$ respectively.

The actual condition of wire electrode after machining at different parametric setting is shown in figure 6. Figure 7 shows the actual image of used electrode at $A_2B_1C_3D_2E_2F_1G_1$ parametric combination. From figure 7, crater depth (d) can be predicted by a relation as:

d = 0.5 (maximum size i.e. diameter of the unused wire electrode – minimum size of wire electrode after used).

Hence, $d = \frac{1}{2}(x_1 - x_2)$. Where, x_1 = maximum size i.e. diameter of the unused wire electrode, x_2 = minimum size of wire electrode after used.

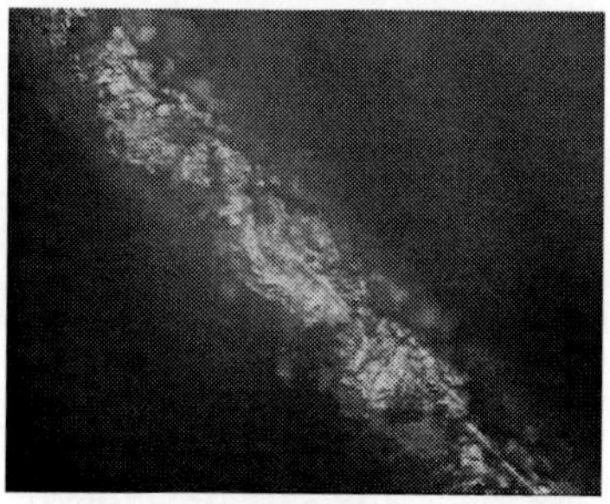

Figure6 (a)

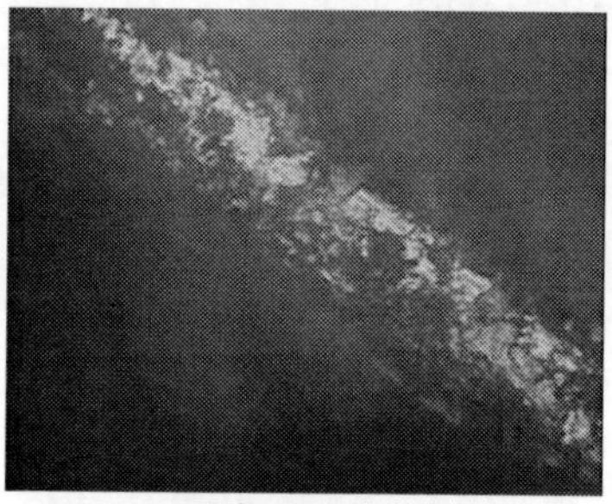

Figure 6 (b)

Figure 6. Actual condition of wire electrode after machining at parametric setting level (a) $A_2B_3C_3D_3E_3F_3G_3$ and (b) $A_2B_2C_2D_2E_2F_2G_2$ respectively.

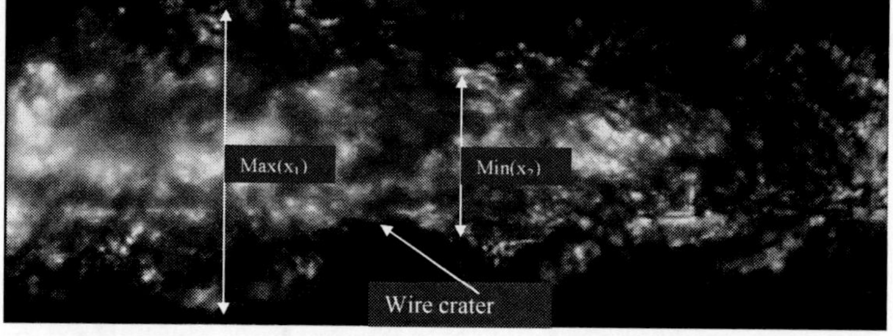

Figure 7. Actual condition of wire electrode at parametric combination $A_2B_1C_3D_2E_2F_1G_1$

4. Validity Test

After selection of the optimal design parameters it is essential to predict and verify the improvement in the quality characteristics for machining of E0300 alloy steel by CNC-Wire cut-EDM operation with respect to the chosen initial i.e. reference parameters setting. The predicted optimal value of S/N ratio can be calculated using the relation (Ross PJ, 1989),

$$\eta_{opt} = \eta_m + \sum_{j=1}^{k} (\eta_j - \eta_m); \quad j = 1 \,\&\, 2, \qquad \text{Eqn. 1}$$

Where,

η_m = Grand mean of S/N ratio

η_j = Mean S/N ratio at optimum level

k = Number of main design parameters that affect the quality Characteristics.

Table 5 represents the results of the improvement of the machining characteristics with respect to the initial cutting parameters setting. Confirmation test results prove that there is a remarkable improvement in the wire crater depth, electrode wear rate and surface roughness characteristics with 26.64%, 31.90%, and 37.52% respectively.

5. Development of Mathematical Models

5.1. Models for WEDM Performance Characteristics

Considering all seven parameters mathematical models for wire crater depth (µm), electrode wear rate (EWR, mm^3/min) and surface roughness, Ra (µm) are developed.
The developed model for wire crater depth (Y_d) is

$$Y_d = 922.25596.VP^{-0.234258}.\ TON^{0.05794136}.TOFF^{0.1446128}.IP^{-0.311153}\ WF^{-0.5779819}.WT^{-0.2521234}.\ SV^{-0.2535048} \qquad \text{Eqn. 1}$$

The developed mathematical model for electrode wear rate (Y_{EWR}, mm^3/min) is

$$Y_{EWR} = 0.0035553.\ VP^{-0.09174}.\ TON^{0.06997869}.TOFF^{0.0889375}.\ IP^{-0.172085}.\ WF^{0.405127}.\ WT^{1.220809}.\ SV^{-0.331405} \qquad \text{Eqn. 2}$$

The developed mathematical model for surface roughness height (Y_{Ra}, µm) is

$$Y_{Ra} = 0.48199.VP^{-0.035925}.\ TON^{0.0955098}.TOFF^{-0.17708}.IP^{-0.03774}\ WF^{0.16568}.\ WT^{0.2858}.\ SV^{0.060429} \qquad \text{Eqn. 3}$$

Table 5. Optimum setting levels of input WEDM parameters for different output parameters

Performance characteristics		Initial cutting parameter	Optimal cutting parameters		% of Improvement of S/N ratio	% of Improvement of machining characteristics
			Predicted	Experimental		
Crater depth (d) in μm	Setting level	$A_2B_2C_2D_2E_2F_2G_2$	$A_1B_2C_1D_1E_3F_2G_3$	$A_1B_2C_1D_1E_3F_2G_3$	-	-
	Results obtained	3.64	--	2. 73	-	26.64%
	S/N ratio	-11.22	-6.67	-8.530	23.97%	
Electrode wear rate (EWR) in mm^3/min	Setting level	$A_2B_2C_2D_2E_2F_2G_2$	$A_1B_2C_1D_1E_1F_3G_2$	$A_1B_2C_1D_1E_1F_3G_2$	-	-
	Results obtained	4.05	--	2.758		31.90%
	S/N ratio	-12.16	-8.41	-8.812	27.53%	-
Surface roughness height (Ra) in μm	Setting level	$A_2B_2C_2D_2E_2F_2G_2$	$A_1B_3C_3D_2E_1F_3G_1$	$A_1B_3C_3D_2E_1F_3G_1$	-	-
	Results obtained	2.497	--	1.56	-	37.52%
	S/N ratio	-7.95	-6.39	-3.362	51.41%	

5.2. Models for WEDM Parameters

Utilizing the experimentally obtained results for the machining performance characteristics e.g. crater depth (d, μm), electrode wear rate (EWR, mm^3/min) and surface finish (Ra, μm) the mathematical models for various WEDM parameters such as open circuit voltage (VP), pulse-on-time (TON), pulse-off-time (TOFF), pulse-peak-current (IP), wire-feed-rate (WF), wire-tension (WT), and spark gap set voltage (SV) are developed. Let, x_1, x_2 and x_3 represent the above three performance characteristics e.g. Y_d, Y_{EWR} and Y_{Ra} respectively then the developed mathematical model for open circuit voltage (VP) is

$$VP = 13.0 + 8.19\, x_1 + 11.1\, x_2 - 42.0\, x_3 - 0.350\, x_1 x_2 - 3.56\, x_1 x_3 - 4.65\, x_2 x_3 + 0.397\, x_1^2 + 0.151\, x_2^2 + 15.3\, x_3^2 \quad \text{Eqn. 4}$$

Similarly, developed mathematical models for Pulse ON Time (TON), Pulse OFF Time (TOFF), Pulse Peak Current (IP), Wire Feed rate (WF), Wire Tension (WT), and Spark gap set voltage (SV) are developed.

The developed model for Pulse ON Time (TON) is

$$TON = 530 - 108\, x_1 - 26.2\, x_2 - 138\, x_3 - 3.55\, x_1 x_2 + 18.4\, x_1 x_3 + 17.6\, x_2 x_3 + 11.8\, x_1^2 - 0.475\, x_2^2 - 1.7\, x_3^2 \quad \text{Eqn. 5}$$

The developed model for Pulse OFF Time (TOFF) is

$$TOFF = 82 + 14.9\, x_1 + 9.7\, x_2 - 60\, x_3 - 0.24\, x_1 x_2 + 3.70\, x_1 x_3 - 6.49\, x_2 x_3 - 3.59\, x_1^2 + 0.769\, x_2^2 + 14.9\, x_3^2 \qquad \text{Eqn. 6}$$

The developed model for Pulse Peak Current (IP) is

$$IP = 32 + 131\, x_1 - 84\, x_2 + 72\, x_3 + 18.7\, x_1 x_2 - 41.0\, x_1 x_3 + 47.0\, x_2 x_3 - 14.0\, x_1^2 - 10.6\, x_2^2 - 23.9\, x_3^2 \qquad \text{Eqn. 7}$$

The developed model for Wire Feed rate (WF) is

$$WF = 41.0 - 5.3\, x_1 + 3.42\, x_2 - 29.2\, x_3 - 0.514\, x_1 x_2 + 3.66\, x_1 x_3 - 0.91\, x_2 x_3 - 0.390\, x_1^2 + 0.081\, x_2^2 + 4.6\, x_3^2 \qquad \text{Eqn. 8}$$

The developed model for Wire Tension (WT) is

$$WT = -7065 - 709\, x_1 - 1476\, x_2 + 10621\, x_3 - 182\, x_1 x_2 + 669\, x_1 x_3 + 566\, x_2 x_3 - 33\, x_1^2 + 69.9\, x_2^2 - 3153\, x_3^2 \qquad \text{Eqn. 9}$$

The developed model for spark gap set voltage (SV) is

$$SV = 213 - 1.2\, x_1 - 59.1\, x_2 - 59.4\, x_3 + 1.06\, x_1 x_2 + 5.12\, x_1 x_3 + 17.7\, x_2 x_3 - 2.67\, x_1^2 + 1.18\, x_2^2 - 3.9\, x_3^2 \qquad \text{Eqn. 10}$$

Utilizing the above mathematical expressions one can predict the parametric setting values in advance by determining the individual parametric value for effective machining of E0300 alloy steel by WEDM process.

5.3. Aditivity of the Developed Models for Machining Performance Characteristic

Table 6 represents the comparative results with percentage of error between actual experimental results and calculated values using developed mathematical models equations 1 - 3 for wire crater depth (Y_d, μm), electrode wear rate (Y_{EWR}, mm^3/min) and surface roughness height, (Y_{Ra}, μm). From the table 6, it is concluded that the developed mathematical models equation 1, 2 and 3 for Y_d (μm), Y_{EWR} (mm^3/min) and Y_{Ra} (μm) respectively bear good agreement with the test results.

Table 6. represents the comparative results with percentage of error

Sl. No.	Cutting parameter setting	Crater depth (Y_d, μm)			Electrode wear rate, Y_{EWR} (mm^3/min)			Surface roughness height, Y_{Ra}, (μm)		
		Exp. Results	Value using Eqn.1	% of Error	Exp. Results	Value using Eqn.2	% of Error	Exp. Results	Value using Eqn.3	% of Error
1	$A_1B_3C_1D_3E_1F_3G_2$	3.675	3.474	5.46	3.859	4.149	6.98	2.397	2.520	4.88
2	$A_1B_2C_3D_3E_2F_1G_1$	3.425	3.529	2.94	4.064	3.702	8.90	2.469	2.244	9.10
3	$A_2B_2C_3D_2E_1F_2G_3$	3.757	3.539	5.80	3.435	3.498	1.80	2.108	2.293	8.06

5.4. Analysis of the Developed Models for Parameters Setting

Parametric setting values are determined utilizing experimentally obtained results and developed equations 4 - 10. Further, determined parametric values are normalized between the ranges of (0 – 1) and thereafter, parametric variations for 18 x 3 i.e. 54 experiments are analyzed graphically. The following relation is used to determine the normalized data

$$Vn = 0.8 \times \left(\frac{Vr - Vmin}{Vmax - Vmin} \right) + 0.1 -$$

Eqn. 11

Where, Vn = Normalized value; Vr = Value to be normalized; Vmax = Highest value; Vmin = Smallest value

Figure 8 shows the individual parametric variation based on the determined normalized value utilizing developed equations 4 – 10 on WEDM output characteristics. Figure 8 scatter diagram represents the 54 experimental results (18 x 3 repetitions for each experiments) based on the performance of the WEDM parameters for different output characteristics. Calculated normalized values for the various parameters e.g. VP, TON, TOFF, IP,WF, WT, SV are 0.512, 0.411, 0.421, 0.713, 0.342, 0.432 and 0.285 respectively for output characteristics such as crater depth (3.78μm), electrode wear rate (4.22 mm^3/min) and surface roughness height Ra (2.48μm) for level 1 (table-2) parametric setting.

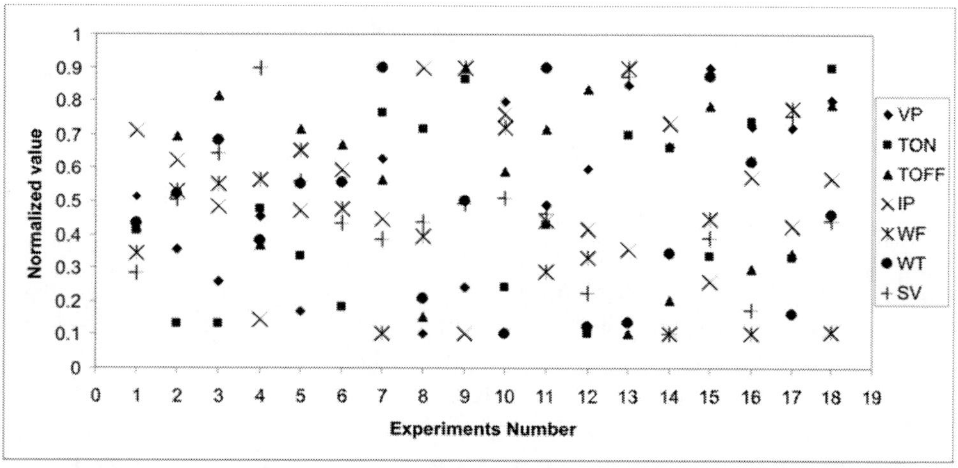

Figure 8. Parametric variations for 18 experiments (table 3).

The figure also represents the individual parametric variation for a particular performance characteristic with different output. The parametric variations suggest that the machine tool WEDM should have more flexibility with provision of wide range of parametric setting to minimize the effect of the above discussed performance characteristics during machining of E0300 alloy steel.

6. Conclusions

Based on the experimental results, ANOVA, F-test, S/N ratio and developed mathematical models during machining of E0300 alloy steel on CNC-Wire-cut EDM the following points are concluded as

For wire crater depth (d, μm), wire feed rate (WF) is the most significant parameter with 40.42 % contribution. For electrode wear rate (EWR, mm^3/min), wire tension (WT) is the most significant parameter with 46.36% contribution. For surface roughness height (Ra, μm), the most significant, significant and less significant parameters are wire tension (WT), wire feed rate (WF) and pulse-on-time (TON) with 26.95%, 22.06% and 15.92 % contribution respectively.

For minimum wire crater depth the optimal parametric combination is $A_1B_2C_1D_1E_3F_2G_3$. For minimum electrode wear rate the optimal parametric combination is $A_1B_2C_1D_1E_1F_3G_2$ and for minimum surface roughness height, Ra, the optimal parametric combination is $A_1B_3C_3D_2E_1F_3G_1$.

The developed mathematical models are of immense important to predict and understand the proper wire crater, electrode wear rate and surface finish for effective machining of E0300 alloy steel in advance.

Confirmation test results also prove that there is a remarkable improvement in the wire crater depth, electrode wear rate and surface roughness characteristics with 26.64%, 31.90%, and 37.52% respectively, when machining is done with the optimal parametric combination.

References

Dauw, D. F. and Beltrami, ETHI. (1994). High Precision wire EDM by online wire position control. *Annals of the CIRP*, Vol 41/1, pp 193-197.

Hewidy, M. S., El-Taweel, T. A. and El-Safty, M. F. (2005) 'Modelling the machining parameters of wire electrical discharge machining of Inconel 601 using RSM' *Journal Material Processing Technology*, Vol.169, pp 328-336.

Kruth, J. P. et. al. (2004). Composites Wires with high tensile core for wire EDM. *Annals of the CIRP*, Vol 53/1, pp 171-174.

Lautre, N. K. and Manna, A. (2005). *A study on wire crater during generation of curved surface in WEDM*, 4th National conference on Precision Engineering COPEN – 2005, Jadavpur University, Kolkata, December 16-17, India.

Manna, A. and Bhattacharyya, B. (2006). Taguchi and Gauss elimination method: A dual response approach for parametric optimization of CNC wire cut EDM of PRAlSiCMMC. *Int. Journal of Advanced Manufacturing Technology*, Vol 28, pp 67-75.

Mahapatra, Swarup S. and Patnaik, A.(2006) "Parametric optimization of wire electrical discharge machining (WEDM) process using Taguchi method", *J. Braz. Soc. Mech. Sci. & Eng.*, Vol.28, No.4, p.422-429.

Ross PJ (1989). *Taguchi technique for quality engineering*. McGraw-Hill, New York.

Saha, Probir; Singha, Abhijit;. Pal Surjya K. and Saha, Partha. (2008) *Soft computing models based prediction of cutting speed and surface roughness in wire electro-discharge machining of tungsten carbide cobalt composite*, Vol 39, No.1-2,pp 74-84

Takino, H. et. al. (2004). Cutting of polished single-crystal silicon by wire electrical discharge machining. *Precision Engineering*, Vol-28, pp 314-319.

Tosun, N., Cogun, C. and Pihtili, H. (2003). The Effect of Cutting Parameters on Wire Crater Sizes in Wire EDM. *Int. Journal of Advanced manufacturing Technology,* Vol. 21, pp 857-865.

Tosun, N. and Cogun, C. (2003). An investigation on wire wear in WEDM. *Journal of Materials Processing Technology*, Vol. 134, pp 273-278.

Taha Ali El-Taweel;(2006) "Parametric study and optimization of wire electrical discharge machining of Al-Cu-TiC-Si P/M composite"; *Int. Journal of Machining and Machinability of Materials,* Vol. 1, No.4 pp. 380 - 395

In: Machining and Forming Technologies, Volume 2
Editor: J. Paulo Davim
ISBN: 978-1-61122-820-5

Chapter 16

OPTIMIZATION OF POWDER MIXED EDM (PMEDM) PROCESS WITH MULTIPLE PERFORMANCE CHARACTERISTICS USING TAGUCHI BASED GREY RELATIONAL ANALYSIS

R. S. Pawade* and P. K. Brahmankar

Department of Mechanical Engineering, Dr. Babasaheb Ambedkar Technological University, Lonere, 402 103 MS, INDIA

ABSTRACT

The paper reports the optimization of process parameters in powder mixed EDM of HSS tool steel using Taguchi based grey relational analysis. Eighteen experimental runs based on orthogonal array were performed. The material removal rate (MRR), relative wear ratio (RWR), and the surface roughness (SR) of the machined specimen were the response variables. By analyzing the grey relational grade, the degree of influence for each controllable process parameters onto individual response variables was estimated. Analysis of variance (ANOVA) of the grey relational grade shows that the pulse current is the most significant parameter influencing PMEDM process according to the weighted sum grade of the MRR, RWR and SR.

Keywords: Powder mixed EDM; Grey relational analysis; Optimization; Taguchi designs; HSS tool steel.

1. INTRODUCTION

Powder mixed electro discharge machining (PMEDM) is a recent variant of the popular EDM, which is extensively used for machining hard materials for making moulds and dies. In

* Corresponding author: E-mail: raju_s_pawade@yahoo.com

powder mixed electro discharge machining (PMEDM), addition of powder into dielectric changes the mechanism of material removal (Klocke et al. 2004). The presence of powder in the electrode-work gap reduces the frequency of abnormal discharges and the discharge channel impulsive force and thus improves the stability of the process (Uno and Okada, 1997; Chow et al. 2004; Mohri et al. 1991). The powder particles modify the plasma channel. The plasma channel becomes enlarged and widened (Chow et al. 2004). The formation of several discharge trajectories within a single input impulse and several discharging spots within a widened plasma channel leads to increase in the MRR. Further uniform distribution of sparking among the powder particles produces shallow craters on the workpiece surface and hence the better surface finish is obtained (Chow et al. 2004; Tzeng and Lee 2001). Literature reveals that the various researchers have performed experiment and analyzed this new modified EDM process to explore the machinability. Few of them have focused on understanding the fundamentals of the process mechanism using numerical modeling (Kansal et al. 2008). Detailed analysis has been made to establish relationships between PMEDM parameters with the MRR, TWR and surface quality using response surface method (Kansal et al. 2005). The improvement in respect of MRR, TWR and surface quality in PMEDM process has been reported in most of these investigations (Klocke et al. 2004; Chow et al. 2004; Mohri et al. 1991; Tzeng and Lee 2001; Kansal et al. 2005; Jeswani, 1981; Zhao et al. 2002; Wong et al. 1998; Pecas and Henriques, 2003; Chow et al. 2009).

PMEDM is a multi-objective, multi-factor die sinking process. Several process parameters interact in a complex manner resulting direct or indirect influence on material removal rate, electrode wear and surface characteristics thereby influencing surface finish of the component. Thus, to produce component with desired accuracy and quality, it is necessary to select the optimal process parametric condition while performing the experiment. The Taguchi method is widely adopted for solving optimization problems in manufacturing engineering. The method utilizes a well-balanced experimental design (allows a limited number of experimental runs) called orthogonal array design, and signal-to-noise ratio (S/N ratio), which serve the objective function to be optimized (maximized) within experimental domain (Phadke, 1989). However, traditional Taguchi design has a major limitation in that it cannot efficiently handle the presence of more than one and conflicting, objectives as there in case of PMEDM process. Thus, optimization needs to be performed in such a way that all the objectives should fulfill simultaneously using multi-response optimization. Such multiple response optimization problems can be solved using grey relational theory developed by Deng in 1989. In grey relational theory, the degree of approximation among the sequences was measured using a grey relational grade. Thus multiple individual optimum values (objectives) will be converted into a single objective in terms of grey relational grade. Many studies have been reported on the application of grey relational theory for optimization of machining problems such as in electro discharge machining (Lin and Lin (2002; Lin et al. 2002; Huang and Lin, 2002).

2. Grey Relational Analysis

The grey relational analysis is a measurement of the absolute value of data difference between the sequences, and can be used to approximate the correlation between the sequences. The various steps in the grey relational analysis are given in the flow chart as shown in Figure 1.

2.1. Data Preprocessing

Data preprocessing can be performed by normalizing the experimental data, which are further used for estimating the grey relational grades. It is necessary to calculate the grey correlation coefficients for determining the grey relational grade. It is required that the quantities required to be processed through grey relational analysis must be dimensionless. Therefore, each series is normalized by dividing the data in the original series by their average values. Let the original reference sequence and sequence for comparison be represented as *xo(k)* and *xi(k)*, *i*=1, 2, ..., *m*; *k*=1, 2, ..., *n*, respectively, where *m* is the total number of experiment to be considered, and *n* is the total number of observation data. Data preprocessing converts the original sequence to a comparable sequence (Deng, 1989).

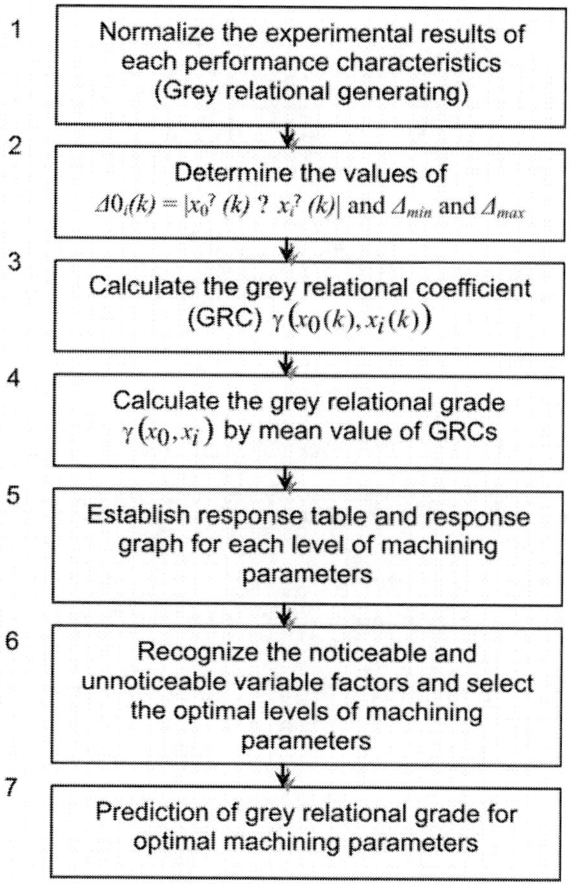

Figure 1. Flow chart for grey relational analysis.

If the quality target (response variable) value of the original sequence is ‘‘the-larger-the-better’’, then the original sequence is normalized using Eq.1 as –

$$x_i^*(k) = \frac{x_i^{(o)}(k) - \min x_i^{(o)}(k)}{\max x_i^{(o)}(k) - \min x_i^{(o)}(k)} \quad (1)$$

If the original sequence is ‘‘the-smaller-the-better’’, then it can be normalized by Eq. 2 as

$$x_i^*(k) = \frac{\max x_i^{(o)}(k) - x_i^{(o)}(k)}{\max x_i^{(o)}(k) - \min x_i^{(o)}(k)} \quad (2)$$

However, if there is ‘‘a specific target value’’, then the original sequence is normalized by the following equation.

$$x_i^*(k) = 1 - \frac{\left|x_i^{(o)}(k) - OB\right|}{\max\left\{\max x_i^{(o)}(k) - OB, OB - \min x_i^{(o)}(k)\right\}} \quad (3)$$

where *OB* is the target value.

Alternatively, the original sequence can be normalized using the simplest methodology that is the values of the original sequence can be divided by the first value of the sequence, *xi (o)(1)*.

$$x_i^*(k) = \frac{x_i^{(o)}(k)}{x_i^{(o)}(1)} \quad (4)$$

where $x_i^{(o)}(k)$ is the original sequence, $x_i*(k)$ the sequence after the grey relational generation, max $xi^{(O)}(k)$ the largest value of $xi^{(O)}(k)$ and min $xi^{(O)}(k)$ the smallest value of $xi^{(O)}(k)$.

2.2. Grey Relational Coefficients and Grey Relational Grades

After data preprocessing, a grey relational coefficient can be calculated using the preprocessed sequences. The grey relational coefficient is determined by Eq. 5 as below –

$$\xi_{ij} = \frac{\min_i \min_j \left|x_i^0 - x_{ij}\right| + \zeta \max_i \max_j \left|x_i^0 - x_{ij}\right|}{\min_i \max_j \left|x_i^0 - x_{ij}\right| + \zeta \max_i \max_j \left|x_i^0 - x_{ij}\right|} \quad (5)$$

where x_i^0 is the ideal normalized S/N ratio for the ith performance characteristic and ζ is the distinguishing coefficient which is defined as $\zeta \in [0, 1]$.

The grey relational grade is estimated by using the grey relational coefficients. A grey relational grade is a weighted sum of the grey relational coefficients, and is expressed by the following equation.

$$\gamma_j = \frac{1}{m}\sum_{i=1}^{m} w_i\, \xi_{ij} \tag{6}$$

where γ_j is the grey relational grade for the jth experiment, w_i the weighting factor for the ith performance characteristic, and m the number of performance characteristics. If the two sequences are identical, then the value of the grey relational grade equals to one. The grey relational grade also indicates the degree of influence exerted by the comparability sequence on the reference sequence. Consequently, if a particular comparability sequence is more important to the reference sequence than other comparability sequences, the grey relational grade for that comparability sequence and the reference sequence will exceed than that of the other grey relational grades.

2.3. Confirmation Tests

Once the optimal level of the process parameters is identified, the following step is to verify the improvement of the performance characteristic using this optimal level. The estimated S/N ratio $\tilde{\eta}$ is calculated as (Phadke, 1989) –

$$\tilde{\eta} = \eta_m + \sum_{i=1}^{\alpha}(\hat{\eta}_i - \eta_m) \tag{7}$$

where η_m is the total mean of the grey relational grade, η_i the mean of the grey relational grade at the optimal level, and α the number of the process parameter that significantly affect the multiple performance characteristics. Based on Eq. (7), the estimated grey relational grade can be obtained.

3. Experimental Details

The experiments were planned to apply grey relational theory to examine how the PMEDM operation influences the machinability indicators of MRR, RWR and surface roughness. An optimal parameter combination was then obtained. Through analyzing the grey relational grade, the most influential factors for individual machinability indicators of PMEDM process was identified. Additionally, the ANOVA was also utilized to examine the most significant factors for the PMEDM process as the MRR, EWR and surface roughness are simultaneously considered.

3.1. Experimental Parameters and Design

In this study, the effects of process, electrode type, and the type of powder mixed dielectric on the response variables were evaluated. The response variables chosen to assess the PMEDM performance were MRR, RWR and surface roughness. These variables were chosen as they are the most representative indicators of the PMEDM process. Further, they are related to the surface quality, the speed of the process and the efficiency of the electrode, respectively. The experiment were designed with seven controllable parameters, in which one is set at two level and the remaining six were set at three levels. Table 1 shows seven controlled parameters that includes Polarity (A) at two level, Electrode (B), Pulse current (C (amp)), Gap voltage, (D (volts)), Pulse duration, (E (sec)), Duty cycle (F) and the Powder type (G) with three levels for each parameter. Considering the importance and widespread use of these parameters by researchers in the die sinking EDM, the selection of these three factors has been made (Sonawane et al. 2008; Pawade et al. 2008). Thus, eighteen experimental runs based on the orthogonal array L_{18} $(2^1 \times 3^7)$ are planned. Table 2 shows the eighteen experimental runs according to the above orthogonal array along with the measured response. Typically, higher values of MRR, smaller RWR as well as surface roughness (SR) are desirable for the improved performance of the PMEDM process.

Table 1. Controllable parameters and their levels

Experimental factors	Levels		
	1	2	3
(A) Polarity	Positive	Negative	
(B) Electrode	Copper	Brass	Aluminium
(C) Pulse current, amp	10	20	30
(D) Gap voltage, volts	50	70	90
(E) Pulse duration, sec	100	200	300
(F) Duty cycle	4	8	12
(G) Powder type	Aluminium	Graphite	Silicon

3.2. Work Material and Electrode

High speed tool steel (HSS) is used in the production of automotive dies, plastic injection molding dies, precision gauge, spindle, and for jigs and fixtures. It was commercially available in the plate form. The chemical composition of the high speed tool steel is 1.55 wt.% C–11.5 wt.% Cr–0.70 wt.% Mo–1.00 wt.% V–0.30 wt.% Mn–0.25 wt.% Si. Electrodes of 12 mm diameter were fabricated from three different metals that include copper, brass and aluminium. These are machined to the required shape and size on CNC turning lathe.

3.3. Machining and Measurement

The experiments were carried out on a fuzzy logic die sinking electric discharge machine (Make - Electronica, Model – Leader ZNC). Figure 2(a-b) shows the experimental set up and the close up view of the powder mixed dielectric EDM operation. Three different electrodes of brass, aluminium and copper were employed for the PMEDM operation. Dielectric mixed with powders of different abrasives were prepared and circulated through tank during the experiments. The powders of silicon, graphite and aluminium and their proportion were varied during the experiments. A jet flushing system was used to assure the adequate flushing of the machining debris from the work-electrode gap. The flushing pressure and the dielectric flow rate are kept constant during the process. The response variables MRR, RWR were measured during the experiment and the surface roughness was examined after the experiments. The volume of material removed from the specimen divided by the machining time, gives material removal rate (MRR) in mm^3/min, and the relative wear ratio (RWR) was determined from the electrode wear rate (EWR). The MRR of the specimen and the RWR were calculated by measuring the weight of the work specimen and the electrode just before as well as just after the process. A PRESSICA make accurate balance with 0.001 gm resolution and 450 gm capacity was used for this purpose. Surface roughness of the machined specimen is the measurement of the small-scale variations in the height of a physical surface. The surface roughness was measured using the surface roughness tester (Make – Mitutoyo, Model - Surftest SJ301), after the process for each specimen. The arithmetic mean deviation of roughness profile (R_a), defined according to ISO 4287:1997 norm were obtained (ISO, 1997). Total sampling length considered while measurement of surface roughness was 6.4 mm (8×0.8 mm).

Table 2. Standard orthogonal array and experimental results

Trial No.	A	B	C	D	E	F	G	MRR (mm^3/min)	RWR	SR (μm)
1	Positive	Cu	10	50	100	4	Al	0.045673	0.031484	10.3533
2	Positive	Cu	20	70	200	8	Gr	0.184075	0.031161	13.0133
3	Positive	Cu	30	90	300	12	Si	0.251058	0.032239	13.9900
4	Positive	Br	10	50	200	8	Si	0.028151	2.311534	3.3600
5	Positive	Br	20	70	300	12	Al	0.036098	1.961946	4.5466
6	Positive	Br	30	90	100	4	Gr	0.148568	0.429715	6.8466
7	Positive	Al	10	70	100	12	Gr	0.102895	0.084416	6.5233
8	Positive	Al	20	90	200	4	Si	0.081942	0.059566	9.5300
9	Positive	Al	30	50	300	8	Al	0.145178	0.050427	9.7133
10	Negative	Cu	10	90	300	8	Gr	0.020592	0.240165	5.7500
11	Negative	Cu	20	50	100	12	Si	0.004059	0.392953	6.6900
12	Negative	Cu	30	70	200	4	Al	0.012738	0.709844	6.9500
13	Negative	Br	10	70	300	4	Si	0.023493	0.515940	10.980
14	Negative	Br	20	90	100	8	Al	0.111000	0.467711	6.6766
15	Negative	Br	30	50	200	12	Gr	0.162745	0.458686	10.050
16	Negative	Al	10	90	200	12	Al	0.034761	9.137222	5.5033
17	Negative	Al	20	50	300	4	Gr	0.016768	10.52743	5.2400
18	Negative	Al	30	70	100	8	Si	0.025899	15.72238	2.9633

a b

Figure 2 (a) Experiemental set up, and (b) close view of the PMEDM process.

Table 3. The data for comparability sequences

Reference/Comparability sequences		$\Delta0_i(k) = \lvert x_0^*(k) - x_i^*(k) \rvert$		
		MRR $\Delta0_1(k)$	RWR $\Delta0_2(k)$	SR $\Delta0_3(k)$
Reference sequence		1.00000	1.00000	1.00000
Comparability sequences	1	0.9400387	0.2169858	0.567151
	2	0.4715479	0.5969856	0.503479
	3	0.2790081	0.2463918	0.413959
	4	0.8390629	0.8149999	0.389243
	5	0.3742855	0.4960664	0.485906
	6	0.7449344	0.4596765	0.723789
	7	0.1208789	0.2042090	0.496667
	8	1.0000000	0.7906489	0.845644
	9	0.0920754	0.1916994	0.000000
	10	0.7322278	0.6846166	1.000000
	11	0.4702294	0.6264497	0.648761
	12	0.1419816	0.7796616	0.424800
	13	0.8343801	1.0000000	0.679939
	14	0.4308420	0.0460815	0.580710
	15	0.2740747	0.6598066	0.387801
	16	0.8594962	0.1812467	0.654832
	17	0.0677055	0.6284077	0.370347
	18	0.000000	0.0000000	0.707115

4. Results and Discussion

4.1. Best Experimental Run

The experimental results for the MRR, RWR and SR are shown in Table 2. Typically, higher values of the MRR, smaller RWR and surface roughness are desirable. Thus, the data sequences have a the-larger-the-better characteristic for the MRR, and the "smaller-the-better" characteristics for RWR and SR. Thus Eq. (1) and (2) were employed for data preprocessing. The experimental results were normalized during preprocessing stage and is shown in Table 3. The maximum MRR, and minimum RWR and SR were set as a reference sequence $x_0^{(O)}(k)$ and the results of the eighteen experiments be the comparability sequences $x_i^{(O)}(k)$, i =1-18, k = 1-3. All the sequences after data preprocessing are denoted as $x_0^*(k)$ and $x_i^*(k)$ for reference sequence and comparability sequence, respectively. The distinguishing coefficient ζ can be substituted for the grey relational coefficient in Eq. (4). If all the process parameters are of equal weighting, then ζ is 0.5. A weighting method is then used to integrate the grey relational coefficients of each experiment into the grey relational grade. The overall evaluation of the multiple performance characteristics is based on the grey relational grade. In this study, all the PMEDM process parameters influence the response variables moreover equally and hence the weighting factor of 0.5 was considered. Thus, the grey relational coefficients are calculated using Eq (5) and the average grey relational grades for each factor level was determined using Eq. (6). The values of GRC and GRG are shown in Table 4. Since the grey relational grades represented the level of correlation between the reference and the comparability sequences, the larger grey relational grade means the comparability sequence exhibiting a stronger correlation with the reference sequence. Based on this study, one can select a combination of the levels that provide the largest average response. In Table 4, the eighth comparability sequence has the largest grey relational grade, i. e. the combination of A1 B2 C2 D_3 E_2 F_1 G_3 shows the largest value of the grey relational grade for the factors A, B, C, D,E, F, and G respectively. Thus, the eighth experimental sequence is the best run giving highest process performance in terms of higher MRR, and lower SR and RWR. Therefore, the PMEDM performed with a positive polarity, a pulse current of 20 amp, a gap voltage of 90 Volts, pulse duration of 200 sec, 4% duty cycle and Si powder added dielectric produced higher MRR and lower SR and RWR.

Table 4. The calculated grey relational coefficient and grey relational grade for eighteen comparability sequences

Experimental runs (Comparability sequences)	Grey relational coefficient $\gamma(x_0(k), x_i(k))$			Grey relational grade γ_j
	MRR	RWR	SR	
1	0.91611	0.45541	0.60203	0.658
2	0.55339	0.61901	0.56873	0.580
3	0.47594	0.46492	0.52770	0.490
4	0.80271	0.77970	0.51740	0.700
5	0.51135	0.56510	0.56018	0.546
6	0.71966	0.54789	0.70332	0.657
7	0.42688	0.45140	0.56539	0.481
8	1.00000	0.75773	0.80923	0.856
9	0.41901	0.44754	0.39569	0.421
10	0.70975	0.67492	1.00000	0.795
11	0.55277	0.63674	0.65086	0.613
12	0.43283	0.74822	0.53235	0.571
13	0.79812	1.00000	0.67168	0.823
14	0.53498	0.40703	0.60963	0.517
15	0.47424	0.65809	0.51681	0.550
16	0.82333	0.44436	0.65482	0.641
17	0.41257	0.63796	0.50978	0.520
18	0.39569	0.39569	0.69094	0.494

4.2. Most Influential Parameter

Grey relational analysis identify the most influencing factors affecting the performance of the process where the multiple criteria is considered. In this study, the values of the parameter level in eighteen experimental runs are set to the comparability sequences for seven controllable parameters. After data preprocessing, the deviation sequences were calculated using the method as mentioned in section 2.1. The deviation sequences and the distinguishing coefficient were then substituted into Eq. 5 to obtain the grey relational coefficients, GRCs. Additionally, the grey relational coefficients are averaged using an equal weighting to obtain the grey relational grade using the procedure outlined in section 2.2.

Table 5. ANOVA results for GRG

Source	DF	Seq SS	Adj MS	F	P	% contribution
Polarity	1	0.00067	0.00067	0.04	0.861	0.257
Electrode	2	0.01368	0.00684	0.40	0.714	5.252
Pulse current	2	0.06901	0.03451	2.02	0.331	26.498
Gap voltage	2	0.02621	0.01311	0.77	0.566	10.064
Pulse time	2	0.01974	0.00987	0.58	0.634	7.579
Duty cycle	2	0.05408	0.02704	1.58	0.387	20.765
Powder type	2	0.03181	0.01591	0.93	0.518	12.214
Polarity×Electrode	2	0.01101	0.00551	0.32	0.757	4.169
Error	2	0.03421	0.01711			
Total	17	0.26043				
S = 0.130788 R-Sq = 86.86%						

In order to determine the influence of process parameters on PMEDM performance, the ANOVA was carried out. Table 5 gives the results of the analysis of variance (ANOVA) for the MRR, RWR and the SR using the calculated values from the grey relational grade given in Table 4. According to ANOVA, the factor C, the pulse current with 26.49% of contribution is the most significant controlled parameters for the PMEDM operation; the duty cycle is with 20.76% contribution, the powder type with 12.21%, and the gap voltage with 10.06%, pulse time with 7.57%, electrode type 5.25% and polarity is 0.25% of contribution if the maximization of MRR, and minimization of RWR and surface roughness are simultaneously considered. This is graphically shown in the Pareto chart in Figure 3.

Further, the ranking of the parameters according to their influence on the MRR, RWR and SR is determined using response analysis. The results are shown in Table 6.

Table 6. The response table for grey relational grade Factors

Levels	A	B	C	D	E	F	G
1	0.6000	0.6167	0.6817	0.5767	0.5700	0.6817	0.5600
2	0.6122	0.6333	0.6067	0.5817	0.6500	0.5833	0.5967
3		0.5683	0.5300	0.6600	0.5983	0.5533	0.6617
Max-Min	0.0122	0.0650	0.1517	0.0833	0.0800	0.1283	0.1017
Ranking	7	6	1	4	5	2	3

Total mean value of grey relational grade is 0.6061

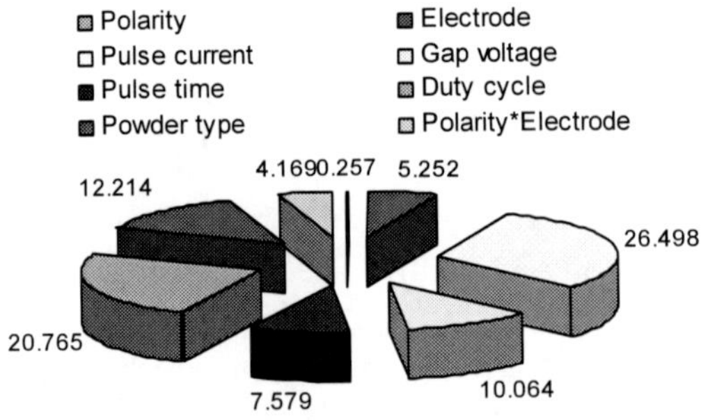

Figure 3. Pareto diagram showing control factors influencing the GRG.

4.3. Effect of Parameters on PMEDM Performance

In order to identify the effects of process parameters on the overall performance of PMEDM process, the means of grey relational grades were analysed. The results of means analysis is shown in Figure 4. It is observed from the plots that the parameters pulse current, duty cycle, powder type, gap voltage and polarity shows linear effects on the machinability of PMEDM of HSS tool steel. At lower pulse current (10 amp), the process gives better performance. However, it reduces at higher values of pulse current till 30 amp. Similarly, lower duty cycle of 4% produce higher process performance and it reduces when the duty cycle increases to 12%. In the case of dielectric medium, silicon powder mixed dielectric gives higher machinability than the other powder (Al and Gr) mixed dielectric during PMEDM process. Further higher gap voltage of 90 Volts shows improved performance.

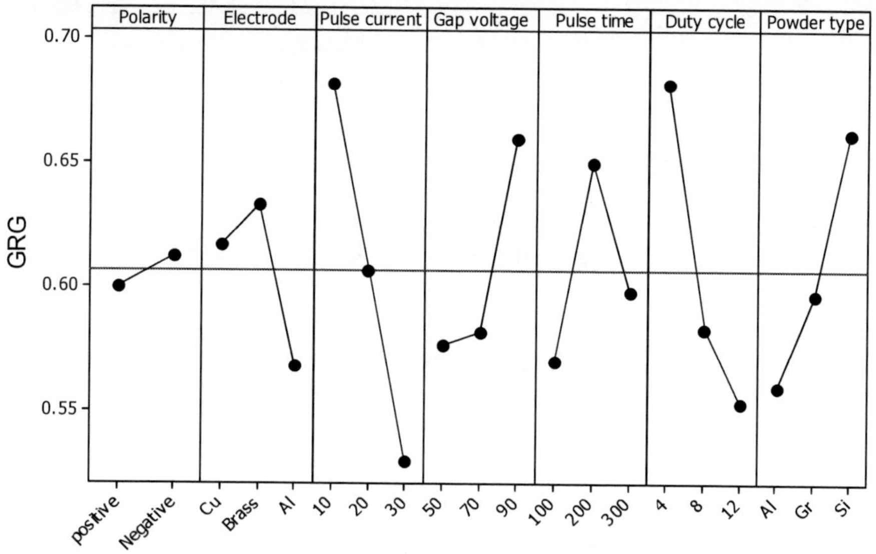

Figure 4. Main effects plots of GRG.

4.4. Comparison between Initial and Optimal Conditions

After obtaining the parametric conditions for optimum grey relational grade, the predicted grey relational grade was determined by conducting the confirmation experiments and using Eq. 7. The A2B2C1D3E2F1G3 is an optimal parameter combination of the PMEDM process via the grey relational analysis. Therefore, the condition A2B2C1D3E2F1G3 of the optimal parameter combination of the PMEDM process was treated as a confirmation test. If the optimal setting with a negative polarity, brass electrode, 10 amp pulse current, 90 Volts gap voltage, 200 sec pulse time, 4% duty cycle with silicon powder added electrolyte is used, the final workpiece gives the MRR of 0.07254 mm^3/min, RWR of 0.01952, and the surface roughness of 8.35 μm. In summary, the result of the confirmation test is better than the experiments on Table 2. Table 10 shows the above values and its comparison with that of the actual parameters by using the optimal machining conditions. It was found that the improvement of 18.7% is observed in the grey relational grade. Good agreement between the two has been observed. This ensures the usefulness of grey relational approach in relation to PMEDM process optimization, where multiple quality criteria has to be fulfilled simultaneously.

Table 7. Estimated grey relational grade for initial and optimal conditions

	Initial parameter setting	Optimal condition	
		Prediction	Experiment
Level	A1B1C1D1E1F1G1	A2B2C1D3E2F1G3	A2B2C1D3E2F1G3
MRR	0.04567		0.07254
RWR	0.03148		0.01952
Surface roughness (μm)	10.3533		8.3500
Grey relational grade	0.658	0.944	0.845
Improvement in grey relational grade 0.187			

5. Conclusions

Powder mixed EDM of HSS tool steel was optimized using a Taguchi based grey relational analysis. It is observed from the response table of the average grey relational grade that the largest value of the grey relational grade was for the eighth experimental condition. Thus, PMEDM process set at 20 amp pulse current, 90 Volts gap voltage, 200 sec pulse duration and 4% duty cycle with silicon powder added electrolyte and aluminium electrode with positive polarity produces the component with lower surface roughness and RWR and higher MRR. It is, therefore, the recommended parameter levels of PMEDM process when the minimization of the surface roughness, and relative wear rate and the maximization of MRR are simultaneously considered. The best performance characteristics were obtained from the means analysis of the grey relational grade. Thus, the optimum parameter settings for improved process performance are A2B2C1D3E2F1G3. Further, the relative significance of the PMEDM process parameters was determined using analysis of variance. ANOVA of the grey relational grade shows that, the parameters pulse current has the highest contribution to the PMEDM process followed by the parameters duty cycle, powder type, pulse time and electrode type. Hence, the pulse current is the most significant input parameter for the improved performance of the PMEDM process.

Acknowledgment

The authors would like to acknowledge the financial support provided by the Government of India, MHRD through TEQIP scheme to Department of Mechanical Engineering for equipment on which the experiments were carried out. Further thanks are due to Graduate students Harshad Sonawane and Rahul Nilangekar for their assistance in conducting the experiments.

REFERENCES

Chow, H. M., Yan, B. H., Huang, F. Y., Hung J. C. (2004). Study of added powder in kerosene for the micro-slit machining of titanium alloy using electro-discharge machining, *Journal of Materials Processing Technology*, 101, 95-103.

Chow, Han-Ming, Yang, Lieh-Dai, Lin, Ching-Tien, Chen, Yuan-Feng (2009). The use of SiC powder in water as dielectric for micro-slit EDM Machining, *Journal of Materials Processing Technology*, Article in Press.

Deng, J. L. (1989). Introduction to grey system theory, *The Journal of Grey Systems*, 1, 1–24.

Huang, J. T. and Lin, J. L. (2002). Optimization of machining parameters setting of die–sinking EDM process based on the grey relational Analysis with L18 orthogonal array. *Journal of Materials Processing Technology,*17, 659–664.

ISO 4287, *Surface Texture: Profile Method - Terms, Definitions and Surface Texture Parameters*, International Standards Organization, 1997

Jeswani, M. L. (1981). Effect of the addition of graphite powder to kerosene used as the dielectric fluid in electrical discharge machining, *Wear,* 70, 133-139.

Kansal, H. K., Singh, Sehijpal, Kumar, Pradeep. (2008). Numerical simulation of powder mixed electric discharge machining (PMEDM) using finite element method, *Mathematical and Computer Modeling,* 47, 1217–1237.

Kansal, H. K., Singh, Sehijpal, Kumar, Pradeep. (2005). Parametric optimization of powder mixed electrical discharge machining by response surface methodology, *Journal of Materials Processing Technology,* 169, 427-436.

Klocke, F., Lung, D., Antonoglou, G. and Thomaidis D. (2004). The effects of powder suspended dielectrics on the thermal influenced zone by electro-discharge machining with small discharge energies, *Journal of Materials Processing Technology*, 149, 191-197.

Lin, J. L. and Lin, C. L. (2002). The use of the orthogonal array with grey relational analysis to optimize the electrical discharge machining process with multiple performance characteristics, *International Journal of machine Tools and Manufacture,* 42, 237–244.

Lin, C. L., Lin, J. L. and Ko, T. C. (2002). Optimization of the EDM process based on the orthogonal array with fuzzy logic and grey relational analysis method, *International Journal of Advanced Manufacturing Technology*, 19, 271–277.

Mohri, N., Saito, N. and Higashi, M. (1991). A new process of finish machining on free surface by EDM methods, *Annals of the CIRP*, 41 (1), 207-210.

Pecas, P. and Henriques, E. (2003). Influence of silicon powder-mixed dielectric on conventional electrical discharge machining, *International Journal of machine Tools and Manufacture,* 43, 1465-1471.

Phadke, M. S. *Quality Engineering using Robust Design*, Prentice- Hall, Englewood Cliffs, NJ., 1989.

Uno, Y. and Okada, A. (1997). Surface generation mechanism in electrical discharge machining with silicon powder mixed fluid, *International Journal of Electrical Machining*, 2, 13-17.

Tzeng, Y. F. and Lee, C. Y. (2001). Effects of powder characteristics on electro-discharge machining efficiency, *International Journal of Advanced Manufacturing Technology,* 17, 586-592.

Wong, Y. S., Lim, L. C., Rahuman, Iqbal and Tee, W. M. (1998). Near–mirror finish phenomenon in EDM using powder-mixed dielectric, *Journal of Materials Processing Technology,* 79, 30-40.

Zhao, W. S., Meng, Q. G. and Wang, Z. L. (2002). The application of research on powder mixed EDM in rough machining, *Journal of Materials Processing Technology,* 129, 30-33.

In: Machining and Forming Technologies, Volume 2
Editor: J. Paulo Davim
ISBN: 978-1-61122-820-5

Chapter 17

PARAMETRIC ANALYSIS ON ELECTROCHEMICAL MACHINING OF AL/SIC$_P$ COMPOSITES

C. Senthilkumar* and G. Ganesan
Department of Manufacturing Engineering, Annamalai University,
Annamalai Nagar – 608 002, Tamilnadu, INDIA

ABSTRACT

Aluminum Matrix Composites (AMC) are difficult to machine due to the presence of hard and brittle ceramic reinforcements. Electro Chemical Machining (ECM) is an important process for machining such materials. In this study four factors: electrolyte concentration, electrolyte flow rate, applied voltage and tool feed rate of ECM, were chosen as variables in order to study the process performance in terms of Metal Removal Rate (MRR) and Surface Roughness (Ra).The Response Surface Methodology (RSM) is used to identified the most influential parameters for maximizing the MRR and minimizing the Ra. This paper attempts to establish a comprehensive mathematical model for correlating the interactive and higher-order influences of various machining parameters through RSM. Analysis of variance (ANOVA) has been performed to verify the fit and adequacy of the developed mathematical models.

Keywords: Electrochemical Machining (ECM), Metal Removal Rate (MRR), Surface Roughness (Ra), Response Surface Methodology (RSM), LM25 Al/20%SiCp.

1. INTRODUCTION

Electrochemical machining (ECM) is an anodic dissolution process where workpiece and tool are respectively anode and cathode, which are separated by an electrolyte. When an electric current is passed through the electrolyte, the anode workpiece dissolves locally, so that the shape of the generated workpiece is approximately a negative image of the tool. The

* Corresponding author: E-mail: senvalli@hotmail.com

electrolyte, which is generally a concentrated salt solution, is pumped at high velocities through the machining gap in order to remove the reaction products and to dissipate the heat generated. (Bhattacharya et al., 2004). The intensive electrolyte flows between electrodes and carries away the metal dissolved from anode workpiece and gaseous products generates in the gap. Since machining is achieved by electrochemical reaction, hard and difficult-to-cut materials can be machined, and there is no residual stress in the workpiece. The advantage of such machining is that there is no wear on the tool electrode, therefore, the cost and time for tool replacement is saved. A required workpiece can be obtained using tool electrode with the shape which is not congruent with the workpiece shape (Domanowskia., 2001). The first process resembling ECM was patented by Gusseff in 1929. Significant advances during the 1950s and 1960s developed ECM into a major technology in the aircraft and aerospace industries (Rajurkar et al., 1995). Metal matrix composites (MMC) are gaining increasing attention for applications in aerospace, defense, and automobile industries. These materials have been considered for use in automobile brake rotors and various components in internal combustion engines, the material used for these applications generally require light weight and greater wear resistance than those of conventional materials. One factor that prevents more manufacturers from embracing MMC technology is difficulty in machining these materials (Mohan et al., 2004).(Teti., 2002) found that the tools coated with diamond using CVD may be superior to PCD in machining the brake drum made of aluminium MMC. According to his findings, the machinability is critically affected by the reinforcement and matrix hardness rather than cutting parameters. (Kannan et al., 2006) observed that the abrasive reinforcements in the metal matrix composites caused rapid tool wear and the results from machining metal matrix composites with conventional tools showed that the main mechanism of tool wear was abrasion. (F. Muller et al., 2000) have studied the machining Particle Reinforced Metal Matrix Composites (PRMMC's) are a class of materials with a wide potential for application in the automotive and aerospace industries. However, the full potential of these materials is hindered by the high manufacturing costs associated with the difficulties experienced in machining these composites. Machining PRMMC's using conventional machining processes such as turning, drilling etc., generally results in excessive tool wear due to the presence of the hard particles which results in a very abrasive nature of this material. Consequently non-conventional machining processes such as Electro Discharge Machining (EDM), laser cutting and Abrasive Water Jet (AWJ) techniques are increasingly being used for the machining of PRMMC's. The result shows that crater-like surface, the size of the crates increases with increased discharge energy during EDM and also produces a relatively small amount of sub-surface damage on the cut surfaces. In Laser machining Striation patterns on the cut surface and burrs at the exit of the laser (dross attachment) were observed. Significant thermal induced microstructural changes were also observed within the PRMMC. AWJ cutting did not result in any thermal damage within the composite. No burr attachment was observed, however, the surface was relatively rough and slotted-edge damage was observed on the top of the cut surface. ECM is a non-traditional process used mainly to cut hard or difficult to cut metals, where the application of a more traditional process is not convenient (Neto et al., 2006). ECM is an extremely prominent machining process among newly developed non-traditional machining techniques. The merit of ECM machining techniques is most apparent in machining MMCs, which have the highest hardness in reinforcement. The surface finish and surface structure is found to be strongly influenced by the current density. (Bhattacharyya et al., 1999) have investigated the influence of various predominant process parameters such as electrolyte flow rate, concentration, the gap thickness and the applied voltage on the machining characteristics metal removal rate and over-cut in electrochemical micro-machining of EN-8 steel. (Munda et al., 2006) have investigated the electrochemical micromachining through response surface methodology approach. They have taken MRR and radial over cut as objective measures and developed mathematics models. Both objectives were dealt separately and analyzed with reference to machining parameters. (El-Taweel., 2008) has investigated the electrochemical turning (ECT) of 6061 Al/Al2O3 (10% wt) composite. The process parameters chosen as magnetic flux density (B), applied voltage (V), tool feed rate (F) and workpiece rotational speed (N) on the material removal rate (MRR) and surface roughness (SR). The material removal rate increases with the increase of magnetic flux density. Magnetic flux density has great influence on improving the surface quality, especially at high values of the tool feed rate due to the higher current density. SR decreases with an increase in applied voltage due to a higher current density, which improves surface quality. Increasing both the applied voltage and the tool feed rate leads to an increase of machining efficiency, and improves the surface roughness significantly. (Liu et al., 2009) studied the effects of applied voltage, current, pulse duration, and the concentration of the electrolyte on the behavior of wire electrochemical discharge machining of Al_2O_3 particle reinforced aluminum alloy 6061. A high applied voltage or long pulse duration reduces the material removal rate. This applies to both 10% and 20% Al_2O_3 reinforced materials. An increase in machining current leads to an increase in MRR. The ECM action becomes more active when the concentration of the electrolyte is increased.(Carinne Coughnowr et al., 1986) have investigated the electrochemical machining of TiC, ZrC, TiB_2, ZrB_2 and TiC/10%Ni composite in NaCl and KNO_3 electrolyte. It was found that the oxide layer is formed on the post ECM surfaces of TiC/Ni and Ni. TiC /10%Ni do not develop a polished surface after ECM in 2M KNO_3. Though a good number of researches on machining of composite materials have been made, no literature reveals the machining of MMC using electro chemical machining. The purpose of the present study, therefore, is to: (a) produce particle-reinforced metal matrix composites by stir casting method; and (b) machining MMC using electrochemical method based on Response surface methodology and developing a mathematical model. The investigation into the influence of ECM process parameters on machining performance criteria has been carried out through the development of mathematical models based on response surface methodology (RSM), utilizing the relevant experimental data as obtained through experimentation. Response surface methodology (RSM) is also used in the present research work for correlating the various machining parameters, such as electrolyte concentration, electrolyte flow rate, applied voltage, tool feed rate on the predominant electrochemical machining criteria metal removal rate and surface roughness. The adequacy of the developed mathematical models has also been tested by the analysis of variance test.

2. Experimental Study

ECM machine has been used in the experiment consists of three major sub-systems:

i. Machining cell
ii. Control panel
iii. Electrolyte circulation

2.1. Machining Cell

The electro-mechanical assembly is a sturdy structure, associated with precision machined components, servomotorized vertical up/down movement of tool, an electrolyte dispensing arrangement. All the exposed components and parts have undergone proper material selection and coating/plating for corrosion protection.

Tool area- 400 mm^2
Cross head stroke- 150 mm
Job holder- 100 mm opening×50 mm depth×100 mm width
Tool feed motor- DC servo type

2.2. Control Panel

The power supply is a perfect integration of high current electrical, power electronics and precision programmable micro-controller-based technologies. Since the machine operates at very low voltage, there are no chances of any electrical shocks during operation.

Electrical output rating- 0–300 Amps. DC at any voltage from 0–20 V.
Tool feed- 0.2 to 2 mm/min.
Supply- 415 v +/− 10%, three-phase AC, 50 Hz.

2.3. Electrolyte Circulation

The electrolyte is pumped from a tank, lined by corrosion resistant coating with the help of corrosion resistant pump and is fed to the job. The reservoir provides separate settling and siphoning compartments.

2.4. Machining Processes

The job to be machined is fixed in the vice. The machining chamber is corrosion resistant. A window allows the operator to view the machining process. The tool is brought near the job with the help of press buttons provided on the control panel and table lifting arrangement, maintaining particular gap. The tool progress is maneuvered vertically by the servo motor and is governed by a micro-controller based programmable drive. In ECM, generally a cathode tool is made out of non-reacting material, such as copper. The process is started in the presence of an electrolyte flow that is circulated with the help of special pump filling the gap between anode (job) and cathode (tool). Electrolyte flow is adjusted by flow control valve. Machining is achieved by sinking the tool forming its replica. During the operation, a sophisticated control panel takes care of any damage to the machine with overload and short circuit protections. After the desired time interval, a hooter gives an indication of completion of the time/ process. The small machining area with given power supply can be machined within 30 min to one hour. The experimentation scheme was designed in such way as to explore the influence of the various predominant machining parameters, based on RSM with the help of MINITAB software, in order not only to obtain the optimal scheme for multi-variable experimentation, but also to perform studies for exploring the interactive and higher-order influences of various parameters considered. MINITAB software can also produce ANOVA tables to test lack of fit of the RSM based model. The workpiece material used in these experiments was LM25 aluminium alloy (7 Si, 0.33 Mg, 0.3 Mn, 0.5 Fe, 0.1Cu, 0.1Ni, 0.2 Ti) reinforced with silicon carbide particles of size 25 μm with 20% volume fractions manufactured through stir-casting technique. The dimensions of the specimens were 30mm diameter and 6 mm height. The experiments were conducted on METATECH Electro chemical machining equipment. The tool was made up of copper with a 20mmX20mm square cross section. The electrolyte used for experimentation was fresh aqueous solution of sodium chloride (NaCl) with varying electrolyte concentration because of its high conductivity and non-passive characteristic. The Workpiece was weighed before and after machining in order to find out the Metal Removal Rate. The specification of weighing machine is as follows: Model: CY-360, Capacity: 0 – 360 gm, Accuracy: ± 0.001 gm. The surface roughness (Ra) of the machined test specimens was measured using a Talysurf tester with a sampling length of 10mm. The specification of Talysurf tester is as follows: Model: Time group ink, serial number of calibration-27105010, Ra 3.29 μm.

3. Response Surface Methodology

The RSM is an empirical modeling approach for determining the relationship between various process parameters and responses with the various desired criteria and searching the significance of these process parameters on the coupled responses (Bhattacharyya et al., 1999). Response surface methodology (RSM) is a collection of mathematical and statistical techniques that are useful for the modeling and analysis of problems in which a response of interest is influenced by several variables and the objective is to optimize this response. It is a sequential experimentation strategy for empirical model building and optimization. By conducting experiments and applying regression analysis, a model of the response to some independent input variables can be obtained. Based on the model of the response, a near optimal point can then be deduced. RSM is often applied in the characterization and optimization of processes (Kansal et al., 2005). The general second order polynomial response surface mathematical model can be considered to evaluate the parametric influences on the various machining criteria as follows:

$$Y_u = b_o + \sum_{i=1}^{k} b_i x_i + \sum_{i=1}^{k} b_{ii} x^2{}_i + \sum_{j>i}^{k} b_{ij} x_i x_j \qquad (1)$$

Table 1. Original values of machining parameters

PARAMETERS	LEVELS				
	-2	**-1**	**0**	**1**	**2**
Electrolyte concentration, X_1(gm/lit)	10	15	20	25	30
Electrolyte flow rate, X_2 (lit/min)	5	6	7	8	9
Applied voltage, X_3 (Volts)	12	13	14	15	16
Tool feed rate, X_4(mm/min)	0.2	0.4	0.6	0.8	1

Table 2. Experimental data

Ex.no	X_1	X_2	X_3	X_4	MRR (g/min)	Ra (μm)
1	-1	-1	-1	-1	0.0112	12.214
2	1	-1	-1	-1	0.0269	11.614
3	-1	1	-1	-1	0.0221	11.121
4	1	1	-1	-1	0.0331	11.754
5	-1	-1	1	-1	0.0246	10.985
6	1	-1	1	-1	0.0338	11.241
7	-1	1	1	-1	0.0263	11.845
8	1	1	1	-1	0.0427	10.287
9	-1	-1	-1	1	0.0334	11.721
10	1	-1	-1	1	0.0416	11.914
11	-1	1	-1	1	0.0356	10.982
12	1	1	-1	1	0.0498	11.141
13	-1	-1	1	1	0.0559	9.217
14	1	-1	1	1	0.0574	9.152
15	-1	1	1	1	0.0562	9.896
16	1	1	1	1	0.0588	9.478
17	-2	0	0	0	0.0201	13.821
18	2	0	0	0	0.0456	11.827
19	0	-2	0	0	0.0262	10.612
20	0	2	0	0	0.0465	9.841
21	0	0	-2	0	0.0150	11.849
22	0	0	2	0	0.0324	9.158
23	0	0	0	-2	0.0415	11.998
24	0	0	0	2	0.1128	9.814
25	0	0	0	0	0.0263	9.621
26	0	0	0	0	0.0285	9.548
27	0	0	0	0	0.0182	9.354
28	0	0	0	0	0.0226	9.721
29	0	0	0	0	0.0245	9.214
30	0	0	0	0	0.0215	8.184
31	0	0	0	0	0.0253	8.162

Where Yu represents the corresponding response. The code values of ith machining parameters for uth experiment are represented by X_{iu}. The values of k indicate the number of machining parameters. The terms bi, bii and bij are the second order regression co-efficient. The second term under the summation sign of this polynomial equation attributes to linear effects, whereas the third term of the above equation corresponds to the higher order effects and lastly the fourth term of the equation includes the interactive effects of the parameters. Using this quadratic model of y in this study is not only to investigate over the entire factor space, but also to locate region of desired target where the response approaches its optimum or near optimal value. The traditional method of experimentation is that varying one factor at a time and studying its effect, was considered time consuming and costly. Furthermore, such experiments cannot predict the presence of interactions between the different factors. With a proper designed experiment such as Central Composite Design (CCD), it is possible to determine, with a much reduced number of experiments, the effect of changing any one variable with the same accuracy as if only one factor has been varied at a time, and the interaction effects between the factors (Cochran et al., 1962). The CCDs are most popular due to the following attributes:(1) CCDs can run sequentially; (2) CCDs are efficient, providing information on experiment variable effect on overall experimental error in a minimum number of runs;(3) CCDs are very flexible (Kuppan et al., 2006).The ECM process was studied with a standard RSM design called CCD. The pertinent process parameters selected for the present investigation are electrolyte concentration (X_1), electrolyte flow rate(X_2), applied voltage (X_3), tool feed rate (X_4) were considered as controlling variables. This design requires five levels of each factor in coded form: - α,-1, 0 1and + α. The α is a function of k, the number of independent parameters. In this design the level interval for each factor depends on the number of factors. The levels of each factor in this CCD design were chosen as -2,-1, 0, 1, and 2 in closed form to have a rotatable design. Table 1 shows the factors and their levels in coded and actual values. For the four variables, the design required 31 experiments with the 16 factorial points, 8 axial points to form a central composite design with α=*2,* 7centre points. The experiment has been carried out according to the designed experiment based on CCD as depicted in Table 2.

4. RESULT AND DISCUSSION

4.1. Metal Removal Rate (MRR)

The model for MRR is carried out using RSM. The fit summary recommended that the quadratic model was statistically significant for analysis of MRR. The ANOVA table for the quadratic model for MRR was shown in Table 3.The value of R^2 and adjusted R^2 are 94.93% and 90.49% which means that the regression model provides an excellent explanation of the relationship between the independent variables and the response (MRR). The associated P-value for the model is less than 0.05 (i.e. α = 0.05, or 95% confidence) indicates that the model is considered to be statistically significant. The standard percentage point of F distribution for 99%confidence limit is 7.87. As shown in Table 4 the F- value (4.34) for lack of fit is smaller than the standard value. Thus the lack-of-fit term is significant. It is also seen that from the p-values, for MRR second order effect of X_4 (tool feed rate) is significant model terms. The other model terms are can be regard as not significant effect due to their "Prob. > *F*" value greater than 0.05. The contour plots were drawn for various combinations. The number represent in the contour plot was MRR. Using the results presented in Table 2, the form of the derived model is as follows:

$$MRR= - 0.1321 + (0.0021X_1) - (0.0162\ X_2) + (0.0325X_3) - (0.3944\ X_4) + (0.000061\ X_1^2) + (0.00238X_2^2) - (0.000773\ X_3^2) + (0.3147\ X_4^2) + (0.00012X_1\ X_2) - (0.00024\ X_1\ X_3) -(0.000161\ X_1X_4) - (0.00095\ X_2\ X_3) - (0.00487\ X_2\ X_4) + (0.01056\ X_3\ X_4) \quad (2)$$

Table 3. Test for significance of metal removal rate

Term	Coefficient	T	P
Constant	- 0.1321	-0.439	0.666
X_1	0.0021	0.429	0.674
X_2	- 0.0162	-0.600	0.557
X_3	0.0325	0.962	0.350
X_4	-0.3944	-3.151	0.006
X_1^2	0.000061	1.357	0.193
X_2^2	0.00238	2.142	0.048
X_3^2	- 0.000773	-0.693	0.498
X_4^2	0.3147	11.285	<0.000
$X_1 * X_2$	0.00012	0.402	0.693
$X_1 * X_3$	- 0.00024	-0.813	0.428
$X_1 * X_4$	- 0.000161	-1.081	0.296
$X_2 * X_3$	- 0.00095	-0.637	0.533
$X_2 * X_4$	- 0.00487	-0.654	0.523
$X_3 * X_4$	0.01056	1.417	0.176

Table 4. ANOVA table for metal removal rate & surface roughness

Source	Degree of freedom	Sum of squares		Mean sum of squares		F-value		P-value	
		MRR	Ra	MRR	Ra	MRR	Ra	MRR	Ra
Linear	4	0.0004	4.829	0.00011	1.207	3.16	3.7	0.043	0.026
Square	4	0.0047	27.07	0.00117	6.769	33.0	20.7	0.000	0.000
Interaction	6	0.0001	3.174	0.00002	0.528	0.81	1.62	0.580	0.205
Lack of fit	10	0.0005	2.569	0.00005	0.256	4.34	0.58		
Error	6	0.00006	2.653	0.00001	0.442				
Total	30	0.01122	53.33						

The Figure 1 shows the effect of applied voltage and tool feed rate. It has been observed from figure that the material removal rate (MRR) increases with applied voltage. It is due to an increase in applied voltage causes greater machining current available in the interelectrode gap (IEG) increases. This result in higher MRR. Hence, maximum MRR was obtained in the range of 14 to 16 volts applied voltage. Faraday's law states that the material removal rate is proportional to the machining current (Bhattacharyya et al., 2003).Increase in tool feed rate IEG becomes smaller. Then the electrical resistance between the cathode and anode decreases. It enhances the electrochemical reaction which results in higher MRR.

The interaction effect of electrolyte flow rate and applied voltage on MRR is shown in Figure 2.The figure shows that the value of MRR increases with increase in applied voltage and flow rate. With increase in applied voltage machining current also increases, thus contributing more noticeable aluminium material debonding from the parent material. This is due to the aluminium having a high electrical conductivity compared to SiC material. Increase in flow rate, more mobility of the ions from the metal to the solution increasing the speed of chemical reactions (Neto et al., 2006). Thus maximum MRR was obtained at high applied voltage range of (14-16 volts) and high flow rate range of (8.5-9 lit/min).

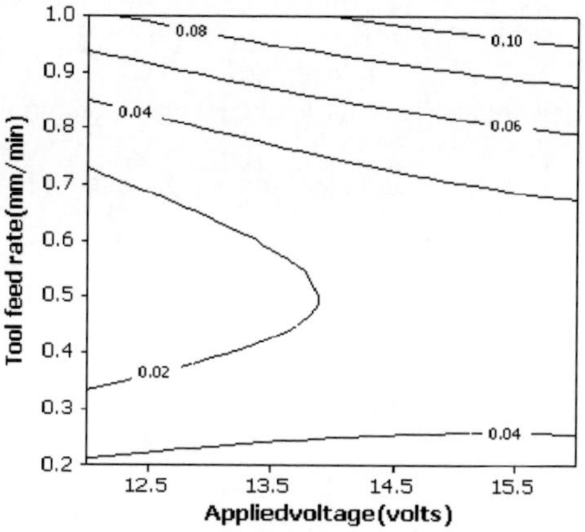

Hold Values
Electrolyte concentration, 20 gm/lit Electrolyte flow rate, 7 lit/min

Figure 1. Effect of applied voltage and tool feed rate on MRR.

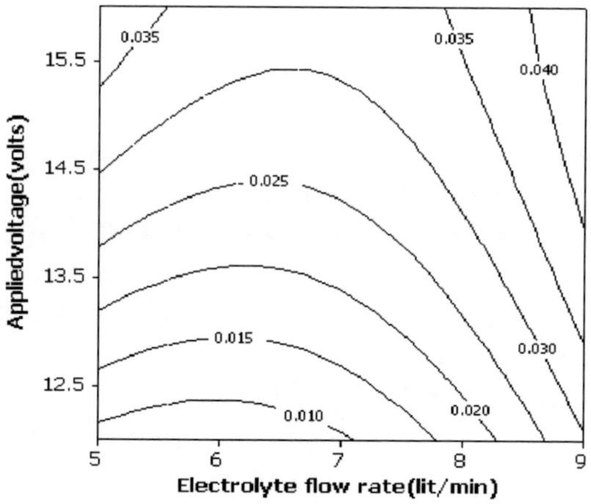

Hold Values
Electrolyte concentration, 20 gm/lit Tool feed rate, 0.6mm/min

Figure 2. Effect of electrolyte flow rate and applied voltage on MRR.

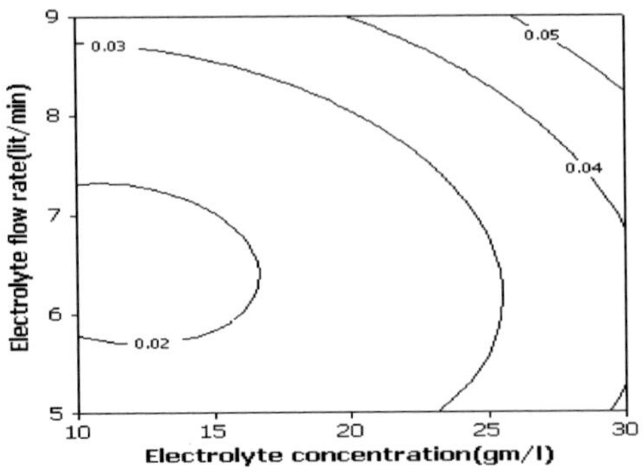

Hold Values

Applied voltage 14 volts Tool feed rate, 0.6mm/min

Figure 3. Effect of electrolyte concentration and electrolyte flow rate on MRR.

Figure 3 illustrates the effect of electrolyte concentration and electrolyte flow rate on MRR. It has been observed from figure that the material removal rate increases with electrolyte concentration. It is due to higher electrolyte concentration, the larger number ions associated with the machining processes, increases the machining current and thus results in higher MRR (Sen et al.,2005).The maximum MRR is obtained at high electrolyte concentration range of (25-30 gm/lit) and high flow rate range of (8-9 lit/min). MRR is proportional to the electrolyte flow rate. This is because increase in electrolyte flow rate removes the reaction products in the IEG effectively and also fresh electrolyte directed into IEG which increases the conductivity of the electrolyte.

4.2. Surface roughness (Ra)

For surface roughness, the fit summary recommended that the quadratic model is statistically significant for analysis. Table 5 presents the ANOVA table for a quadratic model for Ra. The value of R^2 and adjusted R^2 are 90.21% and 81.64 %, respectively. This means that regression model provides an excellent explanation of the relationship between the independent variables (factors) and the response (Ra). The associated p-value for the model is lower than 0.05 (i.e. $\alpha = 0.05$, or 95% confidence) indicates that the model is considered statistically significant.Table 4 shows the F- value 0.58 for lack of fit is smaller than the standard value. Thus the lack-of-fit term is significant. It is also seen that from the p-values, for surface roughness square effect is significant. second order effect X_1 (electrolyte concentration) and X_4 (tool feed rate) are significant model terms. The other model terms are can be regard as not significant effect due to their "Prob. > F" value greater than 0.05. The contour plots were drawn for various combinations. The number represent in the contour plot was surface roughness. Using the results presented in Table 2, the form of the derived model is as follows:

$$Ra = 110.966 - (1.052\ X1) - (6.355X2) - (9.292X3) + (7.308X4) + (0.036\ X12) + (0.244\ X22) + (0.313\ X32) + (10.341\ X42) - (0.012\ X1\ X2) - (0.027X2\ X3) + (0.071X1\ X4) + (0.211X2\ X3) + (0.169\ X2\ X4\ ) - (1.772\ X3\ X4) \tag{3}$$

Table 5. Test for significance of surface roughness

Term	Coefficient	T	P
Constant	110.966	3.853	0.001
X_1	- 1.052	-2.161	0.046
X_2	- 6.355	-2.445	0.026
X_3	- 9.292	-2.872	0.011
X_4	7.308	0.610	0.551
X_1^2	0.036	8.361	<0.000
X_2^2	0.244	2.282	0.037
X_3^2	0.313	2.930	0.010
X_4^2	10.341	3.872	<0.001
$X_1 * X_2$	- 0.012	-0.424	0.677
$X_1 * X_3$	- 0.027	-0.950	0.356
$X_1 * X_4$	0.071	0.498	0.625
$X_2 * X_3$	0.211	1.477	0.159
$X_2 * X_4$	0.169	0.236	0.816
$X_3 * X_4$	- 1.772	-2.481	0.025

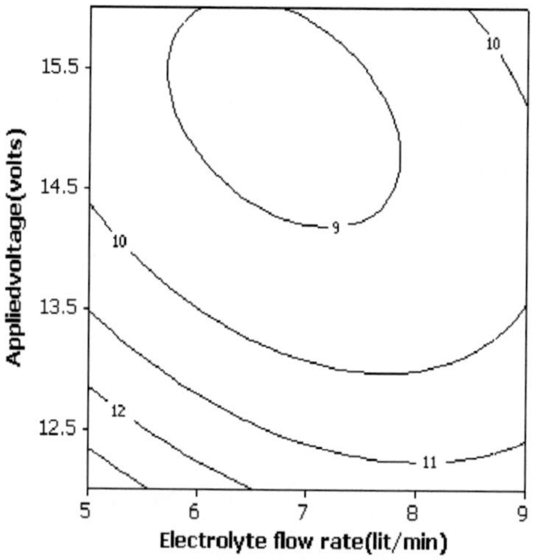

Hold Values
Electrolyte concentration, 20 gm/lit Electrolyte flow rate, 7 lit/min

Figure 4. Effect of applied voltage and tool feed rate on Ra.

The effect of applied voltage and tool feed rate on Ra is shown in. Figure 4. Ra decreases with increase in applied voltage and tool feed rate. At low voltage, current density in the gap is low which reveals crystallographic steps and etches pits, preferred grain boundary attack or finely dispersed microstructure leading to extremely rough surfaces which is shown in Figure 5. Surface roughness of the workpiece decreases as the electrode gap is reduced. In the case of a narrow electrode gap, it is observed that more pits are formed at an electrode gap of 0.2 mm than for one of 1.0 mm.

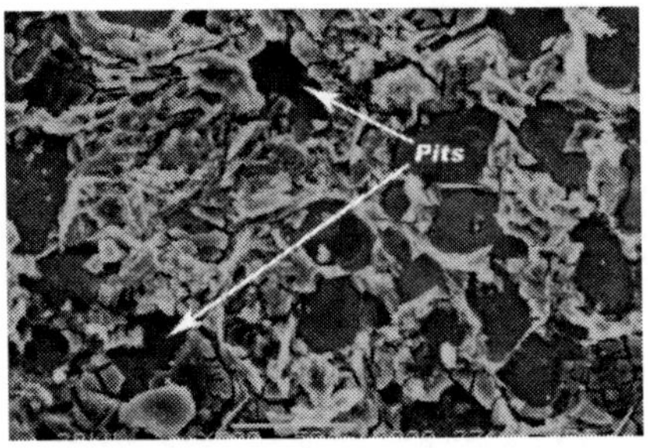

Figure 5. Etch Pits Occurs Below 14 Volts of Applied Voltage.

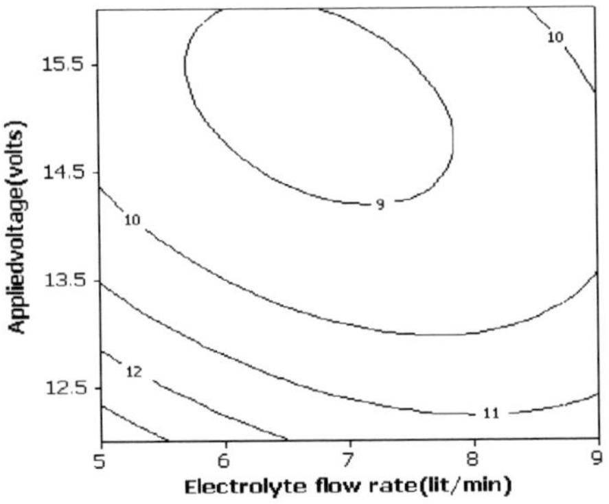

Hold Values
Electrolyte concentration, 20 gm/lit Tool feed rate, 0.6mm/min

Figure 6. Effect of electrolyte flow rate and applied voltage on Ra.

Figure 6 depicts the influence of flow rate and applied voltage on Ra. From the figure, Ra tends to decrease considerably with an increase in applied voltage and electrolyte flow rate. An increase in applied voltage metal dissolution is steady which leads to good surface finish. Increase in electrolyte flow rate result in flushing out the H_2 bubbles, precipitation and sludge's formed in the IEG. This results in reduced surface roughness (Krishnaiahchetty et al., 1981). Figure 7 shows the smoother surface with no pits and flow streaks on the workpiece surface due to high applied voltage and electrolyte flow rate.

Figure 7. No surface damage above 14 Volts of Applied Voltage

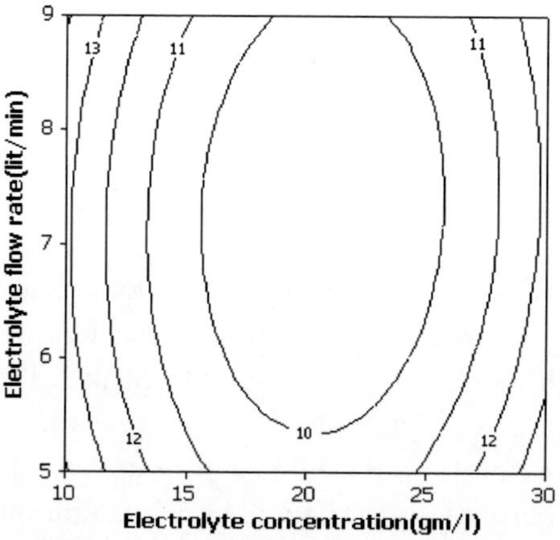

Hold Values
Applied voltage 14 volts Tool feed rate, 0.6mm/min

Figure 8. Effect of electrolyte concentration and electrolyte flow rate on Ra

Figure 8 shows the relationship between electrolyte concentration and electrolyte flow rate on Ra. The figure indicates that Ra initially decreases with increase in NaCl concentration but increases with further increase in NaCl concentration. The reason for this is that the differential dissolution of the phases of the work material, particularly at the higher electrolyte concentration (Sen et al., 2005). Ra decreases with increases of the electrolyte flow rate. This is because during electrochemical machining the atoms along the grain boundaries are attacked first. This leads to the formation of depressions and the surface becomes rough. The thickness of the diffusion layer is affected by the surface roughness and because of the depressions, the thickness of the diffusion layer tends to reduce further. In such case, a rotating eddy is produced since the flow tends to separate from the surface. The metal ions on the upstream side of the valley will be picked up by the eddy and are transported away by the flow. Hence the rate of mass transfer in these areas depends on the velocity of the small eddies. These micro fluid disturbances, termed as 'rotating eddies' remove metal ions from the grain boundaries, thus accelerating the mass transfer and reducing the surface finish. Any effort to increase the velocity reduces the diffusion layer thickness. A direct result of increasing the velocity will be an increase in the level of turbulence. As a consequence to this, the effect of rotating eddies may be reduced resulting in a better surface finish (Krishnaiahchetty et al., 1981).

5. CONCLUSION

In this paper, the electrochemical machining (ECM) of Al/20%SiCp was studied. Four parameters of ECM processes were studied, namely: electrolyte concentration, applied voltage, electrolyte flow rate and tool feed rate. The authors also discuss the influences of process parameters on MRR and Ra.

Mathematical modeling has been developed to evaluate MRR and Ra under various process parameter settings based on response surface methodology. It is also employed to analyze the influence of these machining process parameters on MRR and Ra during ECM.

The influence of different process parameters on machining performance criteria are exhibited though response contour plots.

It is clear from the response contour plot of MRR, the MRR increases with an increase in any of the machining parameters. This is due to an increase in applied voltage and tool feed rate causes high current density in the IEG. This leads to more noticeable aluminium material debonding from the parent material, due to aluminium having a high electrical conductivity compared to SiC material. An increase in electrolyte concentration and electrolyte flow rate the mobility of ions is more and also reaction products and hydrogen bubbles are effectively removed from the IEG which increase the conductivity of the electrolyte. At higher level of machining parameters, except electrolyte concentration, gives lower Ra as shown on the Ra response contour plots. The reason for this could be attributed to the differential dissolution of the phases of the workpiece, particularly at the higher concentration.

References

Bhattacharyya, B., Malapati, M., Munda, J. and Sarkar, A. (1999), "Investigation for controlled electrochemical machining through response surface methodology-based approach", *Journal Materials Processing Technology*, 86, 00-207.

Bhattacharya, B. and Munda, J. (2003), "Experimental investigation on the influence of electrochemical machining parameters on machining rate and accuracy in micromachining domain", *International Journal of Machine Tools and Manufacture,* 43, 1301-1310.

Bhattacharya, B., Munda, J. and Malapati, M. (2004), "Advancement in electrochemical micro-machining", *International Journal of Machine Tools and Manufacture,* 44,1577-1589.

Cochran, G. and Cox, G. M. (1962), "*Experimental Design"*. New Delhi: India.

Coughanow,r C.A., ADissaux, B., Muller, R. H. and Tobais, C. W.(1986), "Electrochemical machining refractory materials", *Journal of Applied Electrochemistry,* 16, 345-56.

Domanowskia, P. and Kozak, P. (2001), "Inverse problems of shaping by electrochemical generating machining", *Journal Materials Processing Technology*, 109,347-353.

El-Taweel, T., A. (2008), "Modelling and analysis of hybrid electrochemical turning – magnetic abrasive finishing of 6061 Al/Al2O3 composite, *Journal advanced manufacturing technology,* 37, 705-714.

Kannan, S., Kishawy, H. A. and Balazinski, M. (2006), "Flank wear progression during machining metal matrix composites," Journal Manufacturing Science and Engineerin,. 128, 787–791.

Kansal, H. K., Singh, S. and Kumar, P. (2005), "Parametric optimization of powder mixed electrical discharge machining by response surface methodology", *Journal Materials Processing Technology*,169, 427-436.

Krishnaiahchetty, O. V. and Radhakrishnan, V. (1981), "A Study on the influence of grain size in electrochemical machining", *International Journal of Machine Tools and Manufacture,* 21, 57–59.

Kuppan, P., Rajadurai, A. and Narayanan, S. (2006), "Experimental investigation into electrical discharge deep hole drilling of Inconel 718 using graphite electrode", In

Proceedings: 17th All India machine tool design and research conference, India, 923–927.

Liu, J. W., Yue, T. M. and Guo, Z. N.(2009), "Wire electrochemical machining of Al_2O_3 Particle Reinforced Aluminum Alloy 6061", *Materials and Manufacturing Processes,* Vol 24, pp. 446–453.

Manna, A. and Bhattacharya, B. (2002), " A study on different tooling sustems during machining of Al/SiC-mmc'"*Journal of Materials Processing Technology* 123, 476-482.

Mohan, B., Rajadurai, A. and Satyanarayana, K. G. (2004), "Electric discharge machining of Al/SiCp metal matrix composites using rotary tube electrode." *Journal Materials Processing Technology,* 153–154, 978–985.

Muller, F. and Monaghan, F.(2000), "Non-conventional machining of particle reinforced metal matrix composite", *International Journal of Machine Tools & Manufacture,* 40,1351–1366.

Munda, J. and Bhattacharyya, B. (2008), "Investigation into electrochemical micromachining (EMM) through response surface methodology based approach", *International Journal of Advanced Manufacturing Technology,* 35, 821-832.

Neto, J. C. D. S., Silva, E. M. D. and Silva, M. B. D. (2006), "Intervening Variables in electro chemical machining", *Journal Materials Processing Technology*, 179, 92-96.

Rajurkar, K. P., Wei, B. and Kozak, J. (1995), "Modelling and monitoring interelectrode gap in pulse electrochemical machining", *Annals of the CIRP,* 44(1),177-180.

Rajurkar, K. P., Zhu, D., McGeough, J. A., Kozak, J. and De Silva, A. (1999), "New developments in electro- chemical machining", Annals of the CIRP, 48(2), 567-579.

Sen, M. and Shan, H. S. (2005), "Analysis of hole quality characteristics in the electro jet drilling process", *International Journal of Machine Tools and Manufacture,* 45, 137–152.

Teti, R. (2002), "Machining of composite materials," *CIRP Annals – Manufacturing Technology,* 51(2), 611-634.

In: Machining and Forming Technologies, Volume 2
Editor: J. Paulo Davim
ISBN: 978-1-61122-820-5

Chapter 18

THE EFFECT OF THE ELASTIC AND PLASTIC ANISOTROPY ON THE ELASTIC-PLASTIC DEFORMATION REGION

J. Prohászka[1], A. G. Mamalis[2*] and Z. Gaál[3]
[1]Research Group for Metals Technology of the Hungarian Academy of Sciences, Bertalan L. u. 7. MT ep., H-1111 Budapest, Hungary
[2]Manufacturing Technology Division, National Technical University of Athens, Athens, Greece
[3]Research Group for Metals Technology of the Hungarian Academy of Sciences, Bertalan L. u. 7. MT ep., H-1111 Budapest, Hungary

ABSTRACT

The elastic-plastic deformation region plays an important role at structural materials operating under low load, e.g. in case of fatigue or continuous load. The elastic-plastic deformation is the most inhomogeneous part of the whole deformation process. The plastic deformation begins in a microstructure which could be described as the softest realizable structure with statistically disordered crystal orientation. The extremely wide deformation options of this structure determine the properties of the intermediate deformation region. These are mainly the elastic and plastic anisotropy: the former provides zero to three times bigger possibility for the start of the plastic deformation, at the latter this value is between 0.23 and 0.5. In this chapter we show that – independently from the accepted or not accepted assumptions of the literature – the start of the plastic deformation always begins in only those few crystals, which have the optimum orientation according to the theoretical assumption. Increasing the stress the plastic deformation extends continuously to the wider fraction of volume. The rate of this phenomenon – till it is extended to all of the crystals – is influenced by the direction dependent Young-modulus. The effects of the anisotropy have to be taken into account in plastic deformation, especially at small deformation rates. In case of fatigue tests, some crystals can take larger part in deformation caused by the cyclic load than other crystals

* Corresponding author: E-mail address: mamalis@central.ntua.gr

do. Similarly to this, only some of the crystals take part in deformation during creep tests, if the load is smaller than a certain value.

Keywords: elastic modulus; elastic-plastic deformation region; anisotropy; shear stress; polycrystalline material.

1. INTRODUCTION

Many papers deal with the topics of elastic and plastic deformation in the literature, but not with the intermediate state between the two phenomena, although the conventional yield strength value can be found in this region, which replaces the hardly definable elastic limit. The work of the design ers is based on this well defined conventional yield strength value; this value is taken into consideration at the dimensioning of the designed structure (ASTM, 1992). The elastic and plastic anisotropy play significant role in this intermediate region. The elastic anisotropy influences the Young-modulus ratio, E111 / E100 in the individual crystals by a value of more than 3 in case of copper and by a value of 2 in case of iron, modifying as well the necessary load for the plastic deformation in the crystals. The Schmid-factor is considered as the indicator of the plastic deformation; the Schmid-factor has almost the same effect like the elastic anisotropy, since the ratio of the maximal and minimal value is 1.8. The chapter deals with the effect of the elastic and plastic anisotropy in the elastic-plastic deformation region, demonstrating it on the examples of often used metals.

Most of the metals are used due to their strength properties in the industry. The strength of the metals can be improved by different technologies, which are widely used in practice. The strength properties of the metals are evaluated by their conventional yield strength value. The value of this parameter falls into a region of the strength values of the metals – the starting region of the elastic and plastic deformation – where the scattering is rather high. The cause of this strong scattering is that the deformation mechanisms change from elastic deformation to plastic deformation in this region.

Table 1. Elastic modulus of some metals in certain direction and mean values. [GPa]

	E_{001}	E_{011}	E_{111}	E_{123}	$E_{2\,9\,20}$	E_{111}/E_{100}	E_{mean}
Aluminium	62,9	72,2	75,9	72,2	68,0	≈1,21	70,5
Copper	67,1	135,8	206,0	135,8	95,2	≈3,07	122,0
Iron	129,0	209,8	265,8	209,8	166,5	≈2,06	212,0
Tungsten	389,5	390,6	391,0	390,6	390,2	≈1,00	388,0

This is the most complex part of the whole deformation process of the steels and this is the cause of the large scattering. In the followings we attempt to summarize the micro and macro mechanisms, which define the deformations in this region and will be resulted in the complexity of the whole intermediate deformation process. The deformation process begins with the elastic deformation which extends to the whole geometry; it is trivial for everybody (Hill, 1948)(Bishop, 1951). But if we are interested in the details, we have to take into consideration some usually neglected facts, namely that the crystals of the polycrystalline materials keep their properties, as they are known from the single crystal. In the polycrystalline bodies, the <uvw> orientation of the crystals show usually statistical distribution, if there is no special reason for the frequenter occurrence of a preferred direction. This anisotropy in the preferred direction can be caused by controlled cooling, large plastic deformation or recrystallization; these are the deformation and recrystallization texture. Certainly, in these textured materials the texture strongly influences the value and scatter of the conventional yield strength. In the followings the chapter deals with the statistically oriented changes. The individual crystals of the polycrystalline bodies consisting of statistically oriented crystals behave differently. In both elastic and elastic-plastic intermediate deformation states only a few crystals are plastically deformed at the beginning; increasing the load, at a certain load rate all crystals show plastic deformation.

2. Basic Considerations

If we take a probe with statistical orientation, one phase material and low impurity content and we apply a significantly lower load than the elastic limit, the deformation extends to the whole geometry. This is valid in the whole elastic region and based on this, we can assume that the deformation rate is constant according to the Taylor assumption (Taylor, 1938). Although the Taylor assumption is often criticized in the literature, this assumption can be accepted in the elastic deformation region. If we accept this assumption to have equal deformation in the individual crystals, the load stress have to be orientation dependent and different according to the elastic anisotropy in the individual crystals; the value of the load stress is equal to the product of the deformation, ε of the whole probe and the orientation dependent elastic modulus, E_{uvw} of the given crystal:

$$\sigma_{uvw} = \varepsilon E_{uvw} \tag{1}$$

Some orientation dependent values of the elastic modulus of various practically used metals are classified in Table 1 using the ratio of the E_{111}/ E_{100} values and the average elastic modulus, E_{mean}. It can be seen in the table that in metals with cubic structure the maximal elastic modulus is parallel to the <111> direction, while the minimal elastic modulus is parallel to the <100> direction.

Although in some cases, like in case of tungsten, the elastic properties are nearly isotropic, the plastic properties are anisotropic. Figure 1 shows the poles of the orientations in the stereographic triangle by the value of the plastic anisotropy and Schmid-factor.

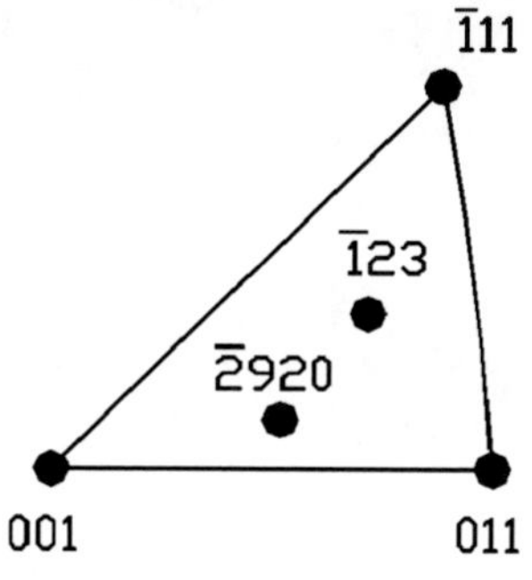

Figure 1. Orientation poles and Schmid-factors from Table 1. in the stereographic triangle.

The same elongation of the crystals in the elastic deformation region results in different stress values in the different crystals; the difference vary in the different materials. The maximal ratio of the different elastic moduli can be hardly displayed at tungsten, it is 1.2 at aluminum, 2.1 in iron and it is bigger than 3.1 in copper. This difference is present till the stress reaches the value of the elastic limit defined by the ASTM (ASTM, 1992). The elastic deformation region is closed at this point, which is seemingly a homogenous deformation process. Assuming that, two facts are not considered: first we don't know how the crystals behave in case of such small deformation, second the reversible movements of the dislocation in the slip planes of the crystals and the reversible deformation of the probe, because the dislocations vary their length under shear stress, which assist the elastic deformation of the whole sample. In our opinion both phenomena can be neglected in case of small deformation, but they have to be mentioned.

3. DISCUSSION

According to Table 1, the tensile stress parallel to the axis of the sample is different almost at all differently oriented crystals, which is also represented in the spatial distribution elastic modulus of the materials introduced in the table (Figure 2). As already mentioned, these properties are very similar in case of cubic structured materials, only the values are different.

The elastic-plastic deformation starts in a probe, where the crystals have significantly different deformation. The deformation is elastic till that limit point; from there the deformation is elastic and plastic. The differences of the tensile stress values of the individual crystals are proportional to the difference of the elastic moduli, given above for the different materials. The ratio of the different stress values is the smallest, 1.1 at the tungsten; the largest, 3.07 is in case of copper, like in case of the elastic modulus values. This is true only in the elastic deformation region.

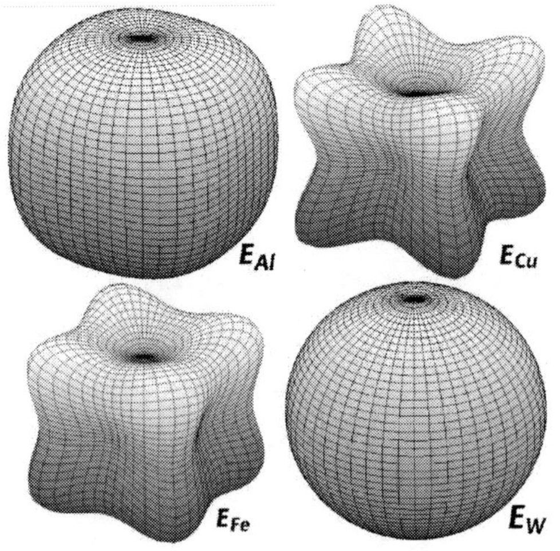

Figure 2. Spatial distribution of the elastic modulus of the metals from Table 1.

At the beginning of the elastic-plastic intermediate deformation period there are homogenously and elastically deformed crystals with strongly different grain size. The plastic deformation begins in the crystal, where in the slip system the shear stress value reaches the critical stress value , τ_{crit}, which is characteristic for the material. It occurs not always in the crystal, in which the stress reaches the maximal value. In the FCC metals there are altogether twelve ⟨1 1 0⟩{1 1 1} slip systems, but simultaneously can be active maximum eight or less systems, but at least one has to be active in the elastic-plastic deformation region, if the stress is enough large. While in face cubic centered mostly one slip system is materialized, in body cubic centered there are three slip systems mentioned in the literature (Smallman, 1985). These are the ⟨1 1 1⟩{1 1 0}, ⟨1 1 1⟩{1 1 2} and ⟨1 1 1⟩{1 2 3} (Dieter, 1988). The slip occurs always in the most dense direction, ⟨1 1 1⟩ in all of the three slip systems and mostly on {1 1 0} slip plane. The reason for that phenomenon can be the not pure metallic bond, which appears at the embrittlement and can cause the slip plane change, too.

At the end of the elastic deformation period there are different stresses present in the individual crystals according to their orientation. The further load let move irreversibly the dislocations, starting the plastic deformation. This process begins in the crystal, in which the stress reaches the material characteristic critical shear stress value, τ_{crit} in its slip systems. This value is defined by Schmid's law:

$$\tau_{crit} = \sigma_{uvw} \cdot \cos\alpha \cdot \cos\beta = \sigma_{uvw} \cdot S \tag{2}$$

where, σ_{uvw} is the stress in the [u v w] oriented crystalline, parallel to the tangential axis of the sample, α is the angle between the [u v w] oriented crystal and the normal direction of the slip plane, β is the angle between σ_{uvw} and slip direction and S is the Schmid-factor of the active slip system of the crystallite (values are represented for the chosen orientations in Figure 1).

If the E_{uvw} values from Table 1. are multiplied by the Schmid values, we obtain the orientation dependent data, which are proportional to the shear stress values in the slip systems of the load stress. It is not exactly known, what kind of elastic deformation occurs when the plastic deformation begins in some crystals. Table 2. gives the Schmid values for the above described slip systems and the sequence of the stress values, τ acting in the slip direction of the given slip systems for the different materials.

Based on the differences between the FCC and BCC metals, the slip direction is ⟨1 1 1⟩ in iron and ⟨1 1 0⟩ in tungsten. As it is already known at single crystals and since the crystals keep their characteristic properties also in polycrystalline bodies; the sequence of the plastically deformed crystals is defined by the maximal Schmid factor, the plastic anisotropy and the orientation dependent elastic modulus. It is interesting that the plastic anisotropy has an opposed effect comparing to the elastic anisotropy, because the plastic anisotropy has largest effect in the direction ⟨1 1 1⟩ and the smallest in direction ⟨1 0 0⟩. Thus, the plastic anisotropy has a more homogenous effect.

Aluminum and tungsten begins the plastic deformation in crystals with ⟨2 9 20⟩ orientation and Schmid factor of 0.499, while iron and copper does it in crystals with ⟨1 2 3⟩ orientation and Schmid factor of 0.466 the same. These plastic deformations are acting in a single slip system at the beginning period of the plastic deformation process. In the other crystals the plastic deformation begins with the irreversible movements of the dislocations in more slip systems at the same time, when the shear stress reaches the critical value. The plastic deformation is started immediately on more slip systems, when the stress reaches the critical shear stress value, necessary for the movement of the dislocations. The critical shear stress value, τ_{crit} is 0.65 MPa in 99.999% pure copper, 0.94 MPa in 99.98% pure copper and 27.5 MPa in 99.96% pure iron (Rosi, 1954)(Cox, 1957). As we mentioned it earlier, the critical shear stress value depends not only on elastic modulus, E_{uvw}, but also on Schmid factor. In aluminum the normal stress has a maximum value in crystal with ⟨1 1 1⟩ orientation in case of uniaxial elastic load, while the shear stress caused by this normal stress has a minimum value. Note that also the point defects, impurities and alloying atoms play an important role. If the stress reaches the critical shear stress, τ_{crit} it activates the residual movement of the dislocations (which generate the plastic deformation). This means different normal and shear stress values at the differently oriented crystals.

We accept that the width of grain boundaries is about 2-3 atomic diameter (Christian, 1965)(Haasen, 1978) and we assume that the distribution of the atoms is not systematic in this width. If a dislocation reaches the grain boundary during its movement, it occupies a layer with a width of b·cosα, where b is the Burgers vector of the dislocation, α is the angle between the slip plane and grain boundary.

We do not have to consider the conditions of Hall–Petch equation (Hall, 1951)(Petch, 1953) at the beginning of the plastic deformation, because there are too few dislocations generated on a slip system to start the movement of the dislocations in the adjacent crystallites. In copper the irreversible movement of the dislocations occurs firstly in crystals with ⟨1 2 3⟩ orientation, secondly in crystals with ⟨1 1 1⟩ orientation. The ratio of the shear stresses which generate the movement of the dislocations is 63,3/56,0 in the above mentioned crystals. This ratio remains constant in the whole elastic deformation region, just like the ratio of the normal stresses which generate the shear stresses above. The ratio of the normal stresses (209,8/265,8 = 0,79) equals to the ratio of the elongation in these two crystals. Consequently, if 10 dislocations reach the grain boundary on ⟨1 2 3⟩ planes, then 12 dislocations reach the grain boundary on ⟨1 1 1⟩ planes. This correlation is valid only in the elastic deformation region and it is true for all slip system.

The plastic deformation starts to develop in two directions on each two slip planes at the same time in crystals with ⟨1 1 0⟩ orientation; simultaneously in two directions on each three slip planes in crystals with ⟨1 1 1⟩ orientation; and simultaneously in two directions on each four slip planes in crystals with ⟨1 0 0⟩ orientation.

The above described plastic deformations extend relatively slow in the whole volume, but the elastic-plastic deformation region stays always active, because only a few slip planes from the possible slip planes of the crystal take part in the plastic deformation; the intermediate region deforms elastically. The material remains in this state till the limit of the plastic deformation capability. If the texture is generated in the material due to the deformation process and the conditions are ideal, the Sachs assumption (Sachs, 1928) can be considered valid in this region: the stress state is homogenous in the whole volume, except of the regions near to the crystal defects. These regions can be neglected comparing to the whole volume. The intermediate region between the two stabile elastic and ideal textured regions is influenced by many other metallurgical conditions.

4. Conclusion

In this chapter the effect of the micro structural elastic and plastic anisotropy is represented on the elastic and plastic deformation properties through some examples. The introduced examples are able the give a good overview about the complexity of the plastic deformation in metals; the picture is even more complicated if the wide range of technological processes are considered, which are applied at certain structural materials before their reach their final use state.

References

ASTM (1992), *Annual Book of ASTM Standards*, Section 01, 01.04, 95.

Bishop, J. F. W. and Hill, R. (1951), "A theory of the plastic dislocation of a polycrystalline aggregate under combined stresses", *Philosophical Magazine,* 42, 414-427.

Cox, G. J., Mehl, R. F. and Horne, G. T., (1957), "Slip, twinning, and fracture in single crystals of iron", *Transactions of the American Society for Metals,* 49, 118.

Christian, J. W. (1965), *The Theory of Transformation in Metals and Alloys*, Pergamon Press.

Dieter, G. E. (1988), *Mechanical Metallurgy,* McGraw Hill Book Co.

Haasen, P. (1978), *Physical Metallurgy*, Cambridge University Press, Cambridge, 46–47.

Hall, E. O. (1951), *"The Deformation and ageing of mild steel"*, Proceedings of the Physical Society, B, 64, 747-753.

Hill, R., (1948), *"A theory of the yielding and plastic flow of anisotropic metals"*, Proc. of the Royal Society of London, A, 193, 281-297.

Petch, N. J. (1953), "The cleavage strength of polycrystals", *Journal of the Iron and Steel Institute,* 174, 25.

Rosi, F. D. (1954), "Stress-strain characteristics and slip-band formation in metal crystals: effect of crystal orientation", Transactions AIME, *Journal of Metals,* 200, 1009-1020.

Sachs, Z. (1928), "Zur Ableitung einer Fliessbedingung", *Zeitschrift des Vereines Deutscher Ingenieur,* 72, 734-736.

Smallman, R. E. (1985), *Modern Physical Metallurgy*, Butterworth, London.

Taylor, G. I. (1938), "Plastic strain in metals", *Journal of the Institute of Metals,* 62, 307.

In: Machining and Forming Technologies, Volume 2
Editor: J. Paulo Davim
ISBN: 978-1-61122-820-5

INDEX

A

B

C

D

E

F

G

H

O

P

T

U

V

W

X

Y